Topics in
Current Physics

32

Topics in Current Physics Founded by Helmut K. V. Lotsch

Superconductivity in Ternary Compounds I

Structural, Electronic, and Lattice Properties

Edited by Ø. Fischer and M. B. Maple

With Contributions by: S. A. Alterovitz O. K. Andersen
S. D. Bader R. Baillif R. Chevrel Ø. Fischer R. Flükiger
W. Klose H.-L. Luo M. B. Maple H. Nohl F. Pobell
D. Rainer B. Renker B. P. Schweiss M. Sergent
S. K. Sinha J. A. Woollam H. Wühl K. Yvon

With 152 Figures

Springer-Verlag Berlin Heidelberg New York 1982

Prof. Øystein Fischer

Département de Physique de la Matière Condensée, Université de Gèneve, 24, Quai Ernest Ansermet, CH-1211 Gèneve 4

Prof. M. Brian Maple

Department of Physics and Institute for Pure and Applied Physical Sciences, University of California, San Diego, La Jolla, CA 92093, USA

ISBN-13:978-3-642-81870-7 e-ISBN-13:978-3-642-81868-4
DOI: 10.1007/978-3-642-81868-4

Library of Congress Cataloging in Publication Data. Main entry under title: Superconductivity in ternary compounds. (Topics in current physics ; v. 32–) Bibliography: p. Includes index. Contents: v. 1. Structural, electronic, and lattice properties. 1. Superconductors. 2. Chemistry, Inorganic. 3. Chalcogenides. I. Fischer, Ø. (Øystein), 1942– II. Maple, M. Brian, 1939– . III. Series. QC612.S8S83 1982 537.6'23 82-5838 AACR2

Softcover reprint of the hardcover 1st edition 1982

2153/3130-543210

Dedicated to the memory of
Bernd T. Matthias

Preface

The structural, electronic and lattice properties of superconducting ternary compounds are the subject of this Topics volume. Its companion volume (Topics in Current Physics, Volume 34) deals primarily with the mutual interaction of superconductivity and magnetism in ternary compounds. These two volumes are the culmination of a project, started nearly two years ago, that was inspired by the intense research effort, both experimental and theoretical, then being expended to explore and develop an understanding of the remarkable physical properties of ternary superconductors. Research activity on this subject has increased in the meantime.

The interest in ternary superconductors originated in 1972, when B.T. Matthias and his co-workers first discovered superconductivity in several ternary molybdenum sulfide compounds that had been synthesized in 1971 by R. Chevrel, M. Sergent, and J. Prigent. The superconducting critical temperature T_c of one of the compounds, $PbMo_6S_8$, was reported to be $\sim$15 K. This value is sufficiently high that there was (and still is) reason to expect that other ternary compounds would be found with superconducting transition temperatures rivaling those of the A15 compounds, of which Nb_3Ge has the record high T_c of 23 K. The interest in ternary superconductors received further impetus when several of the ternary molybdenum sulfides were found to have exceptionally high upper critical magnetic fields, some of them in the neighborhood of 50 Tesla or more.

An immense amount of research on ternary molybdenum chalcogenides then followed. This is described in these volumes by many of the individual researchers involved. Although the original motivation for studying these materials was their superconducting properties, interest quickly spread to other properties such as their electron band structure, phonon spectrum and crystal structure, all of which provide crucial information for developing a detailed understanding of the superconductive behavior of ternary compounds. This volume is to a large extent devoted to these latter properties and is organized in the following way: Chapter 1 gives a brief introduction, historical perspective, and overview of the relatively new subject of ternary superconductors. The chemistry, bonding, and crystal structure of several classes of ternary superconductors are discussed in Chapters 2 and 3. Chapter 4 concentrates on the metallurgical aspects of ternary molybdenum chalcogenides and describes the structural phase transitions that have been observed. Thin films of ternary molybdenum sulfides are considered in Chapter 5, primarily with respect to critical magnetic fields and currents, and normal-state transport properties. Chapter 6 reviews electron band structure calculations for the ternary molybdenum chalcogenides and examines the relationship of band structure to important physical properties of these compounds such as their high superconducting transition temperatures. In Chapter 7, the role of phonons in the extraordinary superconductive behavior of the ternary molybdenum chalcogenides is reviewed. Chapter 8 considers electron-phonon interactions in ternary molybdenum chalcogenides based primarily on isotope effect, electron tunneling measurements, and theoretical considerations thereof.

The discovery of the rare-earth molybdenum chalcogenides and the rare-earth rhodium borides was of particular importance because these compounds were the first superconductors to contain a regular array of magnetic ions. A burst of both experimental and theoretical work on the problem of the interplay of superconductivity and magnetism followed. The result of these investigations has been an ever increasing number of discoveries and a deeper understanding of the problem of the coexistence of superconductivity and long-range magnetic order. The forthcoming volume is devoted to these questions as well as to the superconducting properties of ternary compounds in general.

These two volumes and the Proceedings of the International Conference on Ternary Superconductors [ed. by G.K. Shenoy, B.D. Dunlap, and F.Y. Fradin (North-Holland, Amsterdam 1981)] which was held at Lake Geneva, Wisconsin, September 24-26, 1980, are intended to complement one another and provide the reader with a survey of the present status of this rapidly developing new field and the prospects for its future development, as well as a sense of the excitement that characterizes this research.

No one contributed more to this excitement than B.T. Matthias. It was a great shock to the many researchers in this field and the physics community as a whole when B.T. Matthias died unexpectedly of a heart attack on October 27, 1980. His many contributions to this research area are manifest throughout these volumes. It is clearly a tribute to Matthias's research achievements that a new field in which he was such a guiding and motivating force should have its progress and present status recorded herein.

March 1982

Øystein Fischer
M. Brian Maple

Contents

List of Contributors

Alterovitz, Samuel A.
Department of Physics, Tel Aviv University, Tel Aviv, Israel

Andersen, Ole K.
Max-Planck-Institut für Festkörperforschung, Heisenbergstraße 1,
D-7000 Stuttgart 80, Fed. Rep. of Germany

Bader, Samuel D.
Argonne National Laboratory, Argonne, IL 60439, USA

Baillif, Rémy
Département de Physique de la Matière Condensée, Université de Genève, 24,
Quai Errest Anserment, CH-1211 Genève 4, Switzerland

Chevrel, Roger
Laboratoire de Chimie Minérale B, Faculté des Sciences,
Université de Rennes, Av. Géneral Leclerc, F-35000 Rennes, France

Fischer, Øystein
Département de Physique de la Matière Condensée, Université de Genève, 24,
Quai Errest Anserment, CH-1211 Genève 4, Switzerland

Flükiger, Réné
Kernforschungszentrum, Institut für Technische Physik,
Postfach 3640, D-7500 Karlsruhe 1, Fed. Rep. of Germany

Klose, Wolfgang
Kernforschungszentrum Karlsruhe, D-7500 Karlsruhe 1, Fed. Rep. of Germany

Luo, Huey-Lin
Department of Electrical Engineering and Computer Science,
University of California San Diego, La Jolla, CA 92093, USA

Maple, M. Brian
Department of Physics and Institute for Pure and Applied Physical Sciences,
University of California, San Diego, La Jolla, CA 92093, USA

Nohl, Heinz
Max-Planck-Institut für Festkörperforschung, Heisenbergstraße 1,
D-7000 Stuttgart 80, Fed. Rep. of Germany

Pobell, Frank
Institut für Festkörperforschung, Kernforschungsanlage Jülich,
Postfach 1913, D-5170 Jülich 1, Fed. Rep. of Germany

Rainer, Dirk
Physikalisches Institut, Universität Bayreuth,
D-8580 Bayreuth, Fed. Rep. of Germany

Renker, Burkhard
Institut Laue Langevin, F-38042 Grenoble Cedex, France
and
Kernforschungszentrum Karlsruhe, Institut für Angewandte Kernphysik,
Postfach 3640, D-7500 Karlsruhe, Fed. Rep. of Germany

Schweiss, Bernd P.
Institut für Kristallographie und Mineralogie der Universität
Frankfurt am Main, Senckenberg-Anlage 30, D-6000 Frankfurt am Main 1,
Fed. Rep. of Germany

Sergent, Marcel
Laboratoire de Chimie Minérale B, Faculté des Sciences,
Université de Rennes, Av. Géneral Leclerc, F-35000 Rennes, France

Sinha, Sunil K.
Argonne National Laboratory, Argonne, IL 60439, USA

Woollam, John A.
The University of Nebraska-Lincoln, Department of Electrical Engineering,
W 194 Nebraska Hall, Lincoln, Nebraska 68588, USA

Wühl, Helmut
Kernforschungszentrum Karlsruhe, Institut für Technische Physik
and
Universität Karlsruhe, D-7500 Karlsruhe, Fed. Rep. of Germany

Yvon, Klaus
Laboratoire de Christallographie aux Rayons X, Université de Genève,
24, Quai Ernest Ansermet, CH-1211 Genève 4, Switzerland

1. Superconducting Ternary Compounds: Prospects and Perspectives

Ø. Fischer and M. B. Maple*

With 18 Figures

1.1 Introduction

The remarkable superconducting properties of ternary compounds have stimulated a great amount of interest in these materials within the last decade. Among their many striking properties, three have contributed most decisively to this interest: 1) the superconducting critical temperatures T_c are high, reaching 15 K for $PbMo_6S_8$; 2) the upper critical magnetic field H_{c2} of certain compounds is much higher than in any other class of superconductors; and 3) several structure types of ternary compounds have a regular lattice of magnetic rare-earth (RE) ions and are nevertheless superconducting, thus providing a unique opportunity to study the interplay between superconductivity and long-range magnetic order.

The 15-K superconducting critical temperature of $PbMo_6S_8$, although very high, is still substantially lower than the highest T_c observed to date (T_c = 23 K for the A-15 compound Nb_3Ge). However, if one excludes the A-15 materials from consideration, there are very few binaries that have such high T_c values. Taking into account the rather inadequate knowledge we have of the field of ternary materials at this time, comparing it to the great number of ternary compounds that can be expected to exist, and considering certain favorable factors associated with their crystal structures, it is clear that ternary materials hold great promise for further, and at least modest, increases in T_c.

The upper critical field of $PbMo_6S_8$($H_{c2} \simeq$ 60 Tesla) exceeds the highest value reported for a binary compound, Nb_3Ge, by more than 20 Tesla, in spite of the fact that $PbMo_6S_8$ has a much lower T_c! This result raises the question: What is the maximum critical field that is attainable in a superconductor? It also gives us serious hope for their application in the production of ultrahigh static magnetic fields in the future.

The coexistence of superconductivity and long-range magnetic order has been a long-standing problem in solid-state physics. The apparently contradictory conditions necessary to realize these two collective phenomena have, for more than two decades, challenged both theoreticians and experimentalists alike. The ternary compounds that contain a regular lattice of magnetic moments have brought a completely new situation into the picture. Many old questions have now been answered,

---

*Research supported by the U.S. Department of Energy under Contract No. DE-AT03-76ER 70227

but many new issues have also been raised. The coexistence of long-range antiferromagnetic order and superconductivity is now established in several cases, and it has been found that the competition between ferromagnetic order and superconductivity leads to a long-wavelength oscillatory magnetization in a narrow temperature range below which the superconducting state is destroyed by the ferromagnetism.

It may be worthwhile to reflect for a moment about the distinction between a binary compound and a ternary compound. As the name implies, a ternary compound is comprised of three different elements, each of which occupies one or more special sites throughout the crystal lattice. A ternary compound is distinguished from a pseudobinary compound, which also consists of three elements, but where only two atoms have special sites within the crystal lattice; the third element simply substitutes for one (or sometimes both) of the two elements that form the isostructural binary compounds. This distinction is essential; for several properties of the ternaries, the existence of three distinct sites for the three elements seems to be crucial.

In the field of ternary superconductors, a large number of crystal structures can be obtained by combining three elements. While this contributes to the richness of the field, it also makes it difficult to work with ternary materials. Different compounds with similar compositions may have comparable free energies, and a first attempt to make a ternary compound frequently results in a multiphase material. This can be an advantage in discovering the existence of a compound with a particularly high value of T_c and H_{c2}, but it normally leads to difficulties in identifying the phase with which the superconductive properties are associated. In addition, the phase field of a certain compound may be relatively wide and its properties may depend on the actual composition. A careful chemical, structural, and metallurgical characterization of ternary compounds is therefore necessary.

Most investigations in the field of ternary superconductors have, until recently, been carried out on two classes of materials: 1) the ternary molybdenum chalcogenides of the type $PbMo_6S_8$, first described by CHEVREL et al. [1.1] and often referred to as "Chevrel phases"; and 2) the ternary rhodium borides of the type $ErRh_4B_4$, first reported by MATTHIAS et al. [1.2]. In the following section we shall give a description and a summary of some of the properties of these two classes of compounds as well as other ternary materials.

1.2 The Ternary Materials

1.2.1 The Ternary Molybdenum Chalcogenides (Chevrel Phases)

The ternary molybdenum chalcogenides were first reported by CHEVREL et al. [1.1] in 1971 who pointed out that there existed a whole class of ternary molybdenum sulfides with the approximate formula MMo_nS_{n+1} (n = 3,4,5,6; M: Cr, Fe, Co, Sn,

Pb ...). Crystallographic investigations soon followed [1.3,4], and revealed that the structure of these compounds is basically that of the binary Mo_6Se_8 [1.5] and that the formula should be written $M_xMo_6S_8$ with the possible values of x essentially depending on the size of the M atom. The same kind of compounds can also be formed when S is replaced by Se or Te [1.6,7]. However, the number of known tellurides is considerably smaller than the number of sulfides and selenides. The chemical and structural properties of these materials are discussed in detail in the following two chapters.

The crystal structure of these compounds may be visualized as a slight rhombohedral distortion of the CsCl structure where the cesium atoms are replaced by M atoms and the chlorine atoms are replaced by Mo_6X_8 "pseudomolecules." These Mo_6X_8 (X: S, Se, Te) units resemble somewhat deformed cubes with the 8 X atoms on the cube corners and the 6 Mo atoms at the centers of the cube faces (Fig.1.1). This slightly distorted octahedron of Mo atoms is the most characteristic feature of this crystal structure.

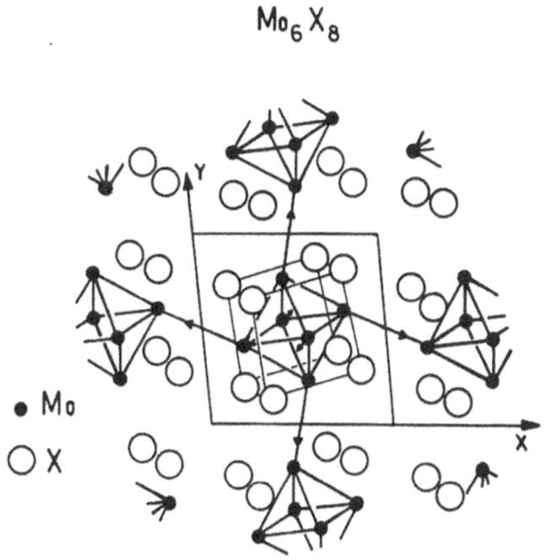

Fig.1.1. Projection of Mo_6X_8 compounds (X: S, Se, Te) on the (001) rhombohedral plane. From Fig.2.3

Three important aspects of this structure should be noted.

1) The size of the Mo_6 cluster is, to a first approximation, independent of the M and X elements.
2) The Mo-Mo intercluster distance depends strongly upon the M and X elements and is typically 20% larger than the intracluster distances.
3) The chalcogen atoms separate the Mo atoms from the M atoms. The M-Mo distances are therefore large, especially when M is a big cation such as Pb, La, or Ca.

One usually distinguishes between compounds where M is a large cation (e.g., M: Sn, Pb) and those where M is a small cation (e.g., M: Cu, Fe). In the latter case, a large homogeneity range for the concentration, x, of M is usually found. This wide range in x values is closely related to a low-temperature structural phase transition to a triclinic structure [1.8,9] that practically all of these

Table 1.1. A selection of superconducting compounds having the $PbMo_6S_8$ structure [1.11]

Compounds	T_c[K]	Compounds	T_c[K]
$PbMo_6S_8$	15	$PbMo_6Se_8$	6.7
$SnMo_6S_8$	14	$SnMo_6Se_8$	6.8
$Cu_2Mo_6S_8$	10.8	$Cu_2Mo_6Se_8$	5.9
$LaMo_6S_8$	7.1	$LaMo_6Se_8$	11.4
$GdMo_6S_8$	1.4	$GdMo_6Se_8$	5.6
$LuMo_6S_8$	2.2	$LuMo_6Se_8$	6.2
Mo_6S_8	1.6	Mo_6Se_8	6.3
$Mo_6S_6I_2$	14.0	$Mo_6Se_7I_1$	7.6

"small M" compounds undergo (Chap.4). Structural phase transitions have also been observed recently in some "large cation" compounds [1.10].

Characteristic of this class of compounds are both the wide variety of M elements which form this structure and the ability to partially substitute other elements for Mo and X. This versatility produces a large spectrum of different physical properties and makes these materials a rich reservoir for a variety of interesting investigations [1.11].

The physics community first became aware of these materials when MATTHIAS et al. [1.12] published a short note in *Science* in 1972, announcing that these compounds were metallic and that many of them were superconducting, with $PbMo_6S_8$ having a critical temperature of 13.7 K. Since then it has been found that $PbMo_6S_8$ can be fabricated with a T_c above 15 K, and that about half of the approximately one hundred compounds in this class are superconducting. In Table 1.1 we list a few examples of superconductors having this crystal structure.

Soon after the discovery of superconductivity in these materials, it was found that many of the compounds have very high critical fields, with $PbMo_6S_8$ giving a new record value of $H_{c2} \simeq 60$ T (at T = 0) [1.13-16]. These anomalously high values of H_{c2} suggested that the electronic structure of these materials must also be rather unusual. Based on simple chemical arguments according to which one would write the valence formula as $M^{2+}(Mo^{2.33+})_6X_8^{2-}$, it was assumed that the conduction electrons would essentially have Mo 4d character [1.17]. This has long since been confirmed by band-structure calculations discussed in Chap.6 and in [Ref.1.18, Chap.6] [1.19-22]. There are several important features of the chemistry and crystal structure that are reflected in the band structure. The existence of clusters shows up in very narrow Mo 4d bands (weak overlap between cluster orbitals). The very high critical fields can be qualitatively understood in terms of these narrow bands since this leads to a small Fermi velocity v_F and consequently a small coherence length $\xi(\xi \sim v_F)$. This, in turn, produces a high critical field because

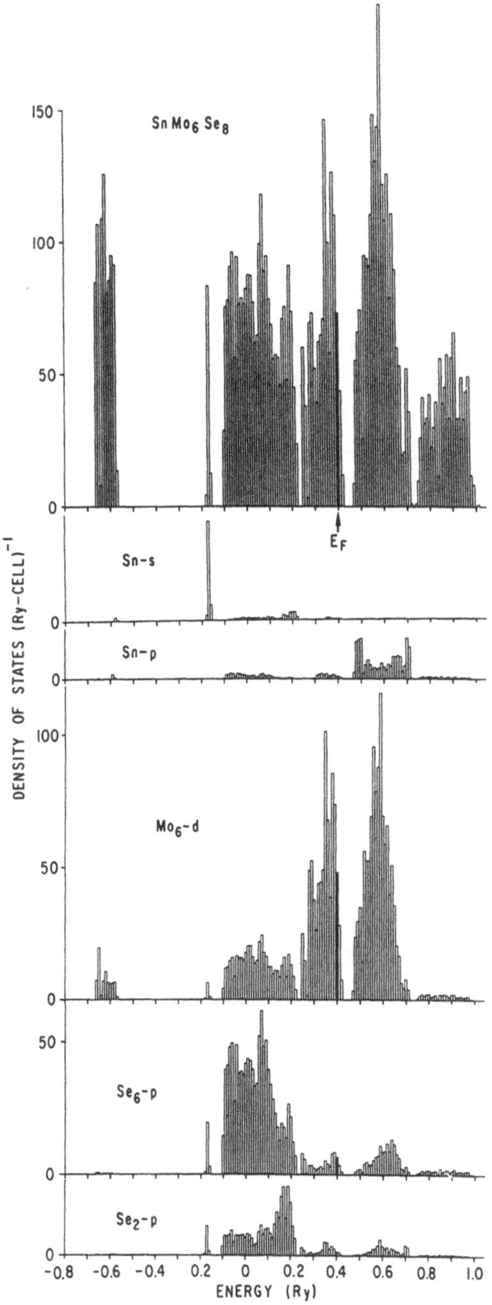

Fig.1.2. Calculated partial densities of states for $SnMo_6S_8$ [Ref.1.18,Fig.6.6]

$H_{c2} \sim \xi^{-2}$. As an example of band-structure calculations we show in Fig.1.2 the calculated partial densities of states for $SnMo_6S_8$. Neither the Sn s and p electrons nor the S p electrons contribute very much to the bands near the Fermi level. However, there is a significant charge transfer from Mo and M to X in the sense of the chemical valence formula. The physical properties depend critically on this charge transfer. Thus one finds that MMo_6S_8 has high values of T_c and H_{c2} when M is divalent, but low values when M is trivalent [1.11]. Similarly, Mo_6S_8 is a low-T_c material whereas $Mo_6S_6I_2$ is a high-T_c superconductor [1.23] (see Table 1.1).

One aspect of the high critical fields which is still not very well understood is the apparent absence of paramagnetic limitation. If the interaction between the external field and the conduction electron spins is taken into account, in addition to the orbital effects, theory predicts critical fields well below the observed ones if "reasonable" values for the parameters are used. This problem is discussed further in [Ref.1.18, Chap.3] where it is shown that anisotropy and multiband effects on H_{c2} may be at least partially responsible for the conflicting results [1.24].

The very high critical fields have caused several groups to study the feasibility of applications of these materials to the generation of high magnetic fields [1.25-28]. The reason for the interest in these materials is best illustrated by comparing H_{c2} for different high-field superconductors (Fig.1.3). We note that at $T = 4.2$ K, H_{c2} for $PbMo_6S_8$ is more than twice that of Nb_3Sn, at present the best superconductor that is commerically available. It has been shown that both thin films (Chap.5) and wires [Ref.1.18, Chap.3] can be made. Critical currents are still about an order of magnitude smaller than those of superconductors currently used in technology, but it is anticipated that high-enough critical currents can be obtained if the microstructure can be better controlled.

The strong coupling character of these materials is important for a detailed understanding of many of their properties. The necessary basic information on phonons and the electron-phonon interaction will appear in the last two chapters. Experimental data reveal that the cluster structure influences the phonon spectrum producing relatively sharp features in the phonon density of states [1.29,30] (Fig.1.4). As a first interpretation of these data a molecular crystal model was invoked [1.29], with 9 external soft modes and 36 internal hard modes giving rise to the broad "hump" above 12 meV in Fig.1.4. This model gives a qualitatively reasonable description, but more detailed investigations show that a strict separation between internal and external modes is not valid, so this model must be used with some caution [1.31]. One of the surprising results of the isotope effect [1.32] and the tunneling [1.33] measurements is that the high-frequency modes seem to contribute more to the electron-phonon interaction than the low-frequency modes. This is contrary to what was initially thought, and is surprising in view of the large value observed for the reduced gap Δ/k_BT_c.

The notion that the Mo 4d electrons were responsible for the superconductivity of these materials soon led to the conclusion that if M was a magnetic element, it would not affect the superconductivity too much [1.17]. The large M-Mo distance would result in a smaller exchange interaction than one normally encounters in bi-

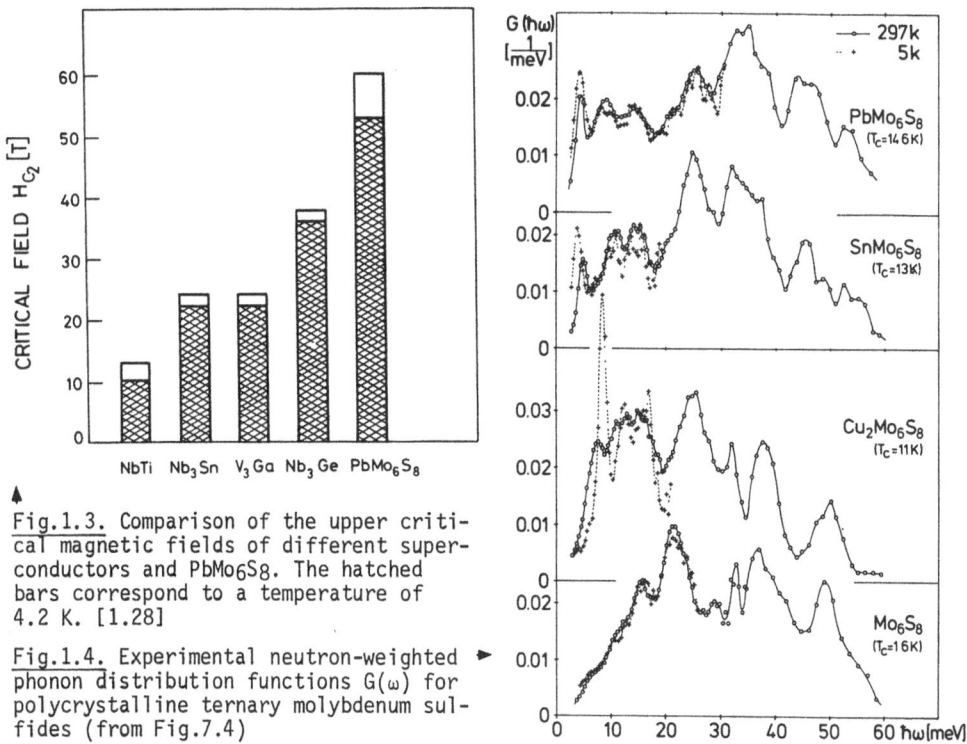

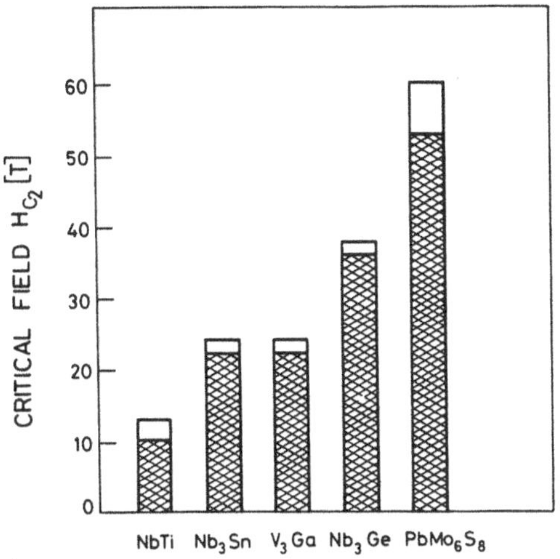

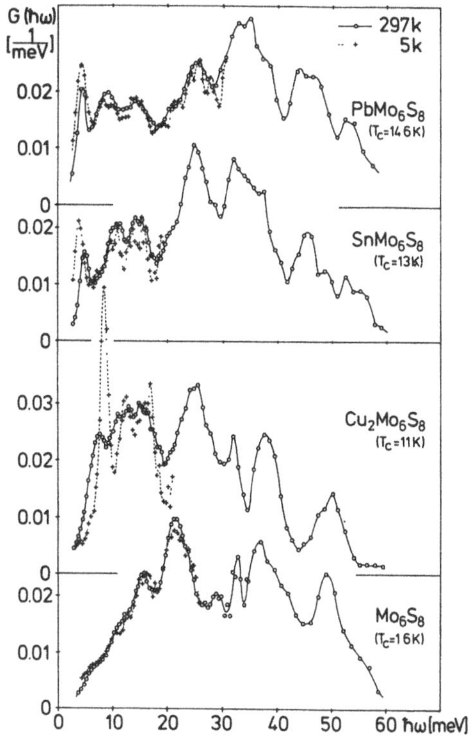

▲ Fig.1.3. Comparison of the upper critical magnetic fields of different superconductors and $PbMo_6S_8$. The hatched bars correspond to a temperature of 4.2 K. [1.28]

Fig.1.4. Experimental neutron-weighted phonon distribution functions G(ω) for polycrystalline ternary molybdenum sulfides (from Fig.7.4) ▶

nary compounds containing magnetic ions. This led to the discovery of the $REMo_6X_8$ superconductors [1.34,35], where RE is a rare-earth element. These compounds, together with the rare-earth rhodium borides, which will be discussed in the next section, are uniquely suited for investigations of the interplay between superconductivity and long-range magnetic order. Since the RE ions in $REMo_6X_8$ compounds form a regular (nearly simple cubic) lattice, these materials are markedly different from the pseudobinary compounds studied earlier, where in the best cases a small percentage of magnetic ions were distributed at random throughout the crystal lattice. As a result of the discovery of these compounds, research activity directed towards solving the problem of the coexistence of superconductivity and magnetism has been intense in recent years.

Most of the $REMo_6X_8$ compounds become superconducting in the proximity of a few degrees Kelvin, well above their magnetic ordering temperatures ($T_M \simeq 0.5$ K). The obvious question was: How would the superconducting state be affected by the magnetic transition? As it happened, one of the compounds, $HoMo_6S_8$, was found to become ferromagnetic; in this case, the superconductivity was destroyed by the onset of the ferromagnetism [1.36,37] (Fig.1.5). A similar effect was observed in the compound $ErRh_4B_4$ at about the same time [1.40,41]. Interest has since shifted to a narrow temperature interval near the ferromagnetic transition temperature within which magnetic order coexists with superconductivity [1.42,43]. However, within

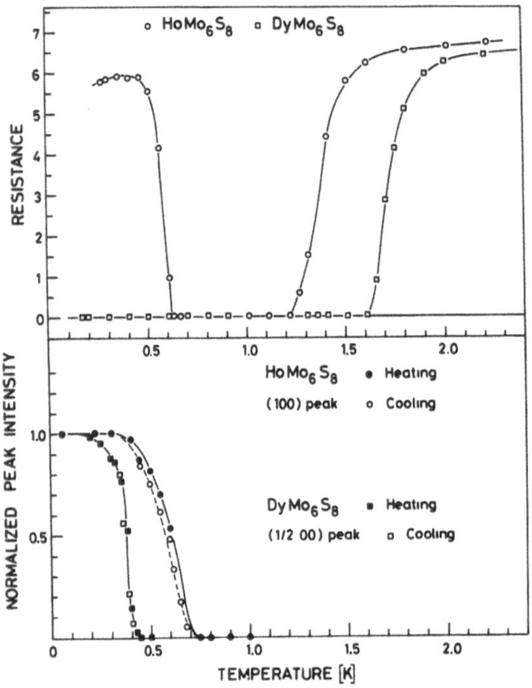

Fig.1.6 ▼

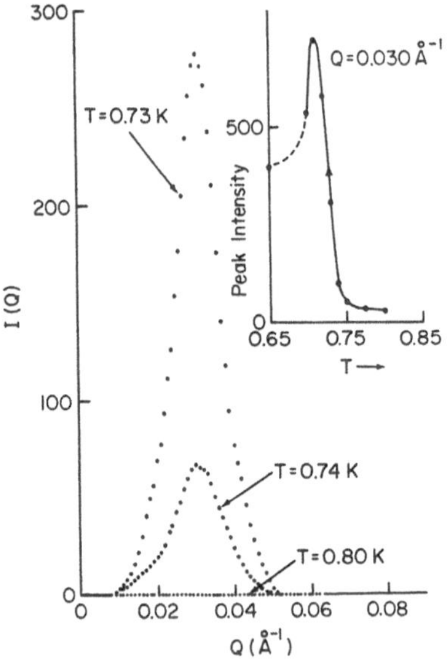

▲

Fig.1.5. Electrical resistance and normalized neutron diffraction peak intensity vs temperature for $HoMo_6S_8$ and $DyMo_6S_8$ [1.36-39]

Fig.1.6. Net scattered intensity, after subtraction of the scattering at high temperatures (2.0 K), for $HoMo_6S_8$, showing the rapid development below 0.75 K of a well-defined peak at a wave vector $Q_c = 0.030$ Å^{-1}. The width of the peak is resolution limited. The inset is a plot of the observed intensity at $Q_c = 0.30$ Å^{-1}. All data were taken while cooling [1.43]

this temperature range, the superconductivity changes the ferromagnetic order into a sinusoidally modulated magnetization state [1.44] with a wavelength of about 250 Å. Figure 1.6 shows recent small-angle scattering results that reveal this oscillatory magnetization state. Most $REMo_6X_8$ compounds are found to order antiferromagnetically, and it is now well established that superconductivity coexists with long-range antiferromagnetism [1.38,39,45-47]. Figure 1.5 shows the electrical resistivity and the intensity of the (1/2, 0, 0) peak from neutron scattering as a function of temperature for $DyMo_6S_8$. Moreover, the critical-field data displayed in Fig.1.7 show that an anomalous reduction of H_{c2} takes place at the Nêel temperature. This unexpected effect has recently been interpreted as the result of a modified pairing and the corresponding change in the electron-electron interaction [1.48]. If this proves correct, it would be the first case of a metal displaying superconductivity with electron pairs built up from non-time reversed states. A detailed account of the experimental results is given in [Ref.1.18, Chaps.5,8], while the theoretical aspects are treated in [Ref.1.18, Chap.9].

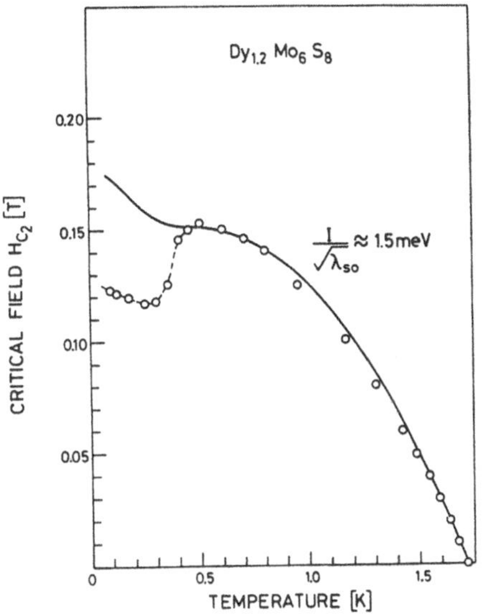

Fig.1.7. Upper critical magnetic field vs temperature for $Dy_{1.2}Mo_6S_8$ [1.38]

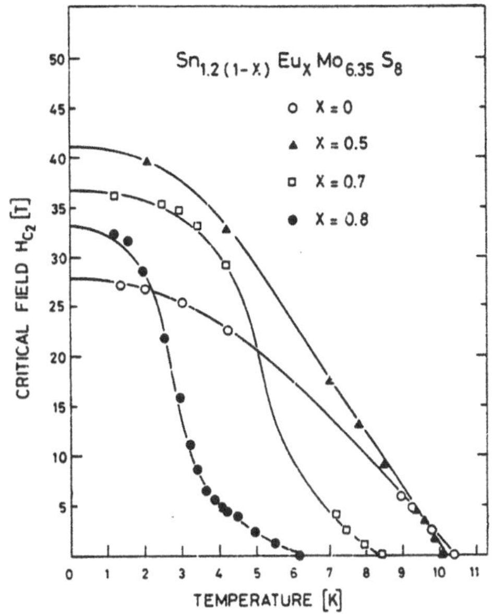

Fig.1.8. Upper critical magnetic field vs temperature for various $Sn_{1.2(1-x)}Eu_xMo_{6.35}S_8$ compounds [1.52]

Related to the questions concerning the coexistence of superconductivity and magnetism is the problem of the absence of superconductivity in the $CeMo_6X_8$ and $EuMo_6X_8$ compounds. In the Ce compounds, this can be understood as a result of a strong antiferromagnetic exchange interaction between the conduction electron spins and Ce^{3+} magnetic moments [1.49,50]. However, for the Eu compounds the question remains open and is presently the object of intense research [1.51].

Early investigations on the $Sn_{1-x}Eu_xMo_6S_8$ system showed that at least for $x \leq 0.5$ the exchange interaction is small [1.3]. In the same concentration domain, a strong increase in H_{c2} with increasing x was observed and has been attributed to the Jaccarino-Peter compensation effect [Ref.1.18, Chaps.3,6,7], [1.52,53]. Figure 1.8 shows the temperature dependence of H_{c2} in this system. Recently, it was found that $EuMo_6S_8$ becomes superconducting under pressure [1.54,55] and, in addition, undergoes a structural transformation at $T \simeq 110$ K [1.10].

The $M_xMo_6X_8$ compounds are not the only ternary molybdenum chalcogenides of interest. Several new compounds with condensed clusters, Mo_9, Mo_{12}, and Mo_∞, have recently been discovered [1.56-61] (Chaps.2,3). We expect that there is a great deal of interesting physics hidden in these new materials. An example is the superconducting compound $Tl_2Mo_6Se_6$ which contains infinite Mo_∞ chains, and shows unusually strong anisotropy [1.62].

1.2.2 The Ternary Rhodium Borides

In 1977, MATTHIAS and co-workers discovered a fascinating set of compounds with the formula MRh_4B_4, where M is Y, Th, or a RE element [1.2]. They found that the compounds with M: Y, Th, Nd, Sm, Er, Tm, and Lu were superconducting, while those with M: Gd, Tb, Dy, and Ho were ferromagnetic. The compounds with M: La, Ce, Pr, Eu, and Yb could not be formed. Later, VANDENBERG and MATTHIAS [1.63] determined that the crystal structure of the MRh_4B_4 compounds is the same as the primitive tetragonal structure of the $CeCo_4B_4$-type compounds investigated earlier by the KUZ'MA and BILONIZHKO [1.64].

The MRh_4B_4 compounds constituted the second series of ternary superconductors and have been studied extensively (Chap.3 and [Ref.1.18,Chaps.2,4,6,7,8]). These materials are similar in many respects to the ternary molybdenum chalcogenides:

1) The compounds $LuRh_4B_4$ and YRh_4B_4 have relatively high superconducting critical temperatures approaching 12 K [1.2].
2) The $CeCo_4B_4$-type crystal structure of the MRh_4B_4 compounds contains Rh_4B_4 "clusters" consisting of interpenetrating Rh and B tetrahedra.
3) The $RERh_4B_4$ compounds have an ordered RE sublattice in which RE ions with partially filled 4f electron shells develop long-range magnetic order via RKKY or dipolar interactions.

The main differences between the MRh_4B_4 compounds and the ternary molybdenum chalcogenides are:

1) Whereas the ternary molybdenum chalcogenides have rhombohedral symmetry, the symmetry of the MRh_4B_4 compounds is tetragonal.
2) The metallic radii of the M atoms are far more critical in forming the MRh_4B_4 compounds than the $M_xMo_6X_8$ compounds.
3) The upper critical magnetic fields of the MRh_4B_4 compounds are rather low, the highest values being ~2.8 T and ~2.1 T for $LuRh_4B_4$ and YRh_4B_4, respectively [1.65].

In Fig.1.9, the crystal structure of the tetragonal RE rhodium borides is depicted in such a way as to emphasize the Rh_4B_4 clusters. The Rh-Rh intracluster distances are nearly the same as in the element and are smaller than the Rh-Rh intercluster distances. VANDENBERG and MATTHIAS [1.67] pointed out that cluster formation occurs in many other high-temperature superconducting compounds and suggested that this may be an important ingredient of many high-temperature superconducting materials. The relative confinement of the Rh 4d electrons within the clusters may be responsible for peaks in the conduction-electron density of states at the Fermi level and, in turn, the relatively high values of T_c, a situation similar to that previously discussed for the $M_xMo_6X_8$ compounds. Band-structure calculations by JARLBORG et al. [1.68] indicate that there is a large Rh 4d partial density of states at the Fermi level in the MRh_4B_4 compounds. The corresponding

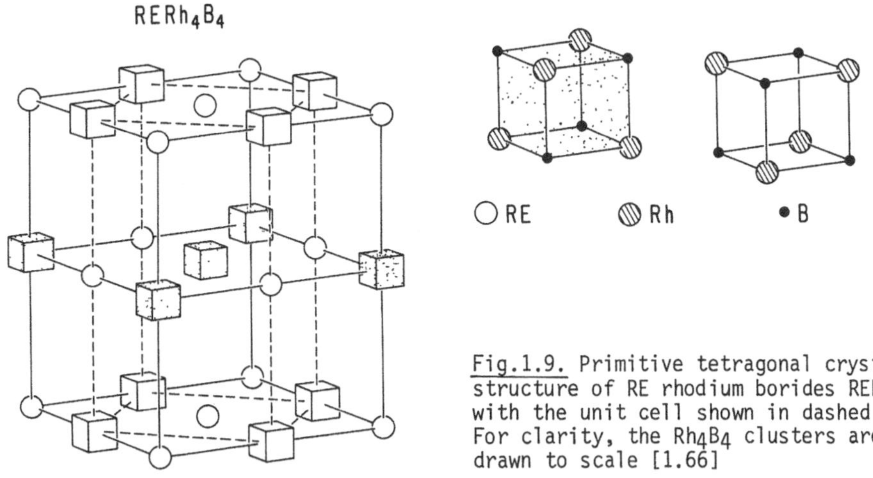

Fig.1.9. Primitive tetragonal crystal structure of RE rhodium borides $RERh_4B_4$ with the unit cell shown in dashed outline. For clarity, the Rh_4B_4 clusters are not drawn to scale [1.66]

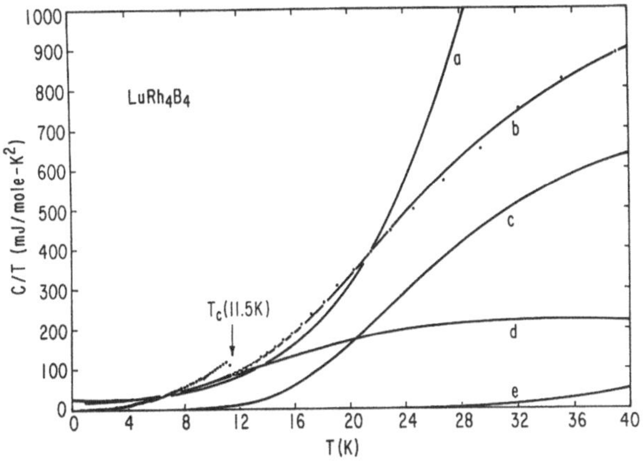

Fig.1.10. Heat capacity C divided by temperature T vs T for $LuRh_4B_4$. The dots indicate the data. Solid lines are calculated contributions to C using (a) a modified Debye model and (b) a molecular crystal model. In the molecular crystal model, contributions from (d) the electronic term and acoustic modes, (c) the soft external Einstein optical modes, and (e) the hard internal Einstein optical modes are shown separately [1.66]

low 4d electron density at the site of the RE ion may account for the small magnitude (~ 0.01 eV) of the exchange interaction between the conduction-electron and the RE magnetic moments in these materials [1.69].

The Rh_4B_4 clusters may also influence the lattice vibrational characteristics of the MRh_4B_4 compounds. This is illustrated by specific-heat C data for $LuRh_4B_4$ [1.66], displayed in Fig.1.10 in the form of a C/T versus T plot. The data could not be fit by a modified Debye model, but were well described by a molecular crystal model similar to that employed by BADER et al. [1.29]for the compounds $PbMo_{5.1}S_8$ and $SnMo_5S_6$, as noted before. A similar molecular crystal analysis provides a good description of the specific-heat data for YRh_4B_4 [1.65].

The moderately high superconducting critical temperatures of $LuRh_4B_4$ and YRh_4B_4 have stimulated investigations of other ternary transition-metal borides as well as other ternary compounds, in general, as potential high-temperature and high-field superconductors [Ref.1.18, Chap.2]. However, the principal interest in the $RERh_4B_4$

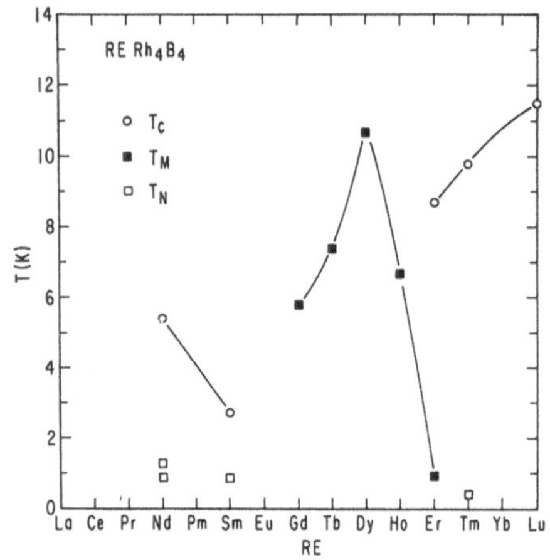

Fig.1.11. Superconducting critical (T_c), Néel (T_N) and Curie (T_M) temperatures of $RERh_4B_4$ compounds. From [Ref.1.18, Chap.4]

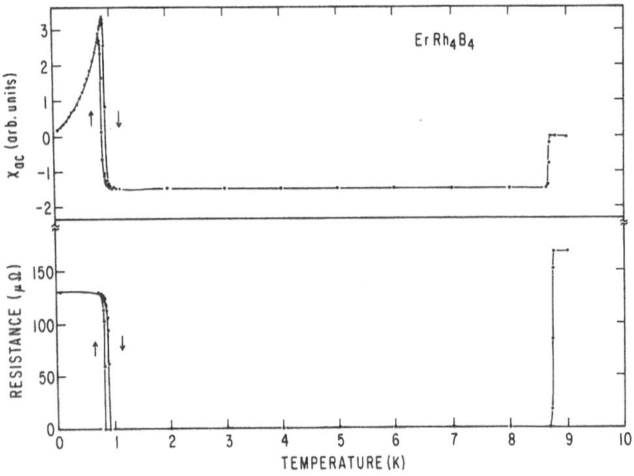

Fig.1.12. Typical ac magnetic susceptibility χ_{ac} and electrical resistance-vs-temperature data for $ErRh_4B_4$ [1.75]

compounds has been the interplay between superconductivity and long-range magnetic order [Ref.1.18, esp. Chap.4]. Like the $REMo_6X_8$ compounds, the $RERh_4B_4$ compounds have an ordered RE sublattice and superconducting and magnetic states with comparable free energies. This latter aspect is apparent in Fig.1.11 in which the superconducting and magnetic critical temperatures of the $RERh_4B_4$ compounds are plotted versus RE. All of the superconducting $RERh_4B_4$ compounds in which the RE 4f electron shell is partially filled undergo some type of magnetic ordering below T_c, at temperatures T_M in the vicinity of 1 K. Whereas $ErRh_4B_4$ becomes ferromagnetic [1.40,41], $NdRh_4B_4$ [1.70,71], $SmRh_4B_4$ [1.72], and $TmRh_4B_4$ [1.73,74] exhibit antiferromagnetic transitions. Investigations at low temperature of $ErRh_4B_4$ by FERTIG et al. [1.40] revealed that superconductivity was destroyed by the onset of long-range magnetic order at a second critical temperature $T_{c2} \sim T_M$. Typical ac magnetic susceptibility χ_{ac}-versus-T and electrical resistance R-versus-T data for $ErRh_4B_4$ [1.75] are shown in Fig.1.12. The thermal hysteresis near T_{c2} in both quantities reflects the first-order nature of the superconducting to normal ferromagnetic transition, which is also evident in the specific heat [1.66] as a sharp spike-shaped feature at T_{c2} due to the latent heat of transformation. Subsequent neutron scattering studies by MONCTON et al. have shown that the long-range magnetic order is ferromagnetic in nature [1.41], and that a sinusoidally modulated magnetic state with a wavelength $\lambda \sim 100$ Å occurs over a narrow temperature interval in the vicinity of T_{c2} [1.42]. The physical properties of $ErRh_4B_4$ and $HoMo_6S_8$ bear a remarkable resemblance to one another, and appear to be general properties of ferromagnetic superconductors.

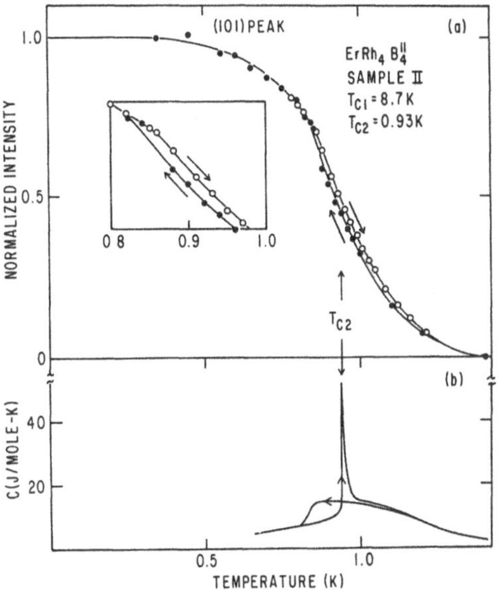

Fig.1.13a,b. Ferromagnetic Bragg intensities for ErRh4B4, compared with the specific-heat data for the identical sample [1.42]

In Fig.1.13, the neutron scattering intensity I and specific heat C of $ErRh_4B_4$ are plotted versus temperature [1.42]. The hysteresis in I and C has been interpreted as evidence for the *macroscopic* coexistence of superconductivity and ferromagnetism [1.42]. It is presumed that the sinusoidally modulated magnetic state develops within the superconducting regions. Whereas the studies on polycrystalline $ErRh_4B_4$ samples by MONCTON et al. [1.42] were interpreted in terms of fluctuations into the spiral magnetization state predicted by BLOUNT and VARMA [1.44] and others [Ref.1.18, Chaps.4,5,8,9], recent neutron diffraction studies on a single crystal specimen of $ErRh_4B_4$ by SINHA et al. [1.76] indicate that a long-range linearly polarized sinusoidally modulated magnetic state, also with $\lambda \sim 100$ Å, and a propagation vector at 45° to the [001] and each of the [100] and [010] axes, coexists with superconductivity.

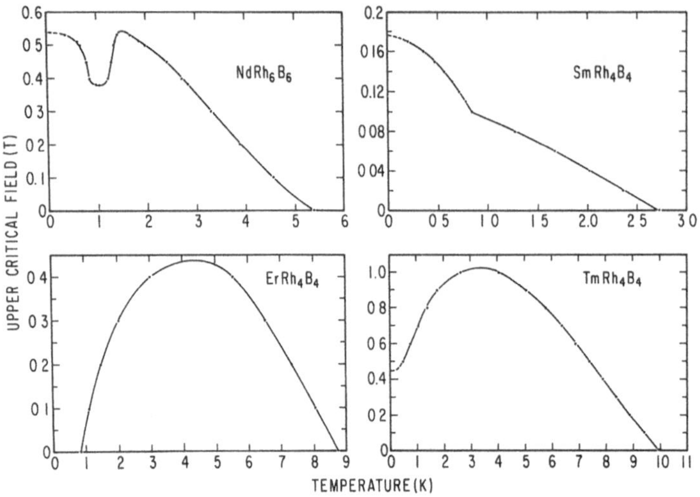

Fig.1.14. Upper critical magnetic field vs temperature for $NdRh_6B_6$ [1.70], $SmRh_4B_4$ [1.72], $ErRh_4B_4$ [1.40], and $TmRh_4B_4$ [1.73]. From [Ref.1.18, Chap.4]

The remaining superconducting $RERh_4B_4$ compounds with partially filled 4f electron shells all exhibit the coexistence of superconductivity and long-range antiferromagnetic order, a property shared by many $REMo_6X_8$ compounds. Figure 1.14 shows the behavior of the upper critical field H_{c2} versus temperature T for the ferromagnetic superconductor $ErRh_4B_4$ and the three antiferromagnetic superconductors $NdRh_4B_4$, $SmRh_4B_4$, and $TmRh_4B_4$. The H_{c2}-versus-T curves display features at the Néel temperatures, similar to those first reported by ISHIKAWA and FISCHER [1.38] for the $RE_{1.2}Mo_6S_8$ compounds. Neutron scattering measurements by MAJKRZAK et al. [1.71,74] have determined the antiferromagnetic structures of $NdRh_4B_4$ (there are two, one at $T \le T_{N1} = 0.9$ K, and the other at $T_{N1} < T \le T_{N2} = 1.2$ K) and $TmRh_4B_4$ ($T_N \simeq 0.6$ K). For example, in $NdRh_4B_4$ the body-centered-tetragonal sublattice of Nd magnetic moments orders antiferromagnetically along the c axis, with a sinusoidal modulation along the [100] direction with $\lambda = 46.5$ Å in the high-temperature magnetic phase, and along the [110] direction with $\lambda = 45.2$ Å in the low-temperature magnetic phase.

Another important aspect of the behavior of the physical properties of the $RERh_4B_4$ compounds is the existence of crystalline electric field effects [Ref.1.18, Chaps.4,7,8]. These are particularly evident in the low-temperature specific heat as Schottky anomalies and reduced entropies of magnetic ordering. The excess specific heat of the compound $ErRh_4B_4$, after the electronic and lattice contributions have been removed [1.66], is shown in Fig.1.15. The solid line represents a Schottky anomaly corresponding to the Er^{3+} energy-level scheme indicated in the figure: a ground-state quartet (or two nearly degenerate doublets, split by $\lesssim$1 K), and excited-state doublets at 12.8, 32.0, 32.4, 33.6, 34.3, and $\gtrsim$100 K above the ground

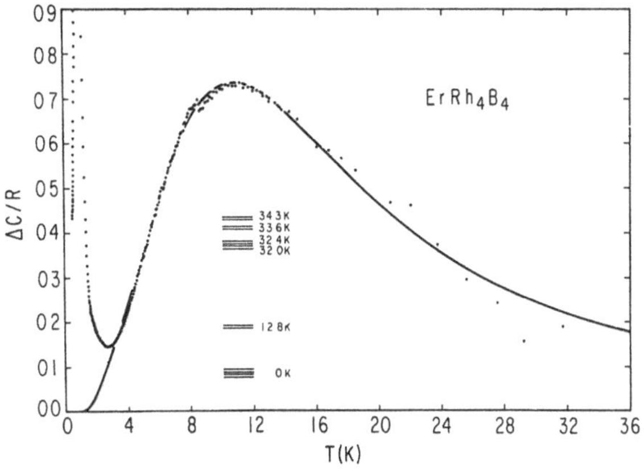

Fig.1.15. Magnetic specific heat ΔC in units of the molar gas constant R vs temperature T for $ErRh_4B_4$. The solid line is a calculated fit to the Schottky anomaly with an Er^{3+} crystal field energy-level scheme consisting of a ground-state quartet and excited-state doublets at temperatures of 12.8, 32.0, 32.4, 33.6, 34.3, and $\gtrsim$100 K [1.66]

state. The Er^{3+} energy-level scheme used to generate the Schottky specific-heat anomaly is not unique, and the correct scheme will have to be established by theoretical considerations and further experimentation. However, these results clearly illustrate the importance of taking crystal-field effects into account in trying to understand and explain the physical properties of the $RERh_4B_4$ compounds. The crystalline electric field is also important in determining the reduced magnetic moment values, the crystallographic directions of magnetization, as well as the magnetic ordering temperatures which do not conform to the expected de Gennes scaling by the factor $(g_J-1)^2J(J+1)$ of the $RERh_4B_4$ compounds [1.75,77]. An interesting "dilemma" [1.78] that has yet to be resolved is the large discrepancy between the values of the ordered Er^{3+} magnetic moment in $ErRh_4B_4$ determined by neutron diffraction ($\sim$5.6 μ_B) [1.41] on the one hand and the Mössbauer effect ($\sim$8.3 μ_B) [1.78] on the other. These and other aspects of the $RERh_4B_4$ compounds are discussed in Chap.3 and in [Ref.1.18, Chaps.2,4,6,7,8].

Some of the most extensive investigations of the $RERh_4B_4$ compounds have been on pseudoternary compounds, formed by substituting a second RE element at the RE sites or some other transition metal at the Rh sites. These investigations allow one to study the interplay between superconductivity, magnetic order, and crystalline electric field effects. For example, it is possible to study systems like $Sm_{1-x}Er_xRh_4B_4$ and $Ho(Rh_{1-x}Ir_x)_4B_4$ in which two types of magnetic order compete (ferromagnetism versus antiferromagnetism) and systems like $Er_{1-x}Ho_xRh_4B_4$ in which ferromagnetic order along different crystallographic axes compete (perpendicular versus parallel to the tetragonal c axis). The competition between ferromagnetic ordering perpendicular and parallel to the c axis in the $Er_{1-x}Ho_xRh_4B_4$ system produces a minimum in the T_{c2}-versus-x curve as can be seen in Fig.1.16 which displays the low-temperature phase diagram of this system [1.79]. A series of experiments [1.79-83] have been carried out on the $Er_{1-x}Ho_xRh_4B_4$ system and have revealed a number of very striking results. 1) There is a critical composition $x_{cr} \sim 0.91$,

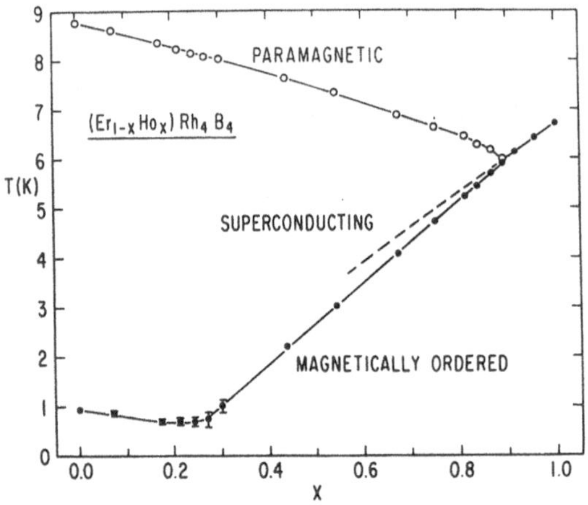

Fig.1.16. Low-temperature phase diagram for the pseudoternary system $(Er_{1-x}Ho_x)Rh_4B_4$ determined from ac magnetic susceptibility measurements. The vertical bars on the T_{c2} data points indicate the observed thermal hysteresis [1.79]

above which ferromagnetism occurs, and below which there is reentrant superconductive behavior with $T_{c2} \sim T_M$. 2) The transition from the superconducting to the normal ferromagnetic state at T_{c2} is first order. The first-order ferromagnetic transition via the superconducting state is depressed in temperature with respect to the second-order ferromagnetic transition that the compound would undergo if it were not superconducting. 3) In some regions of the low-temperature phase diagram, the Er and Ho magnetic moments apparently order independently. Thus, some phase boundaries in the magnetically ordered state are absent in Fig.1.16 and are, in fact, presently in the process of being established. 4) $HoRh_4B_4$, and Ho-rich $Er_{1-x}Ho_xRh_4B_4$ compounds as well, were found to behave as ideal mean-field ferromagnets [1.69,80,84]. This suggests that the interaction responsible for magnetic ordering, presumably the RKKY interaction, has an unusually long-range in these materials, perhaps due to some peculiarity of the $RERh_4B_4$ electron band structure.

The $Ho(Rh_{1-x}Ir_x)_4B_4$ system has also revealed some remarkable types of behavior. The low-temperature phase diagram of this system [1.85] is shown in Fig.1.17. Here, there is a rapid drop in T_c near $x \sim 0.5$, which probably reflects an abrupt change in electronic structure. In addition, the type of magnetic ordering changes from ferromagnetic to antiferromagnetic in the general vicinity of $x \sim 0.2$. The behavior for $x > 0.6$ is especially interesting since superconductivity and antiferromagnetic ordering coexist here, but with Néel temperatures T_N greater than T_c [1.85,86]. The antiferromagnetic structure of the $Ho(Rh_{0.3}Ir_{0.7})_4B_4$ compound has been determined from neutron diffraction studies by HAMAKER et al. [1.87].

TSE et al. [1.88] recently reported ^{11}B NMR and static magnetization measurements on $Y_{1-x}Er_xRh_4B_4$ pseudoternary compounds that suggest a strong exchange interaction between the conduction electron-RE magnetic moments, which is surprising in view of the small depression of the superconducting transition temperature with Er concentration. Large hyperfine interactions below T_c were also observed and attributed to an anomalously large uncompensated conduction electron spin polarization

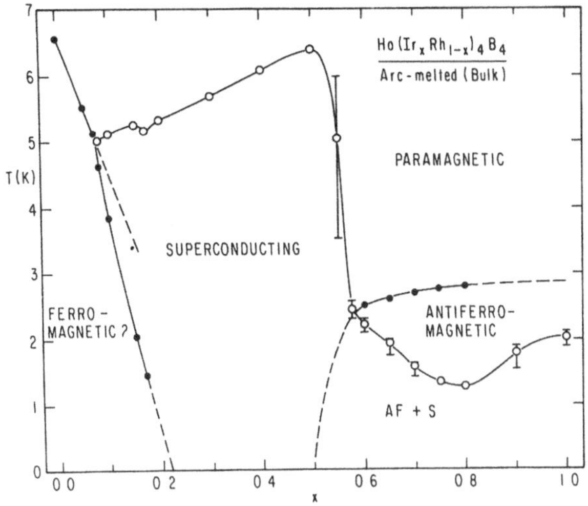

Fig.1.17. Low-temperature phase diagram for the pseudoternary system $Ho(Rh_{1-x}Ir_x)_4B_4$ determined from ac and static magnetic susceptibility measurements on arc-melted samples [1.85]

in the superconducting state. The results were discussed in terms of both itinerant electron antiferromagnetism and non-s-wave Cooper pairing in YRh_4B_4 [Ref.1.18, Chap.7].

Interesting features in the superconducting and magnetic phase boundaries have also been reported for pseudoternary rare-earth molybdenum chalcogenide systems such as $(Eu_{1-x}Ho_x)_{1.2}Mo_6S_8$ [1.89].

1.2.3 Other Ternary Compounds

The striking properties of the $M_xMo_6X_8$ and MRh_4B_4 compounds discussed above have encouraged investigations of other ternary compounds (Chap.3 and [Ref.1.18, Chaps. 2,4,7]). For example, many ternary transition-metal borides which exhibit superconductivity and/or magnetic ordering have been found, in addition to the primitive tetragonal $CeCo_4B_4$-type MRh_4B_4 compounds first reported by MATTHIAS et al. Several of the compounds also have MT_4B_4 stoichiometry (in a chemical formula T denotes transition metal), including some with the body-centered-tetragonal $LuRu_4B_4$-type structure [1.90,91], the orthorhombic $LuRh_4B_4$-type structure [1.92], the primitive tetragonal $NdCo_4B_4$-type structure [1.93], and other variants whose detailed crystal structures have not yet been determined [Ref.1.18, Chap.2]. Whereas many MRu_4B_4 compounds crystallize in the $LuRu_4B_4$ structure, this phase can also be stabilized in MRh_4B_4 compounds by substituting a few percent of Ru for Rh [1.90]. Like the primitive tetragonal $CeCo_4B_4$-type structure, the body-centered-tetragonal $LuRu_4B_4$ phase consists of a slightly distorted face-centered-cubic arrangement of M atoms, separated by Ru_4B_4 "clusters" with two different orientations. However, Ru_4B_4 "clusters" of the same orientation are not arranged in separate sheets perpendicular to the tetragonal c axis, but are distributed equally in a regular array within each sheet. The highest known T_c for this structure type is 9.4 K for $Y(Rh_{0.85}Ru_{0.15})_4B_4$ [1.90].

Superconducting or magnetic MTB_2 compounds with the orthorhombic $LuRuB_2$ structure have been reported [1.94,95]. The compounds $LuRuB_2$ and $YRuB_2$ have relatively high T_c's of 10.0 and 7.8 K, respectively [1.94]. Reentrant superconductive behavior has been observed in the $(Lu_{1-x}Tm_x)RuB_2$ pseudoternary system for $0.52 \leq x \leq 0.62$ [1.96].

Compounds with MT_3B_2 stoichiometry have been found to crystallize in a variety of different structures [Ref.1.18, Chap.2]. The two most common structures are the hexagonal $CeCo_3B_2$ type [1.97] and the $ErIr_3B_2$ type [1.98]. The latter involves a slight monoclinic distortion of the hexagonal $CeCo_3B_2$ structure within the basal plane. The compound $CeRh_3B_2$ has been reported to be an itinerant electron ferromagnet [1.99].

Compounds with the formula MT_6B_4 have been found [1.90,100]. The crystal structure has hexagonal symmetry, but its detailed form is not known [1.100]. The compounds $LaRh_6B_4$ and $Eu^{3+}Rh_6B_4$ have been reported to exhibit itinerant electron ferromagnetism with Curie temperatures of 6 and 19 K, respectively, while YRh_6B_4 and $LuRh_6B_4$ have been reported to show strongly enhanced Pauli paramagnetism [1.100].

Recently, REMEIKA et al. [1.101] reported superconductivity and magnetism in a family of $RERh_xSn_y$ compounds. It is interesting to note that the compounds were discovered while attempting to grow single crystal specimens of $RERh_4B_4$ compounds in a molten tin solvent. Single crystals were indeed produced, but of $RERh_xSn_y$ compounds! Three different phases were identified: phase I, which has a primitive cubic structure with the ideal stoichiometry $RE_3Rh_4Sn_{13}$ [1.101-103], phase III which is face-centered-cubic [1.101,104], and phase II which has a tetragonal structure closely related to that of phase III [1.101,104]. Reentrant superconductive behavior was reported for the phase II compound $ErRh_{1.1}Sn_{3.6}$ with inductively measured values $T_{c1} = 0.97$ K and $T_{c2} = 0.57$ K [1.101]. However, the reentrant transition at T_{c2} in this compound is induced by short-range, rather than long-range, ferromagnetic order. The behaviors of several physical properties with temperature for $ErRh_{1.1}Sn_{3.6}$ are displayed in Fig.1.18. Reentrant superconductive behavior has also been reported for the phase III compounds $ErOs_xSn_y$ and $TmOs_xSn_y$ [1.105,106]. A series of ternary transition-metal germanides $RE_3T_4Ge_{13}$ (T = Ru, Os), isostructural with the phase I $RE_3Rh_4Sn_{13}$ compounds, which exhibit superconductivity or magnetic order have been reported [1.107].

Superconductivity with T_c's up to 6 K in $M_2Fe_3Si_5$ compounds, which have a primitive tetragonal structure, has been observed recently [1.108]. Mössbauer effect experiments indicate that the magnetic moments on both types of Fe sites in this compound are less than 0.03 μ_B [1.109]. The compound $Tm_2Fe_3Si_5$ exhibits reentrant superconductivity with $T_{c1} = 1.3$ K and $T_{c2} = 1.1$ K [1.110]. Another ternary compound

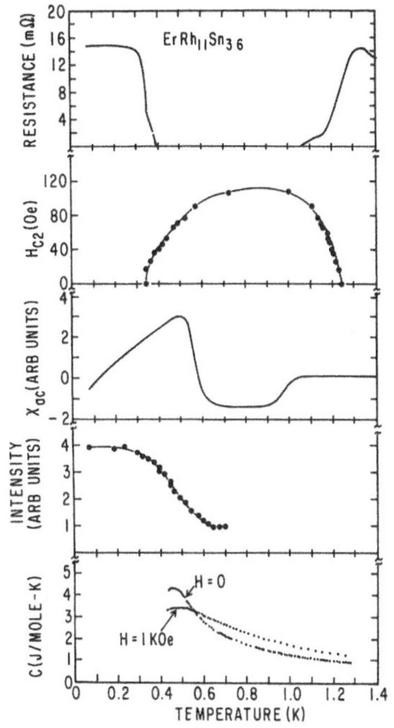

Fig.1.18. Electrical resistance, upper critical magnetic field H_{c2}, ac magnetic susceptibility χ_{ac}, neutron scattering intensity, and heat capacity C vs temperature for $ErRh_{1.1}Sn_{3.6}$ [1.101]

containing Fe has been found to display superconductivity at 4 K, $LaFe_4P_{12}$ [1.111], which has a body-centered-cubic structure [1.112]. The ternary phosphide ZrRuP with the hexagonal ZrRuSi-type structure has a relatively high T_c onset of 13 K [1.113].

Superconductivity or magnetic order has also been observed in ternary transition-metal silicides and germanides of the type $M_5T_4X_{10}$ where M is a trivalent metal atom, T is a transition-metal atom, and X is Si or Ge [1.114,115].

1.3 Concluding Remarks

In this introductory chapter, we have briefly reviewed the research on superconductivity and magnetism in ternary compounds that has been carried out during the past decade. More detailed discussions will appear throughout the various chapters of this treatise. Some of the highlights of this research are summarized below:

1) Certain ternary compounds have relatively high superconducting critical temperatures, as high as 15 K for $PbMo_6S_8$.
2) Some ternary molybdenum chalcogenides have extremely high upper critical magnetic fields, approaching ~60 Tesla or more, again for $PbMo_6S_8$.
3) The coexistence of superconductivity and long-range antiferromagnetic order has been discovered in a number of ternary RE compounds.

4) Several ternary RE compounds exhibit reentrant superconductive behavior due to the occurrence of long-range ferromagnetic order. This is accompanied by a sinusoidally modulated magnetic state that coexists with superconductivity in a narrow temperature interval above the reentrant transition temperature.
5) Large enhancements of the upper critical magnetic field due to the exchange field compensation effect (Jaccarino-Peter effect) have been observed in various superconducting ternary molybdenum chalcogenides doped with Eu ions.
6) Many novel or unusual physical phenomena, in addition to those mentioned above, such as valence fluctuations, the Kondo effect, itinerant electron ferromagnetism, enhanced Pauli paramagnetism, restricted dimensionality effects, giant magnetic moments, metal-insulator transitions, and critical phenomena, have been observed in ternary compounds.

Certain characteristics of the ternary compounds appear to be primarily responsible for their remarkable physical properties. Several of these may be derived from the molecular "clusters" that are found in many of the ternary materials. In the $M_xMo_6X_8$ and MRh_4B_4 compounds, the 4d electrons of Mo and Rh are relatively confined within the Mo_6X_8 and Rh_4B_4 clusters, producing sharp structure in the conduction electron densities of states near the Fermi level. This could account for superconductivity with high values of T_c and H_{c2} and the occurrence of itinerant electron ferromagnetism. In the ternary RE compounds, localization of the conduction electrons within the clusters tends to isolate them from RE ions, leading to a decreased exchange interaction between the conduction-electron spins and the RE magnetic moments. This, of course, is essential since it allows the compounds to retain their superconductivity and produces magnetic ordering at temperatures T_M that are comparable to T_c. For the magnetic superconductors, the other important ingredient is the ordered RE sublattice which insures that the magnetic order has a long range.

In addition to the electronic properties, the clusters also influence the lattice vibrational properties. As a zeroth-order approximation, the $M_xMo_6X_8$ and MRh_4B_4 compounds can be regarded as molecular crystals in which the M atoms are coupled to quasi-rigid Mo_6X_8 or Rh_4B_4 clusters. This leads to low-frequency external modes, associated with vibrations of the coupled M-cluster units and rigid body rotations of the clusters, and high-frequency internal modes associated with nonrigid motions of the clusters. Such "molecular crystal" considerations have been successful in describing the specific-heat data for $M_xMo_6X_8$ and MRh_4B_4 compounds, although they do not adequately account for other aspects of the behavior of these materials.

It should be noted that we have omitted in our discussions certain ternary materials such as the oxides $LiTi_2O_4$, $Ba(Pb,Bi)O_3$, $SrTiO_{3-x}$, and M_xWO_3, as well as intercalated layered materials. Although these materials also have very interesting superconducting properties, they are beyond the scope of this volume as well as of its companion volume [1.18] which is now in preparation. At this point we would like to mention that certain aspects of ternary superconductors are also treated in the recent book by S.V. Vonsovsky, Yu A. Izyumov and E.Z. Kurmaev [1.116].

Finally, it seems clear to us that we have only "scratched the surface" of the field of superconductivity in ternary compounds. The ternary compounds that have been investigated within the last decade have already yielded a rich variety of interesting phenomena; this has stimulated the development of theory and has suggested new and promising directions for experimental research. The extraordinary physical properties of some of these materials are not only of fundamental interest but are also of potential application in technology. Ternary compounds can be expected to fascinate and challenge experimentalists and theoreticians for some time to come.

References

1.1 R. Chevrel, M. Sergent, J. Prigent: J. Solid State Chem. *3*, 515 (1971)
1.2 B.T. Matthias, E. Corenzwit, J.M. Vandenberg, H. Barz: Proc. Natl. Acad. Sci. USA *74*, 1334 (1977)
1.3 O. Bars, J. Guillevic, D. Grandjean: J. Solid State Chem. *6*, 335 (1973)
1.4 M. Marezio, P.D. Dernier, J.P. Remeika, E. Corenzwit, B.T. Matthias: Mat. Res. Bull. *8*, 657 (1973)
1.5 O. Bars, J. Guillevic, D. Grandjean: J. Solid State Chem. *6*, 48 (1973)
1.6 M. Sergent, R. Chevrel: J. Solid State Chem. *6*, 433 (1973)
1.7 R. Chevrel, M. Sergent, Ø. Fischer: Mat. Res. Bull. *10*, 1169 (1975)
1.8 A.C. Lawson: Mat. Res. Bull. *7*, 773 (1972)
1.9 R. Flükiger, A. Junod, R. Baillif, P. Spitzli, A. Treyvaud, A. Paoli, H. Devantay, J. Müller: Solid State Commun. *23*, 699 (1977)
1.10 R. Baillif, A. Dunand, J. Müller, K. Yvon: Phys. Rev. Lett. *47*, 672 (1981)
1.11 Ø. Fischer: Appl. Phys. *16*, 1 (1978)
1.12 B.T. Matthias, M. Marezio, E. Corenzwit, A.S. Cooper, H.E. Barz: Science *175*, 1465 (1972)
1.13 R. Odermatt, Ø. Fischer, H. Jones, G. Bongi: J. Phys. C *7*, L13 (1974)
1.14 Ø. Fischer, H. Jones, G. Bongi, M. Sergent, R. Chevrel: J. Phys. C *7*, L450 (1974)
1.15 S. Foner, E.J. McNiff, E.J. Alexander: Phys. Lett. *49*A, 269 (1974)
1.16 S. Foner, E.J. McNiff, R.N. Shelton, R.W. McCallum, M.B. Maple: Phys. Lett. *57*A, 345 (1976)
1.17 Ø. Fischer: Colloques Int. CNRS No. 242, 79, Grenoble (1974)
1.18 *Topics in Current Physics, Vol. 34*, ed. by Ø. Fischer and M.B. Maple (Springer, Berlin, Heidelberg, New York 1982)
1.19 O.K. Andersen, W. Klose, H. Nohl: Phys. Rev. B *17*, 1209 (1978)
1.20 L.F. Mattheiss, C.Y. Fong: Phys. Rev. B *15*, 1760 (1977)
1.21 D.W. Bullet: Phys. Rev. Lett. *39*, 664 (1977)
1.22 T. Jarlborg, A.J. Freeman: Phys. Rev. Lett. *44*, 178 (1980)
1.23 M. Sergent, Ø. Fischer, M. Decroux, C. Perrin, R. Chevrel: J. Solid State Chem. *22*, 87 (1977)
1.24 M. Decroux, Ø. Fischer, C. Rossel, B. Lachal, R. Baillif, R. Chevrel, M. Sergent: In *Ternary Superconductors*, ed. by G.K. Shenoy, B.D. Dunlap, and F.Y. Fradin (North-Holland, Amsterdam 1981) pp.65-68
1.25 N.E. Alekseevskii, N.M. Dobrovolskii, V.I. Tsebro: JETP Lett. *23*, 639 (1976)
1.26 M. Decroux, Ø. Fischer, R. Chevrel: Cryogenics *17*, 291 (1977)
1.27 S.A. Alterovitz, J.A. Woollam: In *Ternary Superconductors*, ed. by G.K. Shenoy, B.D. Dunlap, and F.Y. Fradin (North-Holland, Amsterdam 1981) pp.113-118

1.28 B. Seeber, C. Rossel, Ø. Fischer: In *Ternary Superconductors*, ed. by G.K. Shenoy, B.D. Dunlap, and F.Y. Fradin (North-Holland, Amsterdam 1981) pp.119-124
1.29 S.D. Bader, G.S. Knapp, S.K. Sinha, P. Schweiss, B. Renker: Phys. Rev. Lett. *37*, 344 (1976)
1.30 B.P. Schweiss, B. Renker, E. Schneider, W. Richardt: In *Proc. 2nd Rochester Conf. on Superconductivity in d- and f-Band Metals*, ed. by D.H. Douglass (Plenum, New York 1976) pp.189-207;
S.D. Bader, S.K. Sinha, R.N. Shelton: In *Proc. 2nd Rochester Conf. on Superconductivity in d- and f-Band Metals*, ed. by D.H. Douglass (Plenum, New York 1976) pp.209-221
1.31 S.D. Bader, S.K. Sinha: Phys. Rev. B *18*, 3082 (1978)
1.32 F.J. Culetto, F. Pobell: Phys. Rev. Lett. *40*, 1104 (1978)
1.33 U. Poppe, H. Wühl: J. Physique *39*, C6-361 (1978)
1.34 Ø. Fischer, A. Treyvaud, R. Chevrel, M. Sergent: Solid State Commun. *17*, 721 (1975)
1.35 R.N. Shelton, R.W. McCallum, H. Adrian: Phys. Lett. *56*A, 213 (1976)
1.36 M. Ishikawa, Ø. Fischer: Solid State Commun. *23*, 37 (1977)
1.37 J.W. Lynn, D.E. Moncton, W. Thomlinson, G. Shirane, R.N. Shelton: Solid State Commun. *26*, 493 (1978)
1.38 M. Ishikawa, Ø. Fischer: Solid State Commun. *24*, 747 (1977)
1.39 D.E. Moncton, G. Shirane, W. Thomlinson, M. Ishikawa, Ø. Fischer: Phys. Rev. Lett. *41*, 1133 (1978)
1.40 W.A. Fertig, D.C. Johnston, L.E. Delong, R.W. McCallum, M.B. Maple, B.T. Matthias: Phys. Rev. Lett. *38*, 387 (1977)
1.41 D.E. Moncton, D.B. McWhan, J. Eckert, G. Shirane, W. Thomlinson: Phys. Rev. Lett. *39*, 1164 (1977)
1.42 D.E. Moncton, D.B. McWhan, P.H. Schmidt, G. Shirane, W. Thomlinson, M.B. Maple, H.B. MacKay, L.D. Woolf, Z. Fisk, D.C. Johnston: Phys. Rev. Lett. *45*, 2060 (1980)
1.43 J.W. Lynn, J.L. Ragazzoni, R. Pynn, J. Joffrin: J. Physique Lett. *42*, 368 (1981)
1.44 E.I. Blount, C.M. Varma: Phys. Rev. Lett. *42*, 1079 (1979)
1.45 R.W. McCallum, D.C. Johnston, R.N. Shelton, M.B. Maple: Solid State Commun. *24*, 391 (1977)
1.46 M.B. Maple, L.D. Woolf, C.F. Majkrzak, G. Shirane, W. Thomlinson, D.E. Moncton: Phys. Lett. *77*A, 487 (1980)
1.47 M.B. Maple: J. Physique *39*, C6-1374 (1978);
M. Ishikawa, Ø. Fischer, J. Müller: J. Physique *39*, C6-1379 (1978)
1.48 G. Zwicknagel, P. Fulde: Z. Phys. B *43*, 23 (1981)
1.49 M. Pelizzone, A. Treyvaud, P. Spitzli, Ø. Fischer: J. Low Temp. Phys. *29*, 453 (1977)
1.50 M.B. Maple, L.E. Delong, W.A. Fertig, D.C. Johnston, R.W. McCallum, R.N. Shelton: In *Valence Instabilities and Related Narrow Band Phenomena*, ed. by R.D. Parks (Plenum, New York 1977) pp.17-28
1.51 M.B. Maple, M.S. Torikachvili, R.P. Guertin, S. Foner: Proc. 15th Rare Earth Research Conf., held June 15-18, 1981, Rolla, Missouri;
C.Y. Huang, W.W. Fuller, D.W. Harrison, H.L. Luo, S.A. Wolf: Proc. 16th Int. Conf. on Low Temp. Physics, Part III, to appear in Physica B+C, 1982
1.52 Ø. Fischer, M. Decroux, S. Roth, R. Chevrel, M. Sergent: J. Phys. C *8*, L474 (1975)
1.53 F.Y. Fradin, G.K. Shenoy, B.D. Dunlap, A.T. Aldred, C.W. Kimball: Phys. Rev. Lett. *38*, 719 (1977)
1.54 C.W. Chu, S.Z. Huang, C.H. Lin, R.L. Meng, M.K. Wu, P.H. Schmidt: Phys. Rev. Lett. *46*, 276 (1981)
1.55 D.W. Harrison, K.C. Lim, J.D. Thompson, C.Y. Huang, P.D. Hambourger, H.L. Luo: Phys. Rev. Lett. *46*, 280 (1981)
1.56 B. Seeber, M. Decroux, Ø. Fischer, R. Chevrel, M. Sergent, A. Grüttner: Solid State Commun. *29*, 419 (1979)
1.57 R. Chevrel, M. Sergent, B. Seeber, Ø. Fischer, A. Grüttner, K. Yvon: Mat. Res. Bull. *14*, 567 (1979)

1.58 A. Grüttner, K. Yvon, R. Chevrel, M. Potel, M. Sergent, B. Seeber: Acta Cryst. *B35*, 285 (1979)
1.59 M. Potel, R. Chevrel, M. Sergent, M. Decroux, Ø. Fischer: C.R. Acad. Sci. (Paris Ser.) *C288*, 429 (1979)
1.60 M. Potel, R. Chevrel, M. Sergent: Acta Cryst. B *36*, 1545 (1980)
1.61 W. Höhle, H.G. von Schnering, A. Lipka, K. Yvon: J. Less Common Metals *71*, 135 (1980)
1.62 J.C. Armici, M. Decroux, Ø. Fischer, M. Potel, R. Chevrel, M. Sergent: Solid State Commun. *33*, 607 (1980)
1.63 J.M. Vandenberg, B.T. Matthias: Proc. Natl. Acad. Sci. USA *74*, 1336 (1977)
1.64 Yu.B. Kuz'ma, N.S. Bilonizhko: Sov. Phys.-Cryst. *16*, 897 (1972)
1.65 L.D. Woolf, H.C. Hamaker, S.E. Lambert, M.B. Maple, H.R. Ott: To be published
1.66 L.D. Woolf, D.C. Johnston, H.B. MacKay, R.W. McCallum, M.B. Maple: J. Low Temp. Phys. *35*, 651 (1979)
1.67 J.M. Vandenberg, B.T. Matthias: Science *198*, 194 (1977)
1.68 T. Jarlborg, A.J. Freeman, T.J. Watson-Yang: Phys. Rev. Lett. *39*, 1032 (1977)
1.69 M.B. Maple, H.C. Hamaker, D.C. Johnston, H.B. MacKay, L.D. Woolf: J. Less Common Met. *62*, 251 (1978)
1.70 H.C. Hamaker, L.D. Woolf, H.B. MacKay, Z. Fisk, M.B. Maple: Solid State Commun. *31*, 139 (1979)
1.71 C.F. Majkrzak, D.E. Cox, G. Shirane, H.A. Mook, H.C. Hamaker, H.B. MacKay, Z. Fisk, M.B. Maple: To be published
1.72 H.C. Hamaker, L.D. Woolf, H.B. MacKay, Z. Fisk, M.B. Maple: Solid State Commun. *32*, 289 (1979)
1.73 H.C. Hamaker, H.B. MacKay, M.S. Torikachvili, L.D. Woolf, M.B. Maple, W. Odoni, H.R. Ott: J. Low Temp. Phys. *44*, 553 (1981)
1.74 C.F. Majkrzak, S.K. Satija, G. Shirane, H.C. Hamaker, Z. Fisk, M.B. Maple: To be published
1.75 M.B. Maple, H.C. Hamaker, L.D. Woolf, H.B. MacKay, Z. Fisk, W. Odoni, H.R. Ott: In *Crystalline Electric Field and Structural Effects in f-Electron Systems*, ed. by J.E. Crow, R.P. Guertin and T.W. Mihalisen (Plenum, New York 1980) pp.533-543
1.76 S.K. Sinha, G.W. Crabtree, D.G. Hinks, H.A. Mook: Proc. 16th Int. Conf. on Low Temp. Physics, Part III, to appear in Physica B+C, 1982
1.77 M.B. Maple: In *Science and Technology of Rare Earth Materials*, ed. by E.C. Subbarao and W.E. Wallace (Academic, New York 1980) pp.167-193
1.78 G.K. Shenoy, B.D. Dunlap, F.Y. Fradin, S.K. Sinha, C.W. Kimball, W. Potzel, F. Probst, G.M. Kalvius: Phys. Rev. B *21*, 3886 (1980)
1.79 D.C. Johnston, W.A. Fertig, M.B. Maple, B.T. Matthias: Solid State Commun. *26*, 141 (1978)
1.80 H.B. MacKay, L.D. Woolf, M.B. Maple, D.C. Johnston: Phys. Rev. Lett. *42*, 918 (1979)
1.81 H.A. Mook, W.C. Koehler, M.B. Maple, Z. Fisk, D.C. Johnston, L.D. Woolf: Phys. Rev. B *25*, 372 (1982)
1.82 L.D. Woolf, D.C. Johnston, H.A. Mook, W.C. Koehler, M.B. Maple, Z. Fisk: Proc. 16th Int. Conf. on Low Temp. Physics, Part III, to appear in Physica B+C, 1982
1.83 H. Adrian, K. Müller, G. Saemann-Ischenko: Phys. Rev. B *22*, 4424 (1980)
1.84 H.R. Ott, G. Keller, W. Odoni, L.D. Woolf, M.B. Maple, D.C. Johnston, H.A. Mook: Phys. Rev. B *25*, 477 (1982)
1.85 H.C. Ku, F. Acker, B.T. Matthias: Phys. Lett. *76A*, 399 (1980)
1.86 L.D. Woolf, S.E. Lambert, M.B. Maple, H.C. Ku, W. Odoni, H.R. Ott: Physica *108B+C*, 761 (1981)
1.87 H.C. Hamaker, H.C. Ku, M.B. Maple, H.A. Mook: To be published
1.88 P.K. Tse, A.T. Aldred, F.Y. Fradin: Phys. Rev. Lett. *43*, 1825 (1979)
1.89 M. Ishikawa, M. Sergent, Ø. Fischer: Phys. Lett. *82A*, 30 (1981)
1.90 D.C. Johnston: Solid State Commun. *24*, 699 (1977)
1.91 K. Yvon, A. Grüttner: In *Superconductivity in d- and f-Band Metals*, ed. by H. Suhl and M.B. Maple (Academic, New York 1980) pp.515-519
1.92 K. Yvon, D.C. Johnston: Submitted to Acta Cryst. B
1.93 Yu. B. Kuz'ma, N.S. Bilonizhko: Dopov. Akad. Nauk Ukr. RSR, Ser. A(3), 275 (1978)

1.94 H.C. Ku, R.N. Shelton: Mat. Res. Bull. *15*, 1441 (1980)
1.95 R.N. Shelton, B.A. Karcher, D.R. Powell, R.A. Jacobson, H.C. Ku: Mat. Res. Bull. *15*, 1445 (1980)
1.96 H.C. Ku, R.N. Shelton: Solid State Commun. *40*, 237 (1981)
1.97 Yu. B. Kuz'ma, P.I. Kripyakevich, N.S. Bilonizhko: Dopov. Acad. Nauk Ukr. RSR, Ser. A (10), 939 (1969)
1.98 H.C. Ku, G.P. Meisner: J. Less-Common Metals *78*, 99 (1981)
1.99 S.K. Dhar, S.K. Malik, R. Vijayaraghavan: J. Phys. C (Solid State Phys.) *14*, L321 (1981);
H. Oesterreicher, F.T. Parker, M. Misroch: Appl. Phys. *12*, 287 (1977)
1.100 I. Felner, I. Nowik: Phys. Rev. Lett. *45*, 2128 (1980)
1.101 J.P. Remeika, G.P. Espinosa, A.S. Cooper, H. Barz, J.M. Rowell, D.B. McWhan, J.M. Vandenberg, D.E. Moncton, Z. Fisk, L.D. Woolf, H.C. Hamaker, M.B. Maple, G. Shirane, W. Thomlinson: Solid State Commun. *34*, 923 (1980)
1.102 J.M. Vandenberg: Mat. Res. Bull. *15*, 835 (1980)
1.103 J.L. Hodeau, J. Chenavas, M. Marezio, J.P. Remeika: Solid State Commun. *36*, 839 (1980)
1.104 J. Chenavas, J.L. Hodeau, A. Collomb, M. Marezio, J.P. Remeika, J.M. Vandenberg: In *Ternary Superconductors*, ed. by G.K. Shenoy, B.D. Dunlap and F.Y. Fradin (North-Holland, Amsterdam 1981) pp.219-224
1.105 S.E. Lambert, Z. Fisk, H.C. Hamaker, M.B. Maple, L.D. Woolf, J.P. Remeika, G.P. Espinosa: In *Ternary Superconductors*, ed. by G.K. Shenoy, B.D. Dunlap, and F.Y. Fradin (North-Holland, Amsterdam 1981) pp.247-250
1.106 Z. Fisk, S.E. Lambert, M.B. Maple, J.P. Remeika, G.P. Espinosa, A.S. Cooper, H. Barz, S. Oseroff: Solid State Commun. *41*, 63 (1982)
1.107 C.U. Segre, H.F. Braun, K. Yvon: In *Ternary Superconductors*, ed. by G.K. Shenoy, B.D. Dunlap, and F.Y. Fradin (Horth-Holland, Amsterdam 1981) pp.243-246
1.108 H.F. Braun: Phys. Lett. *75*A, 386 (1980)
1.109 J.D. Cashion, G.K. Shenoy, D. Niarchos, P.J. Viccaro, A.T. Aldred, C.M. Falco: J. Appl. Phys. *52*, 2180 (1981)
1.110 C.U. Segre, H.F. Braun: Phys. Lett. *85*A, 372 (1981)
1.111 G.P. Meisner: Physica *108B+C*, 765 (1981)
1.112 W. Jeitschko, D. Braun: Acta Cryst. B*33*, 3401 (1977)
1.113 H. Barz, H.C. Ku, G.P. Meisner, Z. Fisk, B.T. Matthias: Proc. Natl. Acad. Sci. USA *77*, 3132 (1980)
1.114 H.F. Braun, C.U. Segre: In *Ternary Superconductors*, ed. by G.K. Shenoy, B.D. Dunlap, and F.Y. Fradin (North-Holland, Amsterdam 1981) pp.239-242
1.115 H.F. Braun, C.U. Segre: Solid State Commun. *35*, 735 (1980)
1.116 S.V. Vonsovski, Yu A. Izyumov, E.Z. Kurmaev: "*Superconductivity of Transition Metals*", in: Solid-State Sciences, Vol. 27 (Springer, Berlin, Heidelberg, New York 1982)

2. Chemistry and Structure of Ternary Molybdenum Chalcogenides

R. Chevrel and M. Sergent

With 37 Figures

The preparation and the structural properties of $M_xMo_6X_8$ (X = chalcogen) compounds are reviewed. The very large facility of substitution of M, Mo, or X elements often involves a drastic change in the physical properties. The new clusters Mo_9, Mo_{12}, $(Mo_{6/2})^1_\infty$ resulting from a condensation of octahedral Mo_6 clusters are exhibited.

2.1 Overview

In recent years, there has been considerable interest in solid-state chemistry in metal-metal bonds: large reviews have appeared on this topic including by COTTON [2.1], BAIRD [2.2], KEPERT [2.3], and KING [2.4]. It is well known that the 4d or 5d orbitals of transition elements are very extended and favour the metal-metal bond. The molybdenum element with the d_5s_1 configuration (metallic radius 1.363 Å) is one of the most favourable among the transition elements.

Many low-oxidation-state molybdenum compounds are characterized by the presence of these metal-metal bonds. They show up in the compounds starting from the IV oxidation state of molybdenum where Mo_2 pairs (i.e., MoO_2 d_{Mo-Mo} = 2.51 Å) [2.5] or Mo_3 triangular clusters (i.e., $Zn_2Mo_3O_8$ d = 2.52 Å) [2.6] are found. As the degree of oxidation decreases there is a tendency to find larger and larger clusters. In the III oxidation state of molybdenum chemistry, discrete pairs are found (e.g., α - $MoCl_3$ d = 2.76 Å) [2.7], but one finds also Mo_4 planar clusters (e.g., $Co_2Mo_4S_8$ d_m = 2.85 Å bounded to each others at an intercluster distance d = 2.96 Å) [2.8], Mo_4 tetrahedral clusters (e.g., $GaMo_4S_8$ d = 2.82 Å or $Mo_4S_4Br_4$ d =2.80 Å) [2.9,10], and Mo_6 octahedral clusters in opposite sharing edges Mo_6O_{12} units (e.g., $NaMo_4O_6$ d_m = 2.80 Å) [2.11].

The II oxidation state of the molybdenum chemistry is essentially characterized by Mo_6 octahedral clusters in Mo_6Y_8 units (Y = halogen) (e.g., Mo_6Cl_{12} d_m = 2.61 Å or $Mo_6Cl_{10}Se$ d_m = 2.62 Å) [2.12,13].

All these compounds consist of clusters of metal atoms connected together by metal-metal bonds in an iono-covalent matrix. In contrast to pure metals and alloys, the metallic coupling between these isolated clusters may be more or less weak. In such well-known materials as Mo_6Cl_{12}, $Mo_6Cl_{10}Se$ there is no coupling between the clusters, and these materials are insulators.

In this chapter, we present the ternary molybdenum chalcogenides containing molybdenum in oxidation state II, and in which a metal-metal coupling between the clusters takes place (d_{Mo-Mo} intercluster = 3.10- 3.60 Å). This coupling leads to a metallic behaviour and is the reason for many of the interesting physical properties to be discussed in the following chapters.

In Sect.2.2, we report the preparation and the characterization of the $M_xMo_6X_8$ compounds and in Sect.2.3 the structural properties. In Sect.2.4, we discuss the relation between structure and other properties of these compounds. In Sect.2.5, we report a new large family of compounds resulting from an extended condensation of the Mo_6 octahedral clusters, with the new Mo_9, Mo_{12}, $(Mo_{6/2})^1_\infty$ clusters.

2.2 Preparation and Characterization

The existence of binary molybdenum chalcogenides is well established: amorphous MoS_3; MoS_2, $MoSe_2$, α-$MoTe_2$ with molybdenite structure; and Mo_2S_3, β-$MoTe_2$ containing Mo-Mo bonds in regular zigzag chains. Only two compounds reported in the literature were not well known. MORETTE claimed these to be the compounds Mo_2Se_3 and Mo_2Te_3 [2.14]. Other authors have thought that these binaries might rather be described as Mo_3Se_4 and Mo_3Te_4 with a large range of non-stoichiometry [2.15].

In order to obtain a lower and lower oxidation state of the molybdenum atom, we need to reduce the existing binaries with the lowest degree of oxidation. This reduction may be carried out by the addition of a metallic element M, able to easily give some electrons to form ternary molybdenum chalcogenides: $M_xMo_yX_z$ (X = chalcogen).

2.2.1 Preparation

The molybdenum was used in the form of powder or chips or in the form of binary chalcogenides. The other starting materials are the chalcogens (sulphur crystals, selenium or tellurium in grains), and the M element. The latter can be used under different forms: powder, lumps, or grains if it is stable in air (3d elements), but only lumps or chips can be used under dry and deoxygenated argon in a glove box or in the form of binary chalcogenides when M is electropositive (rare earth, alkaline, alkaline earth, etc.), or in the form of ternary molybdenum chalcogenides of upper state oxidation molybdenum with alkaline metal, M_2MoS_4 (Se_4).

The mixture with appropriate proportions is prepared in a glove box, pressed into pellets, and sealed in evacuated silica tubes. Often, it was necessary to avoid contact with the silica tube during the reaction. An alumina or a molybdenum crucible is then used.

The time and the temperature of the heat treatment depend on the nature of the M element and the chalcogen, and also of the nature of the ternary molybdenum chalcogenide [Mo_6, Mo_9, Mo_{12}, $(Mo_{6/2})^1_\infty$]. Generally, the first reaction takes place between 800° and 1100° C for twelve hours, and a few further heat treatments are necessary to homogenize the product.

The products are microcrystalline black powders, stable in air. The phase characterization is carried out with the help of X-ray diffraction and density measurements. Often, small brillant black single crystals are obtained in the reacted powder allowing a complete structural study on an automatic 4-circle diffractometer, preceeded by a preliminary examination on Weissemberg, precession and Laüe chambers.

Should the powder not contain single crystals, it is possible to use crystal-growth techniques. The chemical vapor transport technique with halogens (Cl, Br, I) allows the growth of single crystals of 3-mm^3 size when M is a 3d element [2.16]. Another technique which gives the same size of single crystals is to melt the product in sealed Mo tubes [2.17]. We can also obtain single crystals by flux methods with Pb, Sn [2.18]. The adapted Bridgman-Stockbarger technique allows these compounds to be melted under argon pressures up to 100 atomospheres at temperatures of about 1550-1750°C and in some cases to obtain single crystals of an appreciable size (1-3 cm^3) [2.19] (see Chap.4).

Thin solid films have been obtained with M = Pb, Sn, Cu, Ag, La, by sputtering or evaporation [2.20] (see Chap.5). The technique of wire formation is actually in full development [2.20] (see [2.21, Chap.3]).

2.2.2 Characterization: Chalcogenides $M_xMo_6X_8$

We reported in 1971 an investigation of new ternary molybdenum sulfides [2.18] followed by the corresponding selenides in 1973 [2.22]. In the first paper, the compounds were presented with the formula MMo_nS_{n+1} where n could vary from 2 to 6. The non-stoichiometry in the S-Mo ratio and the existence of very large solid solutions of the M element in various systems were the main reasons for this formulation. In fact, if we study, for instance, the Pb-Mo-S compound, we must start from the composition $PbMo_{6.35}S_8$ to obtain a single-phase material [2.23]. If we start from the composition $Pb_xMo_6S_8$, we always end up with some impurity phases (Table 2.1, see page 61).

The exact stoichiometry still remains an open question. Structurally, MAREZIO et al. [2.24] found non-stoichiometry at sulphur and lead defects in $Pb_{0.92}Mo_6S_{7.5}$. Similarly, HAUCK [2.17] showed that the composition $Pb_xMo_6S_{8-y}$ could vary between $0.85 < x < 1.05$ and $0.6 < y < 1.2$ at 1100°C, and is shifted to lower Pb content at higher temperatures. The structure of a lead compound carried out by GUILLEVIC et al. [2.25] and by YVON [2.26] did not exhibit non-stoichiometry in sulphur. Furthermore, our results [2.16,23] and those of FLUKIGER [2.19] related to density measurements, X-ray analysis, chemical analysis, and microprobe analysis on single crystals confirm a formula close to $PbMo_{6.2}S_8$ and exclude the presence of sulphur vacancies. OPALOWSKI et al. and ROUXEL et al. [2. 15] studied the non-stoichiometry of the binary compounds Mo_6Se_8 and Mo_6Te_8. Contrary to these two binaries, the sulfide Mo_6S_8 cannot be prepared by the conventional method described earlier. It was synthesized by an indirect method, described in Sect.2.3. Since all these compounds contain a Mo_6X_8 building block but with possible non-stoichiometry in the M element, we shall use the general formula $M_xMo_6X_8$ (M = alkaline, alkaline earth, 3d element, Ag, Cd, In, Sn, Pb, Sc, Y, rare earth, actinides).

Generally, these compounds crystallize in a rhombohedral hexagonal symmetry, space group $R\bar{3}$. Often, some of these compounds present a triclinic deformation (space group $P\bar{1}$) (for instance, with 3d elements) which usually occurs when the x content increases [2.18,23]. Sometimes, similar distortion also occurs as a function of the temperature in systems with a small M cation [2.16,27].

In selenides and tellurides, it is possible to vary the content x of the M element from x = 0 to an upper limit. On the contrary, in the sulfides, the M element has a stabilizing role and a minimum concentration x is necessary.

The three binaries crystallize in a hexagonal-rhombohedral system (Table 2.2, see page 61). In order to simplify the presentation, we will classify these ternary molybdenum chalcogenides in two groups. The first one includes the phases with a large range of concentration x of the M element; the so-called non-stoichiometric compounds, $M_xMo_6X_8$. The second one includes the phases with a narrow range of concentration ($x \sim 1$) of the M element, the so-called stoichiometric compounds, MMo_6X_8.

The non-stoichiometric compounds $M_xMo_6X_8$ exist for a large number of small cations (M = Li, Mg, Cu, Zn, Cd, 3d elements). Three types of non-stoichiometric compounds are possible at room temperature:

Type I: the compounds $M_xMo_6X_8$ have the rhombohedral symmetry in the whole range of composition.
Type II: the compounds $M_xMo_6X_8$ have the rhombohedral symmetry for lower x content but change into a triclinic symmetry at a higher content of the M element.
Type III: the compounds $M_xMo_6X_8$ have the triclinic structure in the whole range of x content.

Notice that the rhombohedral angle α in the non-stoichiometric compounds is between 92.50° and 95°.

Let us first consider the sulfides.

Type I: M = Cu, Ni, Co, Li. In these compounds, the concentration x may vary continuously between two limits, $1.32 \leqslant x \leqslant 2$ for M = Co, Ni, and between $1.6 \leqslant x \leqslant 4$ for M = Cu (Fig.2.1). For M = Li, the non-stoichiometry is actually less well known. Table 2.3 (see page 62) shows the lattice parameters for the limiting compositions.

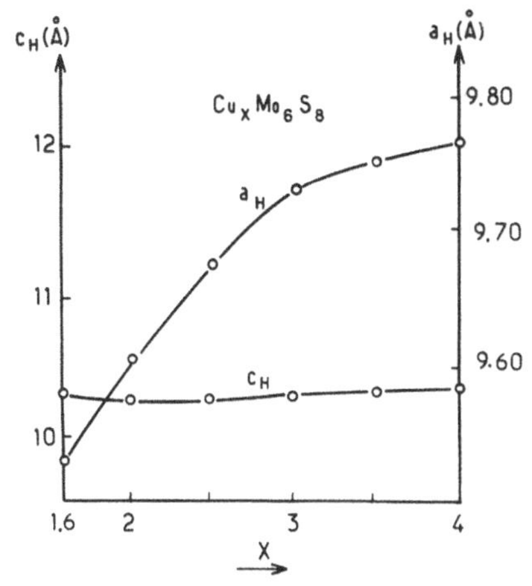

◂ Fig.2.1. Variation of hexagonal parameters of non-stoichiometric $Cu_xMo_6S_8$ compounds $(1.6 \leqslant x \leqslant 4)$ [2.18]

Fig.2.2. X-ray powder diffraction pattern of $Fe_{1.32}Mo_6S_8$ with hexagonal and rhombohedral hkl and $Fe_2Mo_6S_8$ with triclinic hkl [2.16]

▾

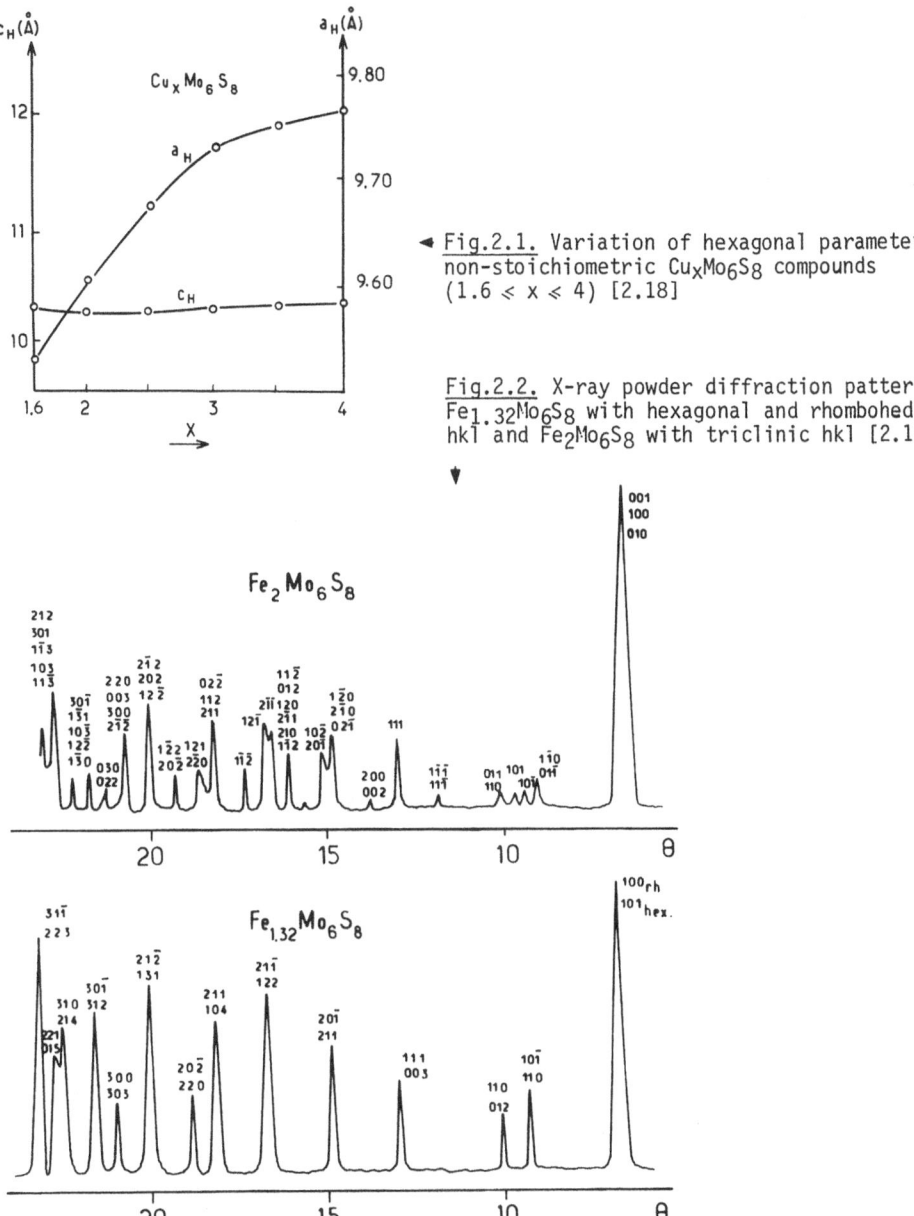

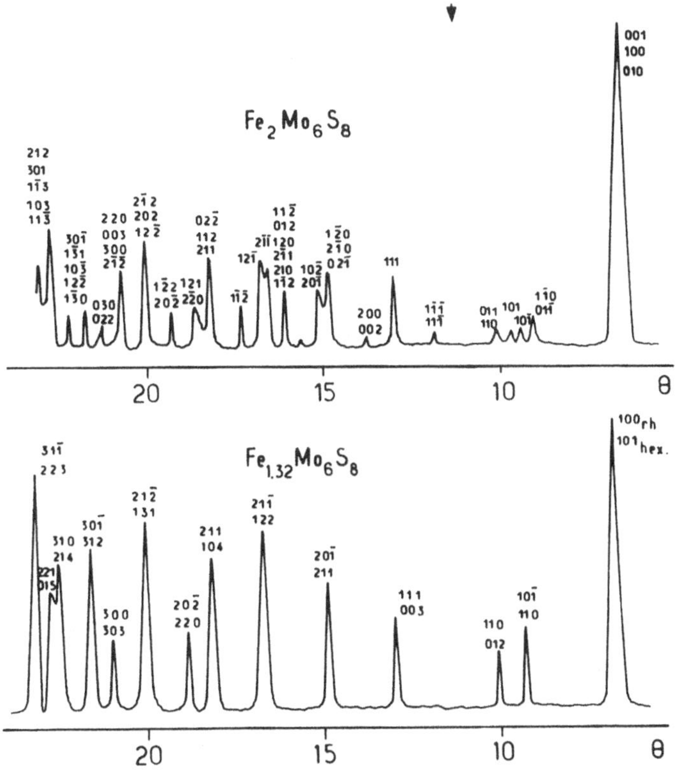

Type II: M = Fe, Mn, Zn, Cd, Mg. Figure 2.2 shows the X diagram for $Fe_{1.32}Mo_6S_8$ (rhombohedral symmetry) and $Fe_2Mo_6S_8$ (triclinic symmetry). The triclinic deformation increases with the concentration of the M element. Table 2.3 (see page 62) shows the lattice parameters for the lower compositions.

Type III: M = Cr. The non-stoichiometric compounds $Cr_xMo_6S_8$ are triclinic for all known values of x. Table 2.4 (see page 62) shows the lattice parameters of these triclinic phases.

In contrast to the sulfides, the existence of the binary Mo_6Se_8 sometimes makes it possible to vary x in the compounds $M_xMo_6Se_8$ continuously from x = 0 up to an upper limit. These upper limits are systematically lower than the ones of the sulfides.

Type I: M = Cu, Co, Li. The phases $M_xMo_6Se_8$ exist in the range for M = Cu $0 < x < 2.8$, and for M = Co $0 < x < 1.4$. For M = Li, the homogeneity domain is still unknown. Table 2.5 (see page 62) indicates the lattice parameters for the compositions at the upper limit.
It is important to note that for M = Ni, two rhombohedral domains are observed between $0 < x < 0.6$ and $1.2 < x < 1.6$ whereas a triclinic domain is observed between $0.6 \leqslant x \leqslant 1.2$,

$a = 6.73$ Å $b = 6.58$ Å $c = 6.75$ Å
$\alpha = 90.61^\circ$ $\beta = 92.17^\circ$ $= 90.95^\circ$ for $Ni_{0.66}Mo_6Se_8$.

Type II. This type does not exist in the $M_xMo_6Se_8$ phase.

Type III. M = Fe, Mn, Cr, V, Ti. These triclinic phases exist for $x \sim 1.2$. Table 2.6 (see page 63) shows the lattice parameters. For M = Mg, the symmetry is triclinic, but the composition has not been well determined yet.

The existence of binary Mo_6Te_8 makes it is possible to vary x in the compounds $M_xMo_6Te_8$ continuously from x = 0 up to an upper limit which, in contrast to the sulfides and the selenides, is very low. Furthermore, the phases exist only with small cations whereas big atoms like Pb do not enter the lattice.

Type I: M = Co, Fe. The phases $M_xMo_6Te_8$ exist in the range of $0 < x < 0.66$ for M = Co, and $0 < x < 2$ for M = Fe. Table 2.7 (see page 63) shows the lattice parameters for the upper limit compositions.

Type II. This type does not exist in the $M_xMo_6Te_8$ compounds.

Type III: M = Cu, Ni. The phases $M_xMo_6Te_8$ are triclinic in the range between $0 < x < 2$ [2.33].

Stoichiometric compounds $M_xMo_6X_8$ exist with many large M cations. All these compounds crystallize in a hexagonal-rhombohedral symmetry with the rhombohedral angle close to 90° (between 88° and 92°). One compound of this series, $SnMo_6S_7$, was reported in 1967 by ESPELUND with cubic symmetry (a = 6.53 Å) [2.34]. No compound of this type exists for X = Te. The range of solubility of the M element is low, but in many cases finite [2.19,23, 24,26,35,36a,b]. It is generally centered around x = 1. Tables 2.8,9 (see pages 64,65) summarize the lattice parameters of these compounds.

2.3 Crystal Structure

As seen in Sect.2.1, nearly all the ternary molybdenum chalcogenides crystallize in a hexagonal-rhombohedral unit cell with $a_R \sim 6.50$ Å and $\alpha_R \sim 90^\circ$. However, several phases of $M_xMo_6X_8$ where M is a small cation crystallize with a slight triclinic distortion [2.16,18]. At room temperature, the sulfides $M_xMo_6S_8$ for M = Fe, Mn, Mg, Zn, Co have a hexagonal-rhombohedral structure for small values of x but become triclinic when x increases. As discussed in Chap.4, this distortion, which is a function of x, also occurs with temperature. For instance, in the phases of type I with M = Cu, Co [2.16,42,43], the hexagonal-rhombohedral structure presents lattice instabilities and becomes triclinic at low temperatures [2.16], whereas $Fe_2Mo_6S_8$ of type II, triclinic at room temperature, becomes rhombohedral above 200°C [2.30].

The first structural investigations on these molybdenum chalcogenides were carried out by BARS et al. who determined the structure of Mo_6Se_8 [2.44], $Ni_{0.66}Mo_6Se_8$ [2.45], and $Ni_2Mo_6S_8$ [2.46]. Then, MAREZIO et al. [2.24] and GUILLEVIC et al. [2.25] determined the structure of $PbMo_6S_8$. GUILLEVIC et al. have also determined the structure of $PbMo_6Se_8$ [2.25], $Ni_{1.4}Mo_6S_8$, $Fe_{1.33}Mo_6S_8$, and $Co_{1.6}Mo_6S_8$ [2.47]. The structure of binary Mo_6S_8 and pseudobinary $Mo_6S_6Br_2$ has been determined by us [2.28,48], and later the structure of several non-stoichiometric compounds $Cu_xMo_6S_8$ was determined by YVON et al. [2.49].

With the knowledge of the phase diagram of $Cu_xMo_6S_8$ determined by FLUKIGER et al. at low temperature [2.42], the triclinic structure of $Cu_{1.8}Mo_6S_8$ was determined at 250 K [2.50]. Later, the triclinic structure of $Fe_2Mo_6S_8$ at room temperature was determined [2.30,31]. The structure of $InMo_6S_8$ with $\alpha \sim 93^\circ$ intermediary to the two classes showed the evolution of the position and of the occupancy of the M element in the interstices of the structure [2.51]. Recently, the structure of $SnMo_6S_8$ compound was determined [2.36,52] and a large structural review on molybdenum chalcogenides was presented on the subject by YVON [2.26].

2.3.1 Structural Parameters of $M_xMo_6X_8$ Compounds

All these investigations give a complete picture of the structure of the ternary molybdenum chalcogenides. All $M_xMo_6X_8$ compounds have a structure which is closely related to that of the binaries Mo_6S_8 and Mo_6Se_8. The structure is hexagonal rhombohedral (space group $R\bar{3}$). The six Mo atoms and six X atoms are located at 6f positions around the ternary axis and two X atoms are in 2c positions on this axis. The refined parameters given in Table 2.10 (see pages 66,67) are the three positional parameters x, y, z of the Mo atom and of the chalcogen atoms X(1) in 6f positions, and the x parameter of the axial chalcogen atoms X(2) in 2c positions. In Table 2.11 (see page 68), the three positional parameters for the two 6f positions of the M element, the anisotropical vibration parameters, and the occupancy factors are given.

2.3.2 Crystal Structure

The structure can be well described as a stacking of Mo_6X_8 units (Fig.2.3). Each Mo_6X_8 building block in one unit cell is a slightly distorted cube where the chalcogen atoms sit at the corners; the Mo atoms are located slightly outside of the middle of the faces of the X-atom cube. The Mo atoms in one unit form a distorted octahedron ("cluster"). This involves short Mo-Mo intracluster distances. The symmetry of the Mo_6X_8 unit is hexagonal C3i or $\bar{3}$ (ideal symmetry D3d). Building blocks of the same geometry and composition are known to occur in transition-metal halides such as in $MoCl_2 \rightarrow (Mo_6Cl_8)Cl_{4/2}Cl_2$ [2.12] where each unit is separated from each other by additional ligands. In our compounds, the clusters are arranged in a more compact manner. As can be seen from Fig.2.3, the faces of the Mo_6X_8 pseudo-cubes are not parallel to the (100) planes of the rhombohedral unit cell; we have to rotate each of these units by about 27° around the ternary axis or [111] rhombohedral axis in order to obtain the true structure. So, six corners of each Mo_6X_8 cube lies directly opposite the face centers of the adjacent cubes which means that there are close contacts between the six Mo atoms of one unit and the chalcogen atoms of the six other units. The Mo-X interunit bonds are grouped in Table 2.12 (see page 69).

Each Mo atom of a Mo_6 octahedral cluster is surrounded by five chalcogen atoms constituting a square base pyramid. This square base of chalcogen atoms is a face of a Mo_6X_8 cubic unit and the top is a chalcogen atom belonging to an opposite Mo_6X_8 unit. The six 6f chalcogen atoms of one Mo_6X_8 unit belong to square faces of a unit and so are simultaneously the summits of six pyramids. In contrast, the two 2c chalcogen atoms belong only to the square faces of the pyramid.

This particular arrangement of the Mo_6X_8 units leaves a certain number of cavities in the chalcogen atom network. The largest one has approximately a cubic shape and is formed by eight chalcogen atoms (site 1) belonging to eight different Mo_6X_8 units (Fig.2.4). This cavity is situated at the origin of the rhombohedral unit cell with point symmetry $\bar{3}$. Much smaller holes (site 2) are found close to the middle of the rhombohedral axes between two sites 1. They are more irregularly shaped and have a chalcogen environment which appears to be intermediate between a tetrahedral and a trigonal prismatic configuration.

These cavities are all interconnected and form infinite channels which run along the rhombohedral axes. They are empty in the binary Mo_6X_8 compounds and are filled by the M atoms in the $M_xMo_6X_8$ ternaries. The type of filled cavity and its occupancy depend mainly on the size of the M atom: big atoms (e.g., Pb) occupy exclusively the large hole at the origin (site 1) and yield stoichiometric compound ($x \sim 1$), whereas the small atoms (e.g., Cu) simultaneously occupy the small interstices of both sites 1,2. This partial occupancy leads to a non-stoichiometric compound ($x > 1$).

The particular arrangement of the Mo_6X_8 units leads to short distances between tow Mo atoms belonging to two neighbouring units. However, this intercluster distance

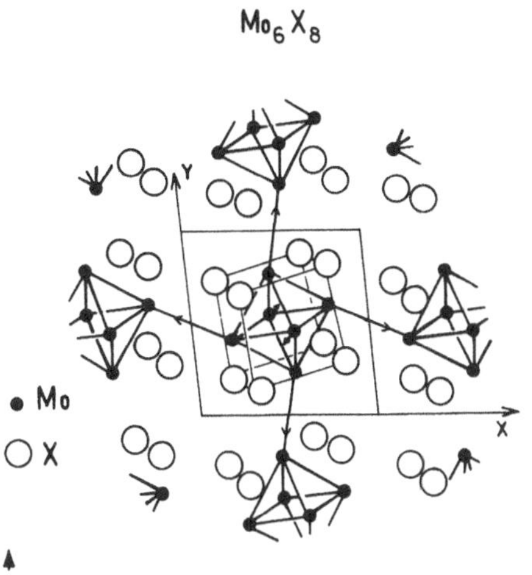

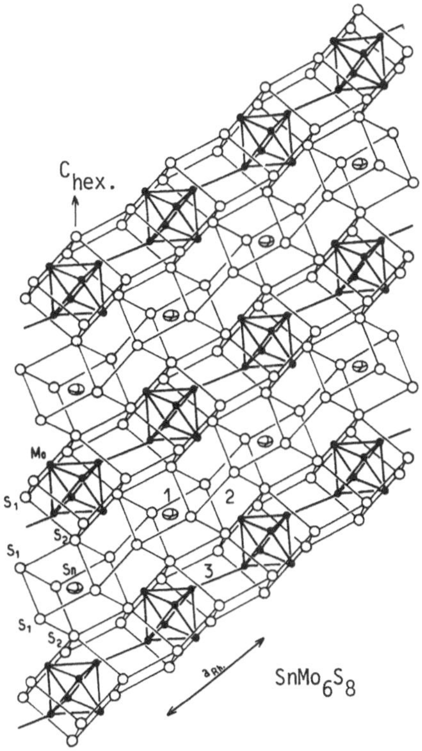

Fig.2.3. Projection of Mo_6X_8 compounds (X = S, Se, Te) on the (001) rhombohedral plane [2.28]

Fig.2.4. View of the projection of the $SnMo_6S_8$ structure on the hexagonal (11$\bar{2}$0) plane. One can see a certain number of cavities in the chalcogen atom network: sites 1,2 are partly filled by M atoms (M_1, M_2) in the $M_xMo_6X_8$ ternary compounds; the intercluster bond takes place through site 3 [2.53b]

is larger (3.10-3.60 Å) than the intracluster distance (2.65-2.80 Å). The latter is close to that of pure molybdenum (2.72 Å) (Table 2.12, see page 69). Thus, the Mo_6 octahedra form a network of intersecting linear chains running in three approximately perpendicular directions along the rhombohedral axes (Fig.2.4). In view of these results, these chalcogenides resemble another class of superconducting materials, the A15- or (β-W)-type phases which contain non-intersecting linear chains of transition-metal atoms in three perpendicular directions.

As can be seen in Table 2.12, the Mo-Mo intercluster distances vary considerably from one compound to another as a result of the clustering of the Mo atoms in spatially separated Mo_6 octahedra. Their electronic 4d wave functions are thought to be rather localized and the distance between these clusters are therefore of primary importance in the interpretation of their physical properties [2.53c-f]. As can also be seen in Table 2.12, the Mo-Mo intercluster distances are about the same in $M_xMo_6S_8$ and $M_xMo_6Se_8$, and they are about 20%-30% larger than the Mo-Mo intracluster distance.

On the other hand, the Mo-Mo intracluster distances are also important because they reflect the degree of "filling" of the conduction band due to charge transfer of the M element. These distances vary mainly as a function of the valence of the M element (see also Chap.3).

For instance, in $Cu_xMo_6S_8$, the unit cell volume increases when x increases, however the intracluster distance decreases, i.e., the Mo_6 cluster contracts. The addition of monovalent copper atoms does increase the number of electrons on the Mo_6 cluster. Thus, the resulting oxidation state of the Mo atom decreases from 2.36 ($Cu_{1.8}Mo_6S_8$) to 2.06 ($Cu_{3.66}Mo_6S_8$). The Mo_Δ-Mo_Δ) distance (Table 2.12) is always longer than the (Mo-Mo)$_\Delta$ distance, i.e., the Mo_6 octahedron is always elongated along the ternary axis. As can be seen in Table 2.12, as the Cu concentration increases, the Mo_Δ-Mo_Δ distance decreases by about 0.2 Å in the sulfides and by about 0.1 Å in the selenides, whereas d(Mo-Mo)$_\Delta$ remains practically unchanged. Thus, the Mo_6 octahedron contracts mainly along the ternary axis, becoming more regular, and the Mo atoms which were slightly above the face centres of the chalcogen cube in the $Cu_xMo_6X_8$ compounds with a low Cu concentration become almost coplanar with these faces in the Cu-rich compounds.

In the same way, the Mo_Δ-Mo_Δ intracluster distance decreases when the valence of the M cation increases. For example, the (Mo_Δ-Mo_Δ) distance is equal to 2.80 Å for $AgMo_6S_8$ (Ag^+), as compared to 2.74 Å for $SnMo_6S_8$ (Sn^{2+}) and 2.71 Å for $REMo_6S_8$ (RE^{3+}).

The Mo_6 octahedron in Mo_6Se_8 is smaller than the one in Mo_6S_8 and the Mo_6 octahedron of Mo_6Te_8 is even smaller than the one in the selenide. It seems that in these compounds sulphur has a valence -2, whereas the selenium has a valence ~ -1.8 and the tellurium has a valence ~ -1.7 [2.54]. This trend can be also seen in the fact that the insertion possibilities of a third element M are much reduced in the tellurides as compared to the sulfides.

The essential structural difference among the different compounds arises from the distribution of the M^{n+} cations in the X-atom network along the channels. The largest site (site 1) in the chalcogen atom array is located on the 3-fold axis between two Mo_6X_8 block units (Fig.2.5). This site has a pseudo-cubic environment of point symmetry $\bar{3}$. If we look at the MMo_6X_8 compounds of the second class, for example, $SnMo_6S_8$ [2.36], the tin atoms are located at the origin of the rhombohedral unit cell which is the centre of this site (Fig.2.6). This site is formed by eight sulphur atoms belonging to eight different block units: six sulphur atoms S_1 in 6f positions and two sulphur atoms S_2 on the ternary axis. This cube is already compressed along the 3-fold axis (direction S_2-Sn-S_2). The distance SnS_2 is much shorter than the distance SnS_1 (Sn-S_2 = 2.741 Å, Sn-S_1 = 3.094 Å). The environment of the Sn atom is thus 6+2.

In Tables 2.13,15 (see pages 71,72), if we compare the typical distances and angles of the MS_8 cube in $PbMo_6S_8$ and $SnMo_6S_8$, we see that the SnS_8 cube is more compressed than the PbS_8 cube. Indeed the fractional deformation of the cube can be defined by $(MS_1 - MS_2)/MS_2$ (M = Pb, Sn). The deformation percentage is 0.107 in the PbS_8 cube and 0.114 in the SnS_8 cube.

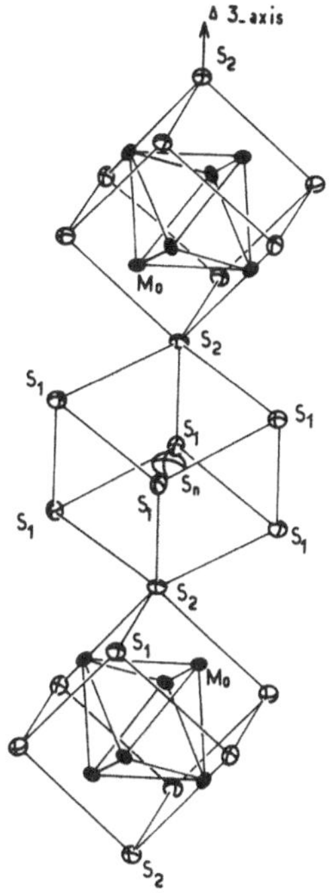

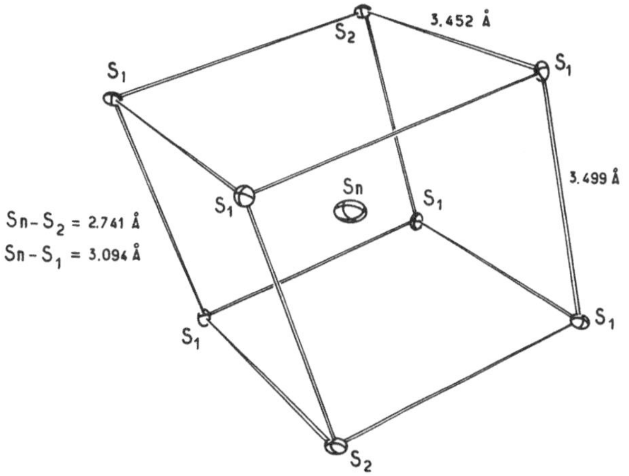

▲ Fig.2.6. A perspective view of the SnS_8 cube in $SnMo_6S_8$. The 3-fold axis goes through S_2-Sn-S_2. The thermal ellipsoids are drawn on a 50% probability scale [2.51]

◄ Fig.2.5. The stacking of Mo_6X_8 units and X_8 cubic sites along the three fold axis [2.53b]

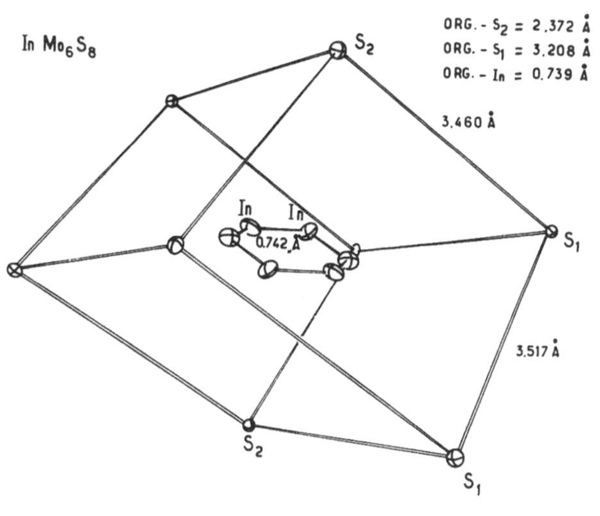

Fig.2.7. A perspective view of the InS_8 cube in $InMo_6S_8$ compound. The thermal ellipsoids are drawn on a 50% probability scale [2.51]

Furthermore, the root-mean-square displacement (RMSD) shows a more anisotropic thermal motion for tin atoms (Table 2.14, see page 71). We shall write $(\chi_\perp)$ for the RSMD perpendicular to the ternary axis and $(\chi_{||})$ for the RSMD parallel to it. We thus obtain $\Delta\chi^2 = (\chi_{||}^2) - (\chi_\perp^2) = 5.8 \times 10^{-2} > 0$ for lead and $\Delta\chi^2 = -3.16 \times 10^{-2} < 0$ for tin; the negative value for tin atoms means that the vibration ellipsoid of a tin atom is oriented perpendicular to the ternary axis (Fig.2.6).

The value of 2.3 for the ratio $(x_{\perp}^2)/(x_{\parallel}^2)$ calculated from our X-ray data at T = 300 K fits well to the Mössbauer measurements of BOLZ et al. [2.55] and differs by a factor of two from the results of KIMBALL et al. [2.56].

If we now consider the crystal ionic radii for S^{-2} ($r \simeq 1.70$ Å) and Sn^{2+} $r \simeq 1.26$ Å), the larger interatomic distance Sn-S_1 (3.094 Å) of the SnS_8 cube is still greater than the expected value by 0.134 Å. In the lead compound, there is no difference. This again accounts for the greater delocalization of the tin atom. Whether this delocalization is a true delocalization or only thermal motion cannot be determined from X-ray data. In fact, if the delocalization is less than about 0.6 Å, it is difficult to distinguish in the two phenomena. However, as we shall see later, on comparing $SnMo_6S_8$ with other $M_xM_6S_8$ compounds with smaller cations, it seems reasonable to interprete the data as a true delocalization.

Now, if we consider the S_8 cube in the $InMo_6S_8$ structure, one indium atom is located on one 6-fold equivalent crystallographic position closely grouped around the inversion centre (origin) with the occupancy of 0.166. The cube is more and more compressed and the distance origin-S_2 is equal to 2.372 Å against 2.741 Å in the case of tin. The environment of the indium atom is five sulphur atoms. Notice that the rhombohedral angle α is 93°.

The $M_xMo_6X_8$ materials (M = small atom) present a non-stoichiometry in M atoms. Their rhombohedral angle is about 94°-96°. The structure of several compounds of the first class has been determined. Site 1 (origin site), previously occupied by a big atom (Sn, Pb), is here occupied in a different way by the small atoms. In fact, the small atoms are delocalized in two 6- fold positions in two different sites: site 1 and site 2, and partially occupy both of these sites. The Table 2.15 (see page 72) indicates the occupancy of M_1 on site 1 and of M_2 on site 2. The occupancy factor depends upon the natur of the small M element.

For instance, let us take the case of the non- stoichiometric $Cu_xMo_6S_8$ compounds, where the pseudo-cube (site 1) is very distorted (origin-S_2 = 2.06 Å and origin-S_1 = 3.28 Å) (Fig.2.8). This contraction means that the rhombohedral cell becomes contracted along the 3-fold axis forcing the Cu cation to sit in a potential minimum removed from the centre of inversion symmetry. There is a first ring of six positions [Cu(I)] around [0, 0, 0] forming the inner sites. A second ring of six positions referred to as Cu(II) (outer sites) are located close to the second very distorted site centered about [0, 0, 1/2]. The interstices of type 1 that belong to different cubes are interconnected by interstices of type 2 (2.44 Å) showing a possible ionic conductivity at room temperature. In $Cu_xMo_6S_8$, only the first site is occupied at the lower limit of x ($x \simeq 1.8$). Increasing x results in an increase in the occupancy of the second site and leaves the occupation of the first site nearly unchanged (Fig.2.9). This increasing occupancy of site 2 does increase the origin-S_1 distance of site 1 with respect to the unchanged origin-S_2 distance. This distortion of site 1 can be easily seen by inspecting the angles of this pseudo-cube given in Table 2.13.

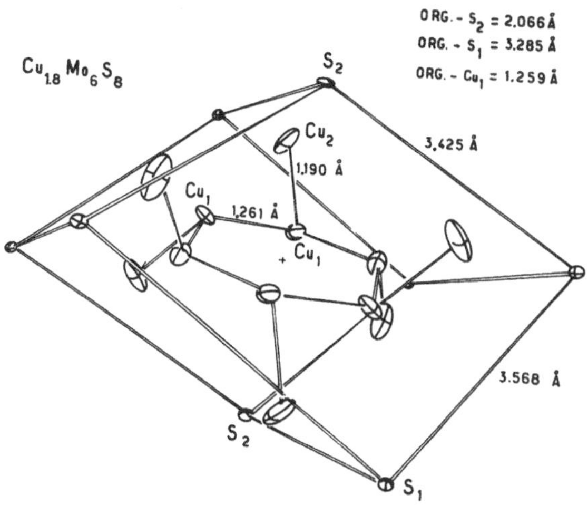

Fig.2.8. A perspective view of the six innermost positions of the site 1 and the six positions further out belonging to site 2 in $Cu_{1.8}Mo_6S_8$ compound. The thermal ellipsoids are drawn on a 50% probability scale [2.52]

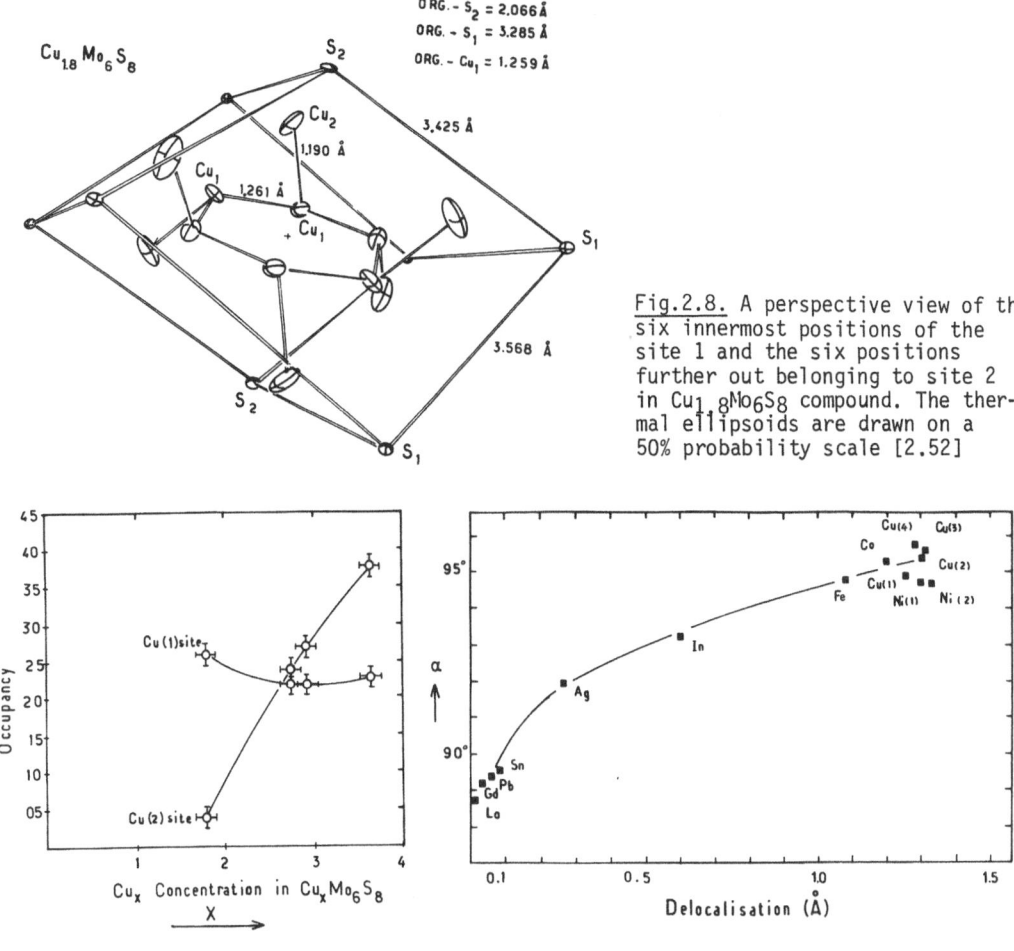

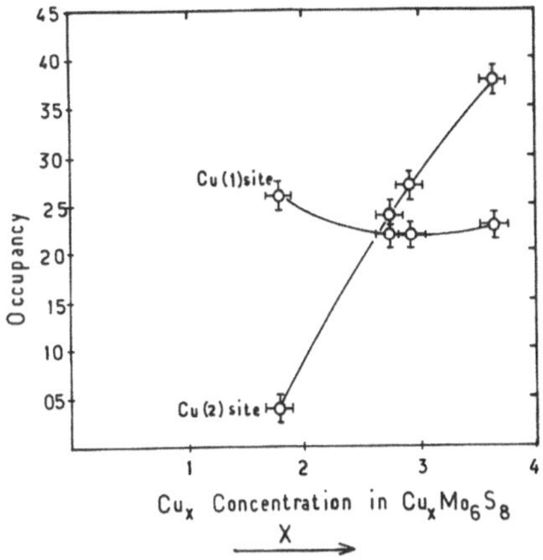

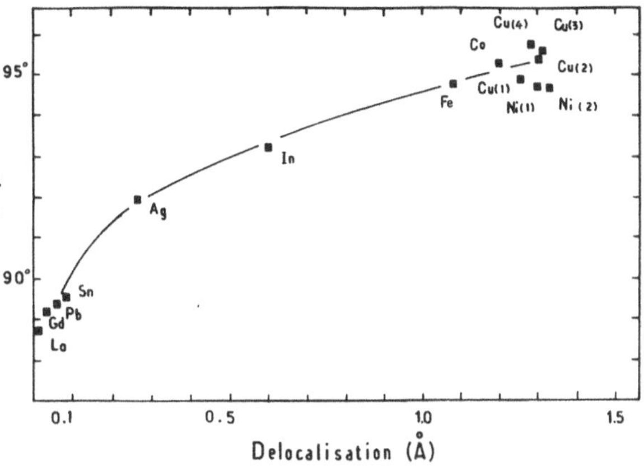

Fig.2.9

Fig.2.10

Fig.2.9. Site occupancy of Cu in site 1 and Cu in site 2 as a function of the Cu concentration in $Cu_xMo_6S_8$ [2.49]

Fig.2.10. The delocalization of the cation M as a function of the rhrombohedral angle α in $M_xMo_6S_8$ (M = Ni, Cu, Co, Fe, In, Ag, Sn, Pb, Gd, La) [2.57]

The structures of the compounds $PbMo_6S_8$ and $Cu_xMo_6S_8$ are two limiting situations. In fact, the structures yield evidence for a systematic delocalization of the cation M from the inversion centre, and YVON found a relation between the rhombohedral angle and the delocalization of the M element [2.57] (Fig.2.10).

The larger the delocalization is, the more the origin pseudo-cube is compressed and distorted along the 3-fold axis and the larger the rhombohedral angle. Consequently, the delocalization can be directly related to a simple geometrical argument involving the rhombohedral angle. The least delocalized cations are the big ones like RE, Pb, Sn. Then come silver and indium, and then the 3d elements (small cations). Since the structural instabilities occur with compounds containing the latter cations, we expect that these instabilities are correlated to the amount of delocalization.

The structure of the $Pd_{1.6}Mo_6S_8$ compound has recently been carried out [2.31]. The compound has a rhombohedral angle $\alpha = 92.40^\circ$ inferior to one of the $InMo_6S_8$ compound. The delocalization of the Pd atom is much more important. The origin-Pd distance is equal to 1.527 Å and the occupancy is equal to 0.27. Thus, the formula of the compound is $Pd_{1.6}Mo_6S_8$ [2.31,33].

2.3.3 The Triclinic Phase

As we saw in Sect.2.1, the great majority of $M_xMo_6X_8$ chalcogenides have trigonal symmetry at room temperature. However, some $M_xMo_6X_8$ compounds, for M = small cations, have a triclinic symmetry. This triclinic structure is merely a simple distortion of the rhombohedral symmetry. Three triclinic structural phases are actually well known: $Cu_{1.8}Mo_6S_8$ [2.50], $Fe_2Mo_6S_8$ [2.30], and $Ni_{0.66}Mo_6Se_8$ [2.44].

The rhombohedral positions 6f (Mo, X_1) split into three positions of 2-fold multiplicity in the space group $P\bar{1}$. The positional parameters of the M, Mo, and S atoms and their occupancy are given in Table 2.16 (see page 74). The structure of the triclinic $Cu_{1.8}Mo_6S_8$ compound was determined at 250 K, whereas the two other structural investigations were carried out at room temperature.

In Table 2.17 (see page 75), the mean Mo-Mo intracluster distances, the mean Mo-Mo intercluster distances, the mean Mo-X intercluster distances, and the mean M-Mo distances are given. These distances are close to the corresponding values found in the rhombohedral compounds.

Of the two sites for the M atoms in the rhombohedral structure, site 1 remains in the triclinic structure, although slightly distorted. In $Cu_{1.8}Mo_6S_8$ (below 250 K), it contains (occupancy = 0.9) pairs of copper atoms at d_{Cu-Cu} = 2.58 Å, a distance close to that found in the corresponding rhombohedral phase (2.52 Å) (Fig.2.11a). Each copper atom is surrounded by four sulphur atoms forming a distorted tetrahedron.

The same site 1 is occupied in the structure of the $Fe_2Mo_6S_8$ phase at room temperature. The iron atoms form pairs at 2.51 Å and have an approximately square-pyramidal configuration (Fig.2.11b). In the $Ni_xMo_6Se_8$ triclinic phase, the "filling" of the channels is completely different. In the $Cu_xMo_6S_8$ compounds, the three equivalent sites (2) are equally occupied by the copper atoms (Cu 2). In the $Ni_{0.6}Mo_6Se_8$ compound, one of these distorted sites (2) is occupied by the nickel atoms, which are coordinated by four selenium atoms in a distorted tetrahedral configuration (Fig.2.11c).

The reversible structural transition thus appears to be a "freezing" of small cations in the channels of the structures of $Cu_xMo_6S_8$ and $Fe_xMo_6S_8$ with a slight distortion of the octahedral cluster. The temperature of the phase transition varies with the nature and the content x of the M small cation.

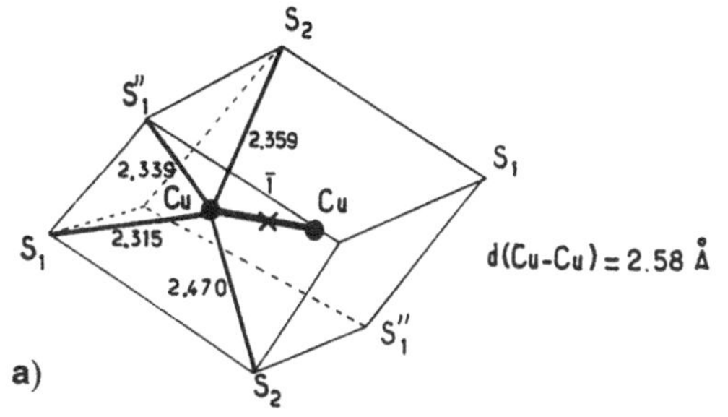

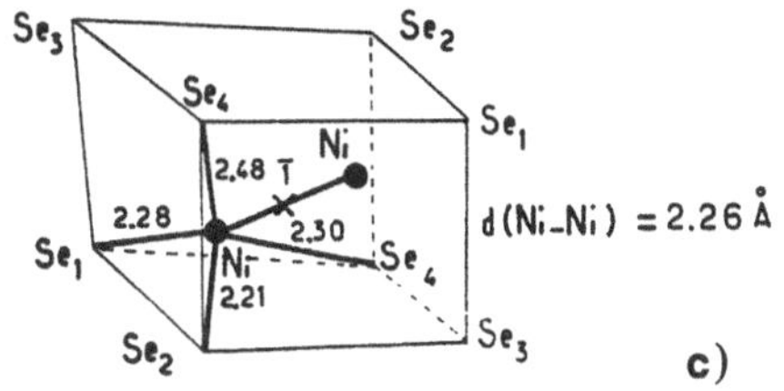

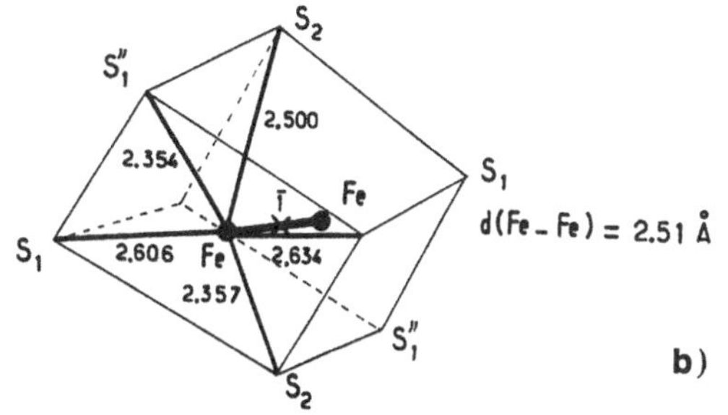

Fig.2.11a-c. Configuration of small cations (Cu, Fe, Ni) on the sites 1 and 2 of the triclinic compounds [2.30,45, 50]. (a) The Cu in origin site (000) (site 1) of the low-temperature phase $Cu_{1.8}Mo_6S_8$; (b) The Fe in origin site (000) (site 1) and (c) Ni in site 2 (001/2) of the room-temperature phases, $Fe_{1.68}Mo_6S_8$ and $Ni_{0.66}Mo_6S_8$

2.4 Relations Between Structure and Properties

The results of the band structure calculations fit well with the electronic and chemical properties of new materials found of the same type structure (see Chap.6 and [Ref.2.21, Chap.6]).

In fact, it had been suggested that the superconductivity in the $M_xMo_6X_8$ compounds is due to the 4d electrons of molybdenum atoms and that electronically one may consider these compounds as being formed by weakly coupled Mo_6X_8 rigid units and by M atoms. Moreover, band calculations for $PbMo_6S_8$ and other chalcogenides made by ANDERSEN [2.58], MATTHEISS [2.59], BULLET [2.60] and JARLBORG [2.61] confirm that the band structure near the Fermi level is essentially composed by narrow d bands corresponding roughly to the molecular levels of the Mo_6X_8 units. The reason why the narrowness of the bands occurs near E_f is the non-bonding $d_{x^2-y^2}$ character of the corresponding orbitals.

One result of these calculations is that the sulphur p band is filled, i.e., the sulphur atom has the valence -2. So the position of the Fermi level, or the number of d electrons on the cluster Mo_6, is related to the charge transfer from the M element to the Mo_6 cluster. Since sulphur is found have the valence -2, the eight sulphur atoms take 16 electrons from the six molybdenum atoms, and 20 electrons per Mo_6 cluster remain in the d band in Mo_6S_8. A characteristic property of these compounds is that the Mo_6 octahedron is formed with less than the 24 valence electrons found in the insulator $MoCl_2$, and that this number may vary continuously from 20 to almost 24 electrons. The deficiency of bonding electrons is greatest for Mo_6S_8 ($4e^-$)

and the octahedron is found to be most distorted. The addition of Cu atom to Mo_6S_8 increases the number of valence electrons for bonding, through the charge transfer from Cu to Mo and this is a reason why the Mo_6 octahedron contracts and becomes more regular (as in $MoCl_2$).

In $PbMo_6S_8$, the Pb atom is divalent, giving its two 6p e^- to the Mo_6 4d bands, resulting in 22 e^- on the Mo_6 cluster or 3.66 e^-/Mo atom, the so-called cluster VEC introduced by YVON [2.26]. Thus, the effect of the third element M in $M_xMo_6S_8$ is not only to push the Mo_6S_8 units apart but also to modify strongly the charge on the Mo_6 cluster, i.e., to modify the position of the Fermi level in the 4d band. So, one is led to believe that a necessary condition for the formation of the $M_xMo_6S_8$ compound is to have a number of d electrons on the Mo_6 cluster superior to 20 electrons.

2.4.1 Stabilization of Mo_6S_8

As we have seen in Sect.2.2, the two binaries Mo_6Se_8 and Mo_6Te_8 can be obtained by direct synthesis, by reaction of the elements at 1150°C and 860°C, respectively. On the other hand, the M elements entering the channels of $M_xMo_6S_8$ stabilize Mo_6S_8. The direct synthesis of Mo_6S_8 at 1150°C gives a mixture of Mo_2S_3 and Mo metal. The binary Mo_6S_8 has so far only been produced by indirect methods: by "leaching" in diluted hydrochloric acid [2.28] or by electrochemical oxidation [2.62-64] of the $M_xMo_6S_8$ compound with the M atom as a small cation (M = Mg, Fe, Co, Ni, Cu, Li, etc.). The oxido-reduction reaction may be described by:

$$M_x^{n+}Mo_6S_8 + nx(H^+, Cl^-) \rightarrow Mo_6S_8 + x\,(M^{n+}, n\,Cl^-) + nx/2\;H_2\uparrow .$$

solid	in aqueous hydrochloric acid	solid	liquid	gas

Thus, $(Mo_6S_8)^{xn^-} + nx\,H^+ \rightarrow Mo_6S_8 + nx/2\;H_2\uparrow$.
The oxidation state of the M^{n+} ion remains unchanged:

$$nx\,H^+ + nx\,e^- \rightarrow nx/2\;H_2\uparrow$$
$$(6\,Mo)^{(16-xn)^+} \rightarrow nx\,e^- + (6\,Mo)^{16+}$$

For example, for $MgMo_6S_8$ compound, n = 2, x = 1 (Mg^{2+}).

a) $2H^+ + 2\,e^- \rightarrow H_2\uparrow$

b) $6(Mo^{2.33+}) \rightarrow 2\,e^- + 6(Mo^{2.66+})$ (octahedral Mo_6 cluster) .

Specifically, the oxido-reduction reaction is an oxidation of the molybdenum atoms with the escape of hydrogen gas. The redox topotactic reaction (b) is reversible by electrochemistry. The reduction reaction takes place with the insertion of small cations.

The oxido-reduction reaction in an aqueous hydrochloric acid has been carried out on a single crystal of $Ni_2Mo_6S_8$ compound. The initial single crystal broke into several crystals. The structure of Mo_6S_8 was determined using one of these crystals [2.28]. The crystallographic data are given in Tables 2.10,12. The increasing Mo_Δ- Mo_Δ distances of Mo_6S_8 compared to $Ni_2Mo_6S_8$ verify the oxidation of the Mo_6 cluster. This compound decomposes at 470 C. Before decomposition, it forms a very unstable new variety β-Mo_6S_8 with the following parameters [2.16]:

a_R = 6.46 Å, α_R = 89°46, a_H = 9.10 Å, c_H = 11.30 Å, V_H = 810 Å^3

The α-Mo_6S_8 form has the crystalline parameters:

a_R = 6.432 Å, α_R = 91°34, a_H = 9.20 Å, c_H = 10.88 Å, V_H = 797 Å^3.

The metastable α-Mo_6S_8 form is very reactive. In fact, it is possible to make new compounds by inserting new M elements [2.28,65] such as aluminium, in a low-temperature (~300° C) reaction:

$Al_{x\simeq 2}Mo_6S_8$: a_R = 6.48 Å, $_R$ = 95°83, a_H = 9.62 Å, c_H = 10.02 Å,
V_H = 803 Å^3.

In the same way, it is possible to modify the range of concentration x of the $M_xMo_6S_8$ compounds. In $Cu_xMo_6S_8$, the x range now begins at x = 0 and goes to x = 4.

It is possible, by the same method, to obtain the binaries Mo_6Se_8 or Mo_6Te_8. These compounds, obtained by indirect synthesis, have the same lattice parameters as the ones obtained by direct synthesis. The solid solution Mo_6Se_8-Mo_6S_8 is possible by direct synthesis from Mo_6Se_8 up to the limit composition $Mo_6Se_4S_4$: a_H = 9.10 Å, c_H = 10.61 Å, V_H = 827 Å^3 (Fig.2.12, solid lines). The indirect synthesis gives, for the composition $Mo_6Se_4S_4$ different crystalline parameters: a_H = 9.39 Å, c_H = 10.89 Å, V_H = 831 Å^3 (Fig.2.12, dotted lines). At high temperatures (~700°C), this second phase (dotted lines) changes into the first phase (solid lines). From the phases $Mo_6Se_{8-x}S_x$ obtained by indirect synthesis for x > 4, it is possible to obtain a new series of metastable phases (dashed line) by annealing. The tranformation temperature varies from 700°C to 470°C for the $Mo_6Se_4S_4$ and Mo_6S_8 compounds, respectively. Figure 2.12 shows the variations of the hexagonal parameters of these solid solutions.

2.4.2 Mixed Anion: Substitution of Halogen for Chalcogen

The binary Mo_6S_8 with 20 electrons per Mo_6 cluster is not stable (decomposition at 470°C). As we suggested above, a way of stabilizing this compound may be to add electrons to the Mo 4d band. This can be done by substituting halogens for the sulphur atoms. The substitution of one sulphur by a halogen will add one electron to the d band. The compounds $Mo_6S_6Br_2$ and $Mo_6S_6I_2$ were synthetized in silica tubes using MoY_2 (Y = Br, I), Mo and S as starting materials (Table 2.18, see page 75).

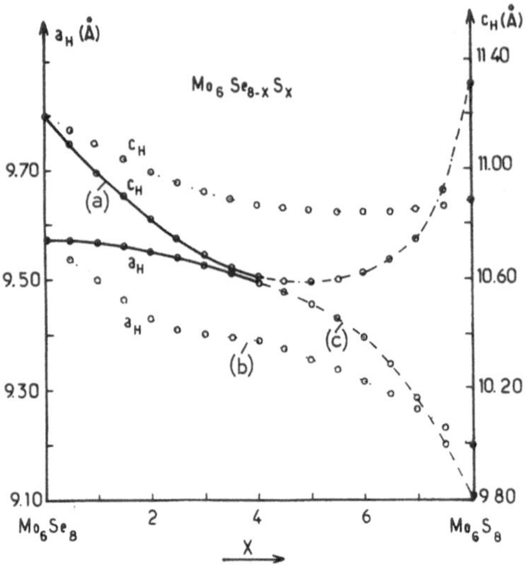

Fig.2.12a-c. Solid solutions in the pseudo-binaries $Mo_6Se_{8-x}S_x$: variations of the hexagonal crystalline parameters with x; (a) binaries made by direct synthesis (——); (b) binaries made by indirect synthesis at room temperature (····); (c) binaries obtained by annealing of the indirect synthesis phase (—·—·—·)

These compounds have 22 d electrons per Mo_6 cluster, the same number as found in $PbMo_6S_8$. The superconducting critical temperature is equal to 13.8 K and 14.0 K in $Mo_6S_6Br_2$ and $Mo_6S_6I_2$, respectively, and is very close to the critical temperature of $PbMo_6S_8$ (14 K) [2.54].

The structure of $Mo_6S_6Br_2$ [2.48] was determined from single crystal data. This compound is isostructural with Mo_6S_8 (space group $R\bar{3}$), but also has crystallographic order between the sulphurs and the bromines. Indeed, the six sulphur atoms are located on the 6f site whereas the two bromine atoms are located on the 2c site on the ternary axis (Table 2.19, see page 75). Inspection of Table 2.20 (see page 76) shows furthermore that the Mo_6 octahedral cluster is the most regular which has been found so far.

Why are these two compounds stable at high temperature? The answer is given in Fig.2.13. The pseudo-cube origin (S_6Br_2) is the most distorted one found in the $M_xMo_6S_8$ phases. It is so compressed along the 3-fold axis that the two bromine atoms touch one another along the ternary axis leading to a van der Waals bond (3.914 Å) (Fig.2.13). This distortion of this pseudo-cube looks like the one found in the $M_xMo_6S_8$ compounds (M = small cation) [2.48]. Here, the stability of the phase is related to the delocalization of the M cations, i.e. , to the M-S bonds inside this pseudo-cube origin. Likewise, the stability of the Mo_6Se_8 and Mo_6Te_8 binaries is explained by the close contacts Se_1-Se_1 and Te_1-Te_1 in the pseudo-cube origin. Similarly, the Mo_6S_8 compound is unstable because of the weak contacts S_1-S_1 (Fig.2.13). The "$Mo_6S_6Cl_2$" compound has not yet been synthetized. This would involve a very big distortion of this pseudo-cube due to Cl-Cl contacts.

The insertion of small cations has been realized in the host structure $Mo_6S_6Br_2$ and $Mo_6S_6I_2$ [2.48,67]. For instance, $Cu_{1.5}Mo_6S_6Br_2$ has been synthetized with the parameters a_R = 6.53 Å, α_R = 94°68, a_H = 9.60 Å, c_H = 10.34 Å, V_H = 825.3 Å^3.

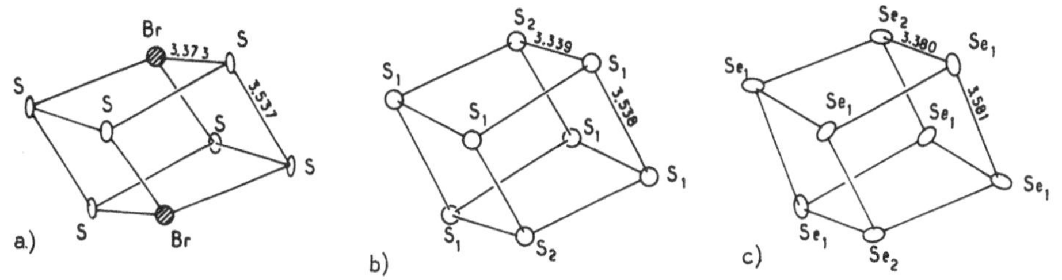

Fig.2.13a-c. Origin cubic site for (a) $Mo_6S_6Br_2$; (b) Mo_6S_8; (c) Mo_6Se_8

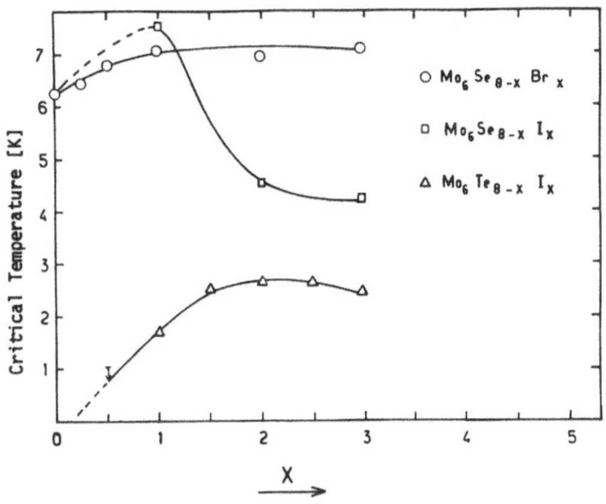

Fig.2.14. Critical temperature T_c, as a function of halogen concentration x [2.54]

In the selenides and the tellurides the situation is different since the $Mo_6X_{8-x}Y_x$ solid solutions exist on both sides of x = 2 ($0 \leqslant x \leqslant 3$). However, the fact that the variation of the lattice parameters saturates above x = 2 may indicate that the halogens also preferentially occupy the 2c sites here [2.54]. The substitution of bromine or iodine for selenium in Mo_6Se_8 results in an increase in T_c up to 7.1 K for Mo_6Se_7Br and 7.6 K for Mo_6Se_7I (Fig.2.14). The compound Mo_6Te_8 is not superconducting above 1 K and so far no compound in the tellurides has been found to be superconducting. However, the substitution of iodine for tellurium in Mo_6Te_8 makes the compound $Mo_6Te_6I_2$ superconducting with a maximum T_c equal to 2.6 K. The lattice parameters and critical temperatures for binary and pseudo-binaries are reported in Table 2.18 (see page 75).

2.4.3 Substitution on the Octahedral Cluster $(Mo, Me)_6$ (Me = Nb, Ta, Re, Ru, Rh)

Another interesting feature of the ternary molybdenum chalcogenides is the possibility of making new compounds with mixed octahedral $(Mo, Me)_6$ clusters. These new materials are all isostructural to the Mo_6X_8 type structure (Table 2.21, see page 76). The three first materials in Table 2.21 are semiconducting and are stoichiometric compounds. In fact, one finds 24 electrons per octahedral cluster corresponding to a gap predicted by the electronic band-structure calculations. The other

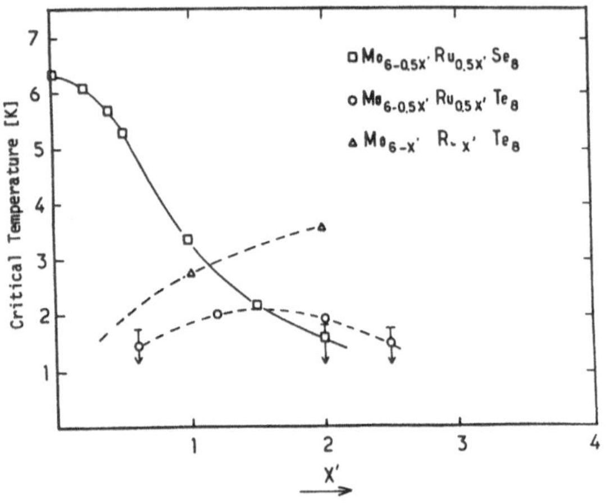

Fig.2.15. Superconducting critical temperature for $Mo_{6.0.5x'}$ $Ru_{0.5x'}X_8$ (X = Se, Te) and $Mo_{6-x'}Re_{x'}Te_8$. x' represents the number of electrons added to the Me_6 cluster by the substitution of Re or Ru for Mo. Note the relations x' = 2x [2.66]

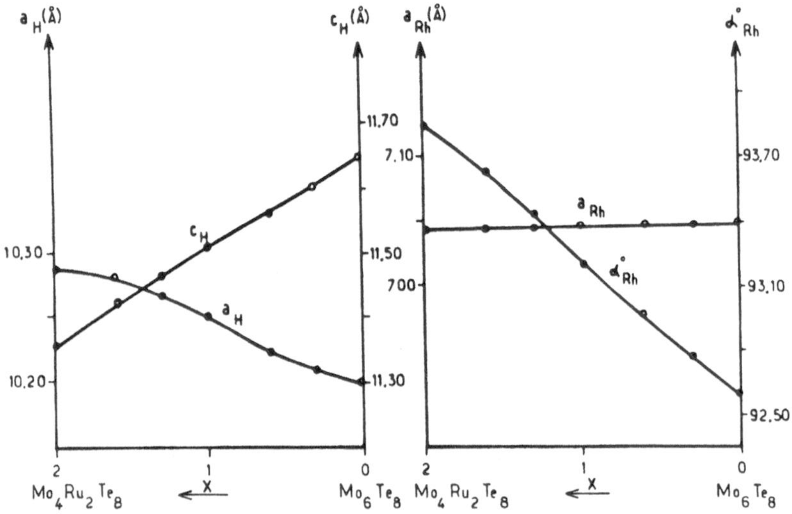

Fig.2.16. Hexagonal and rhombohedral lattice parameters for the solid solution $Mo_{6-x}Ru_xTe_8$ [2.66]

phases in Table 2.21 have a metallic behaviour, and the $Mo_4Re_2Te_8$, Mo_5RuTe_8 phases are even superconducting (Fig.2.15), having formally 22 electrons per cluster.

The metallic solid solution $Mo_4Ru_2Te_8$-Mo_6Te_8 on the octahedral cluster is continuous (Fig.2.16) from Mo_6Te_8 to $Mo_4Ru_2Te_8$. This example can be extended to others; it seems possible to make a continuous metallic solid solution from substitution on the octahedral cluster with two metallic limits.

For the solid solution between semiconducting $Mo_4Ru_2Se_8$ and metallic Mo_6Se_8, the situation is different. The diagram of the variations of lattice parameters (Fig.2.17) shows two single-phase regions with an inhomogeneous domain between them. A region with a large homogeneity range near Mo_6Se_8 exists because of its metallic behaviour. In this range, the rhombohedral parameter a_R is practically

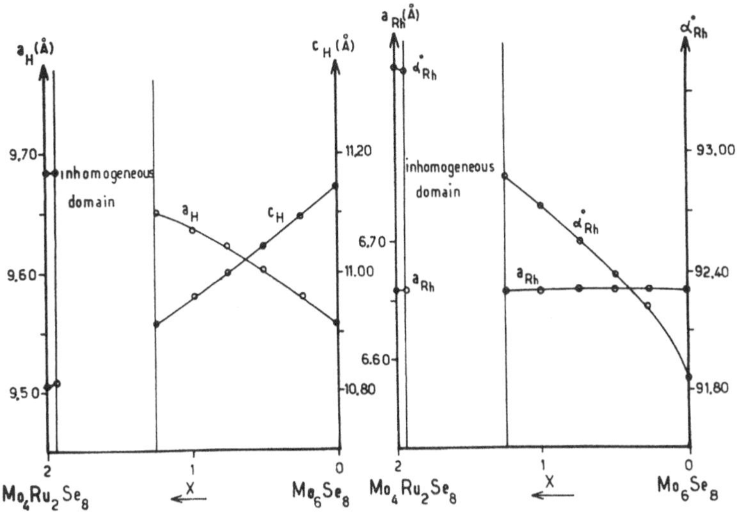

Fig.2.17. Hexagonal and rhombohedral lattice parameters for the solid solution $Mo_{6-x}Ru_xSe_8$ [2.66]

constant whereas the rhombohedral angle α_R increases as we replace Mo by Ru. This is consistent with the assumption that α_R is correlated with the charge transfer.

It does not look possible to make a continuous solid solution from substitution on the octahedral cluster with two different limits, semiconducting and metallic.

Since $Mo_4Ru_2Se_8$ is semiconducting and $Mo_4Ru_2Te_8$ is metallic, it seemed to us interesting to study the solid solution ($Mo_4Ru_2Se_{8-x}Te_x$) thus substituting for the chalcogens in keeping the composition of the octahedral cluster constant. It turns out that the solid solution is continuous in the whole domain $0 \leq x \leq 8$. The resistance measurements show a transition from a semiconducting to a metallic phase between $1 < x < 1.3$ (Fig.2.19).

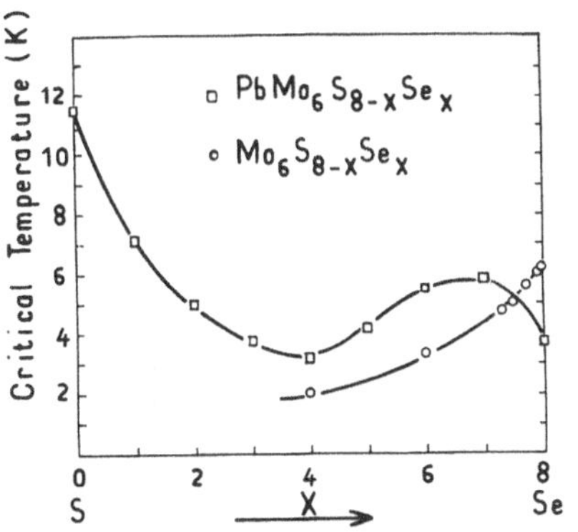

Fig.2.18. T_c vs the concentration of x in the solid solutions $PbMo_6S_{8-x}Se_x$ and $Mo_6S_{8-x}Se_x$ [2.69]

2.4.4 Chalcogen-Chalcogen Substitution (S, Se, Te)

The solid solutions of substitution chalcogen-chalcogen have been carried out in $PbMo_6(X, X')_8$ [2.69,70] and $SnMo_6(X, X')_8$ [2.71] compounds. These investigations indicate that there exists a correlation between the critical temperature and the rhombohedral angle α (or c/a). In general, one observes a sharp drop in T_c when one chalcogen is replaced by another and T_c is minimum around the composition $MMo_6X_4X'_4$. It is generally thought that this behaviour is related to a substitution of one chalcogen for another on the site 6f. That should destroy the 3-fold point symmetry except, for example, in $PbMo_6Se_6S_2$ in which an order may take place between selenium atoms in 6f positions and the last remaining sulphur atoms in 2c special positions, corresponding to a maximum in the curve of T_c (Fig.2.18).

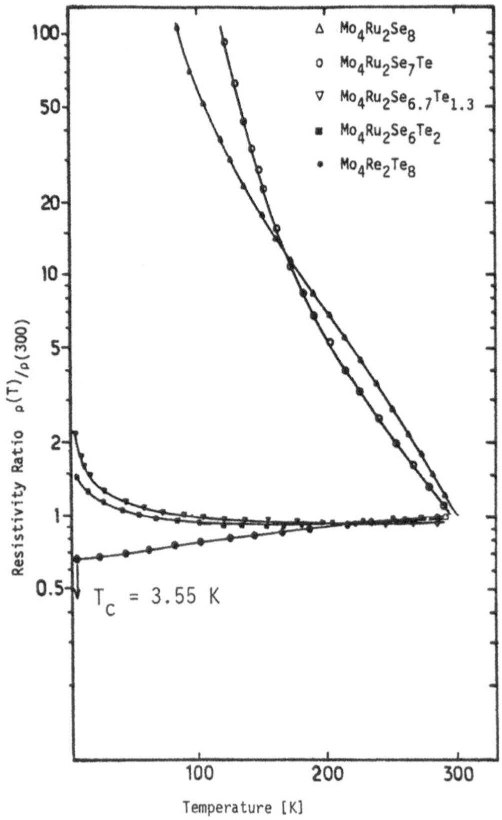

Fig.2.19. Resistance ratio $\rho(T)/\rho(300)$ for $Mo_4Ru_2Se_{8-y}Te_y$ (y = 0, 1, 1.3, 2) and $Mo_4Re_2Te_8$ [2.66]

2.4.5 Substitution of Oxygen for Sulphur

Very recently, UMARJI et al. claimed the synthesis of new materials $M_xMo_6S_6O_2$ ($M_x = Co_2, Ni_2, Cu_2$, Pb), $M_xMo_6S_{8-x}O_x$ with $x < 2.0$ of the same type structure [2.72]. These compounds were synthetized by the high-temperature solid-state reaction. The pseudo-binary $Mo_6S_6O_2$ was obtained in the same way as Mo_6S_8, by the acid dissolution from $M_xMo_6S_6O_2$ (M = Cu, Ni). The lattice parameters and some T_c values are presented in Table 2.22 (see page 76).

2.4.6 Addition of MM'_x and Substitution of (M, M') on an M Element in the Channels

The compound $PbCu_xMoS_8$ [2.23], $SnZn_xMo_6S_8$ [2.73] and $SnFe_xMo_6S_8$ [2.73,74] have been studied so far. As seen in Table 2.23, the x insertion of small cations in the big cation compounds (Pb, Sn) increases the rhombohedral angle. In fact, it has been proved that the small cations (Fe) occupy the site 2 analogous to the sites occupied by the Cu(2) atoms [2.74]. The iron ions in the $SnFe_{0.4}Mo_6S_8$ compound form an octahedron around site 1, but are external to site 1 (d_{Fe-Fe} = 2.98 Å, 3.07 Å). The distance Fe-Fe is equal to 2.61 Å between two equivalent interstices of type II, i.e., between two octahedra. The iron ion has a square face pyramidal environment of sulphur atoms (4x $Fe-S_1$ = 2.43, 2.52, 2.75 and 2.61 Å and $Fe-S_2$ = 2.25 Å). The Sn atom sits at the cubic origin site (cf. Fig.2.4) at the centre of the Fe octahedron.

Many results have been reported about the insertion of big cations (RE^{3+}) in $PbMo_6S_8$ and $SnMo_6S_8$ compounds [2.23,36]. For instance, the $PbLu_xMo_6S_8$ compound is shown in Fig.2.20. When the x content increases, the critical temperature increases. In fact, we believe that this insertion is the result of increasing the occupancy of the origin site up to the optimal value of unity.

Some of the substitutions big cation-big cation (Fig.2.20) and small cation-small cation have been carried out [2.23,36,71,73]. In constrast, it has not yet been possible to realize the substitution small cation-big cation. This can be explained by their different occupancy in the channels of the structure.

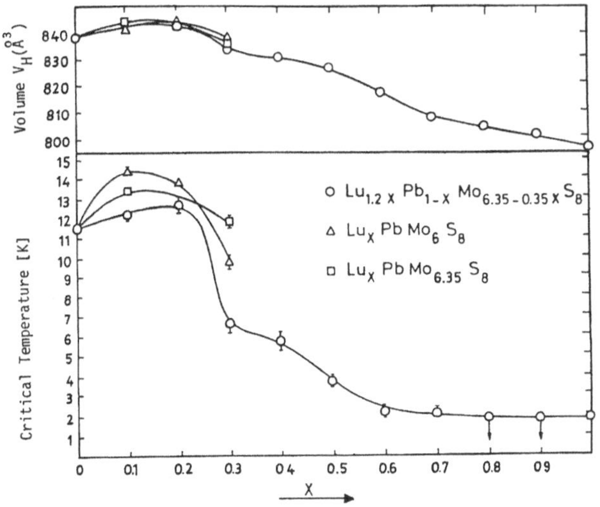

Fig.2.20. Critical temperature and volume vs the compositions $Lu_xPbMo_6S_8$ and $Lu_{1.2x}Pb_{1-x}Mo_{6.35-0.35x}S_8$ [2.23]

2.5 New Ternary Compounds Resulting from Linear Condensation of the Octahedral Mo_6 Clusters

The low valence chemistry of molybdenum is characterized by the presence of metal-metal bonds in the form of clusters. Here we present the preparation and the characterization of compounds containing new Mo_9X_{11}, $Mo_{12}X_{14}$ or $(Mo_{6/2}X_{6/2})^1_\infty$ units with new Mo_9, Mo_{12} and $(Mo_{6/2})^1_\infty$ clusters resulting from the linear condensation of octahedral Mo_6 clusters [2.37].

2.5.1 A New Cluster Mo_9 in a Mo_9X_{11} Unit (X = S, Se)

The cluster Mo_9 can be considered as the condensation of two octahedral Mo_6 clusters (Fig.2.21). The X_{11} chalcogen-polyhedron of the Mo_9Se_{11} unit is made up of six square faces like the ones found in the Mo_6X_8 units and six triangular faces. The molybdenum atoms form a bioctahedron, i.e., two octahedra which are fused by sharing a face along the 3-fold axis leading to a new Mo_9 cluster. Each Mo atom of the Mo_3 medium plane has six neighbouring Mo atoms of the metal cluster and faces four chalcogen atoms forming two edge-sharing triangles. Each extremal Mo atom forming the lower and higher levels of the Mo_9 cluster has an environment like one of the Mo atoms of the $M_xMo_6X_8$ compound; it has four Mo neighbours on the Mo_9 cluster and four chalcogen neighbours belonging to a square face of a Mo_9X_{11} unit and one external chalcogen atom belonging to a neighbouring unit forming the top of the square-base pyramid.

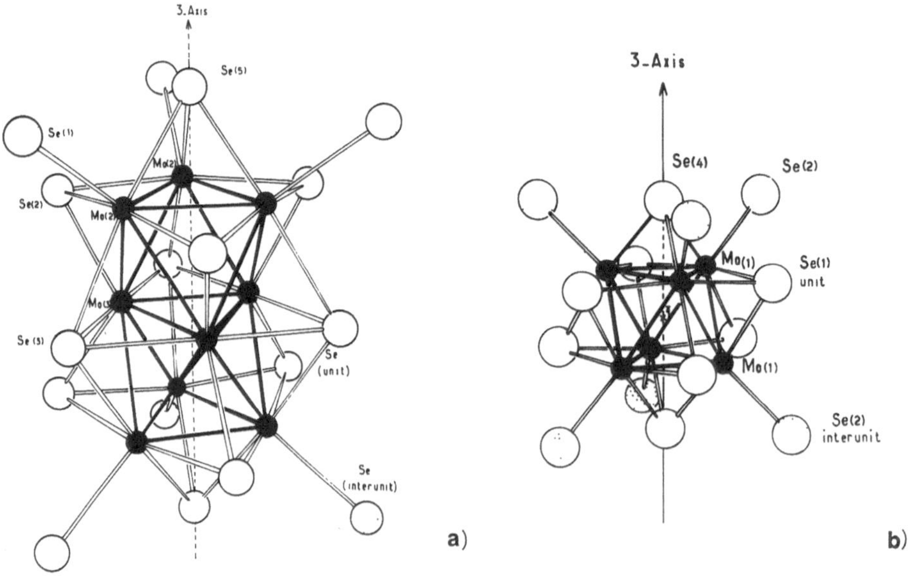

Fig.2.21a,b. The new Mo_9 cluster in Mo_9Se_{11} unit (a) compared with the Mo_6Se_8 unit (b) in the $M_xMo_{15}Se_{19}$ compounds [2.37,75]

The synthesis of the $In_{\sim 3}Mo_{15}Se_{19}$ phase was carried out using In and Se grains and Mo powder. The reaction took place in an evacuated silica tube at 1100°C for 25 hours followed by several heat treatments. Special care was taken in order to minimize contamination of the sample by the silica: the reactions were carried out in alumina or molybdenum crucibles [2.76]. The compound $In_{\sim 3}Mo_{15}Se_{19}$ decomposes into $In_2Mo_{15}Se_{19}$ and impurities like InSe above 1250°C. New materials with this latter formula have also been produced by direct synthesis [2.77].

The results of X-ray investigation and density measurements show the existence of a single-phase region in the In-Mo-Se system for the composition $In_xMo_{15}Se_{19}$, with 2.9 < x < 3.4. This new phase crystallizes with a hexagonal unit cell, space group $P6_3/m$. In Fig.2.22, the X-ray powder diagrams of $InMo_6S_8$ and $In_{3.33}Mo_{15}Se_{19}$ are compared.

The $M_2Mo_{15}X_{19}$ phases crystallize in a hexagonal-rhombohedral symmetry. The space group is $R\bar{3}c$. The X-ray powder diffraction data of $In_2Mo_{15}Se_{19}$ are shown in Fig.2.22. In Table 2.24 (see page 77) the lattice parameters of these compounds are given.

In the structure of $In_{\sim 3}Mo_{15}Se_{19}$ [2.75] there are two different crystallographic positions for the In atoms. One position In(1) is occupied by one In atom in a large seven-coordinated cavity (Fig.2.23a), the other position In(2) corresponds to a triangular group of six-coordinated cavities like the ones of the rhombohedral $InMo_6S_8$ phase. These cavities are only partially occupied (0.9 and 1.3 In atoms per triangular group for $In_{2.9}Mo_{15}Se_{19}$ and $In_{3.3}Mo_{15}Se_{19}$, respectively). In the structure of $In_2Mo_{15}Se_{19}$ [2.82] there exists only one position for the In atoms in a large eleven-coordinated cavity (Fig.2.23b). In Table 2.25 (see page 78) the fractional atomic coordinates of these two series of compounds are reported.

The new phases are characterized by the presence of Mo_6Se_8 units and new Mo_9Se_{11} units.

The phase $In_xMo_{15}Se_{19}$ may be described with the formula: $In_xIn_2(Mo_6Se_8)(Mo_9Se_{11})$ with 0.9 < x < 1.4. The projection onto the $(11\bar{2}0)$ hexagonal plane is shown in Fig.2.24a. Two kinds of units can be seen. The stacking of these units turns out to be constituted by two distinct columns along the hexagonal c axis: one column is formed by the stacking Mo_6Se_8-In(2)-Mo_6Se_8. The indium atom In(2) is delocalized and occupies statistically three equivalent positions in three one-edge-sharing octahedral sites; a second column is constituted by a stacking Mo_9Se_{11}-In(1)-In(1)--Mo_9Se_{11}. Both indium atoms In(1), grouped together in pairs, are located in a site of symmetry 3/m with 14 neighbouring selenium atoms belonging to six Mo_6Se_8 and three + two Mo_9Se_{11} units (Fig.2.23a,24a).

The $In_2Mo_{15}Se_{19}$ compound may be described by the structural formula $In_2(Mo_6Se_8)(Mo_9Se_{11})$. As can be seen in Fig.2.24b, the stacking of the Mo_6Se_8 and Mo_9Se_{11} units is quite different here. We find, in the (001) direction, a stacking sequence Mo_6Se_8-In-Mo_9Se_{11}-In. The In^{1+} cation is situated on the ternary axis and

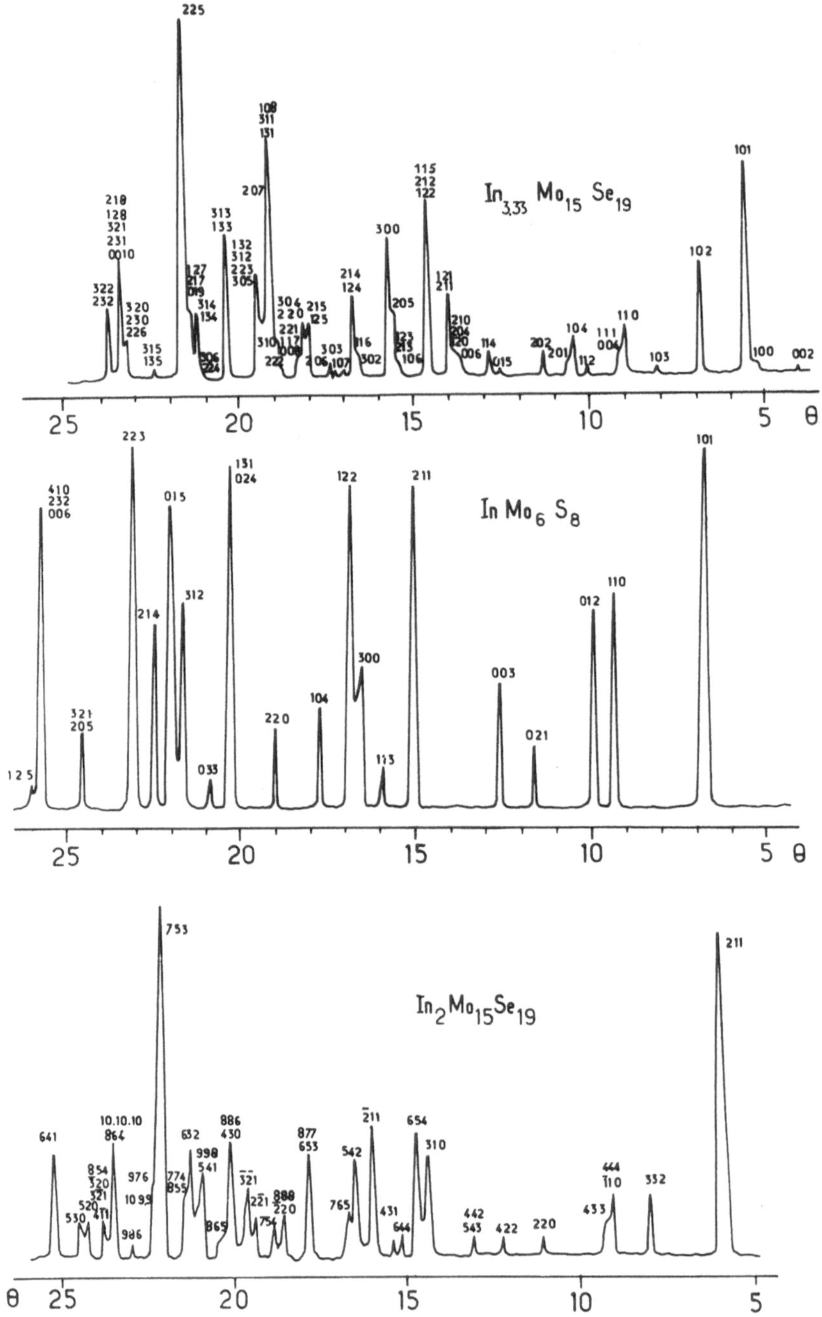

Fig.2.22. X-ray powder diffraction patterns for $In_{3.33}Mo_{15}Se_{19}$, $InMo_6S_8$ with hexagonal hkil and $In_2Mo_{15}Se_{19}$ with rhombohedral hkl ($CuK\bar{\alpha}$ = 1.541 Å) [2.78]

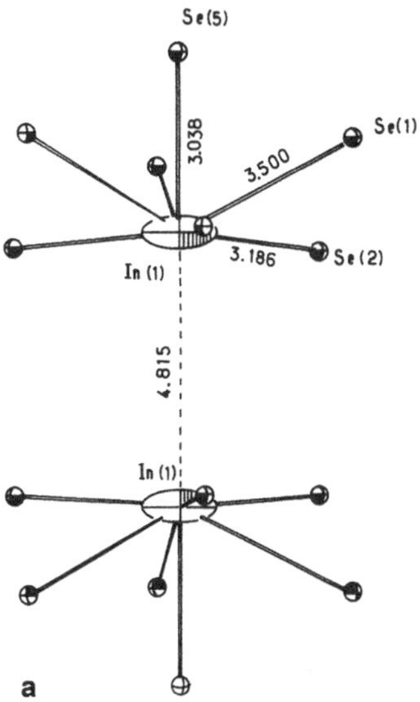

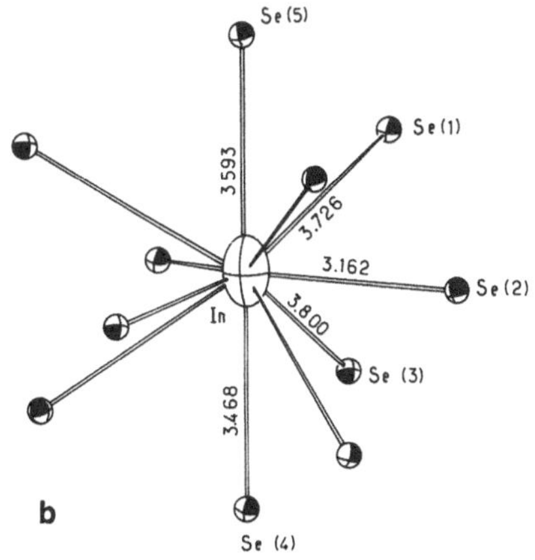

Fig.2.23a,b. Environment of indium cations in $In_3Mo_{15}Se_{19}$ (a) and $In_2Mo_{15}Se_{19}$ (b) compounds [2.79]

is surrounded by eleven selenium atoms belonging to three Mo_9Se_{11} units and three Mo_6Se_8 units on both sides, and by one Mo_6Se_8 unit and one Mo_9Se_{11} unit at the upper and the lower levels (Fig.2.23b,24b). Here, the indium atoms are situated in a three-dimensional network of channels, similar to the ones found in the MMo_6S_8-type compounds as in Fig.2.24c.

The intercluster Mo-Mo bonds in these phases are always realized between two Mo atoms belonging to adjacent square faces of selenium atoms. One Mo_9Se_{11} unit is bonded by six intercluster Mo-Mo bonds to six neighbouring Mo_6Se_8 units and conversely. However, the symmetry of this arrangement is different in the two series of compounds. As can be seen in Fig.2.25, the symmetry of the intercluster bonds in $In_2Mo_{15}Se_{19}$ is 32 whereas in $In_3Mo_{15}Se_{19}$ it is 3/m. Indeed the Mo_9Se_{11} building block has the 32 symmetry in the $In_2Mo_{15}Se_{19}$ compound and is bonded by the six intercluster Mo-Mo bonds equal to 3.39 A, whereas the Mo_9Se_{11} unit has the 3/m symmetry in $In_{\sim 3}Mo_{15}Se_{19}$ compounds and is bonded by the six intercluster Mo-Mo bonds equal to 3.50 Å (cf. Table 2.26, see page 79).

In the $M^{n+}Mo_6X_8$ compounds, it has been established that the Mo-Mo intertriangle distance in the Mo_6 clusters decreases as the number of electrons per cluster increases [2.49]. Comparing the Mo_6Se_8 clusters in $In_{\sim 3}Mo_{15}Se_{19}$ and $In_2Mo_{15}Se_{19}$, one can see that the same behaviour is true for the new compounds (Table 2.26). Now, if one looks at the Mo_9 clusters, it is found that an increasing charge transfer on this cluster leads mainly to an increase in the size of the median triangle. When one goes from $In_2Mo_{15}Se_{19}$ to $In_{3.3}Mo_{15}Se_{19}$, the Mo-Mo distance of the median triangle increases from 2.681 to 2.768 Å showing that the same tendency is found here. The overall effect of an increased number of electrons on the Mo clusters is to reduce c_H (contraction of the Mo_6 cluster) and to increase a_H (expansion of the

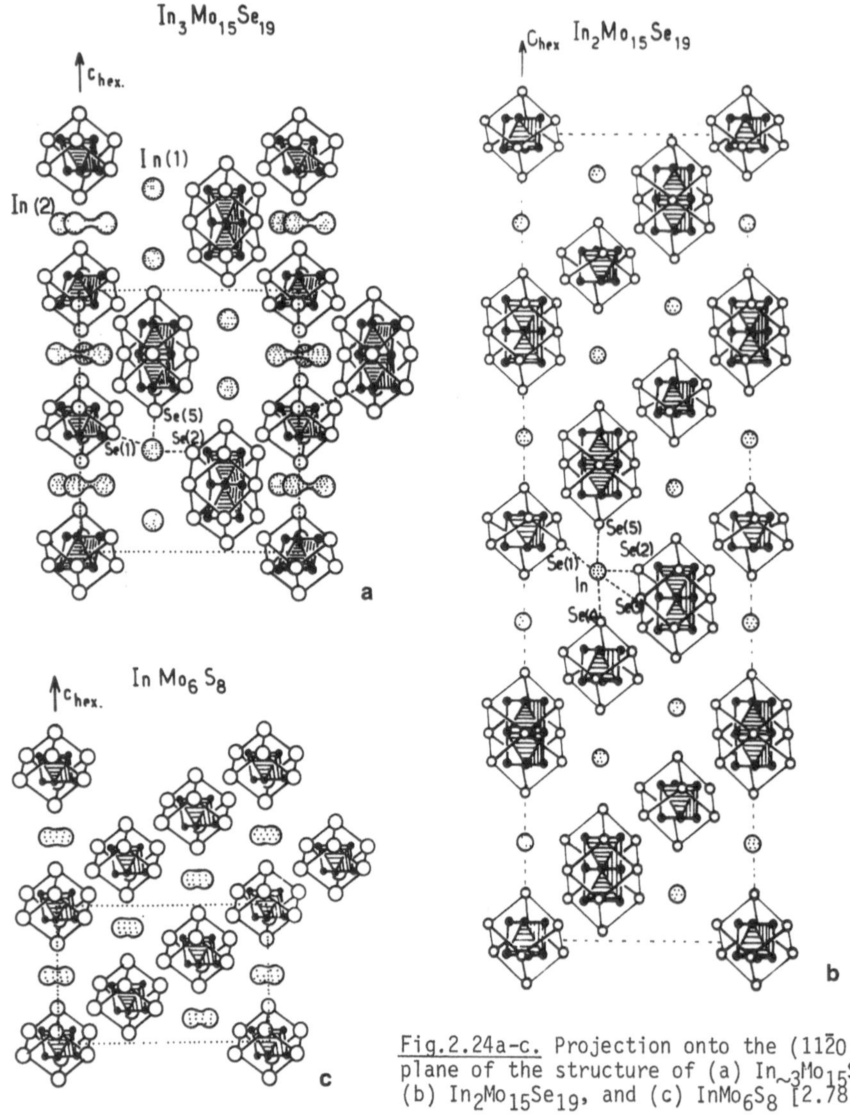

Fig.2.24a-c. Projection onto the $(11\bar{2}0)$ hexagonal plane of the structure of (a) $In_{\sim 3}Mo_{15}Se_{19}$, (b) $In_2Mo_{15}Se_{19}$, and (c) $InMo_6S_8$ [2.78]

Mo_9 cluster). This type of variation can be seen in Table 2.24 where Ba is the only divalent M element in $M_2Mo_{15}Se_{19}$.

The substitution of sulphur for selenium is possible in a large domain in $In_{\sim 3}Mo_{15}Se_{19}$, but has not been studied in $In_2Mo_{15}Se_{19}$. Variation of the lattice parameters for the $In_{3.33}Mo_{15}Se_{19-x}S_x$ solid solution is shown in Fig.2.26. A phase of the type $In_{\sim 3}Mo_{15}Se_{19}$ exists in the domain $0 < x < 12$ (Fig.2.26). For $X = 12$, a narrow two-phase region containing the $In_{\sim 3}Mo_{15}Se_{19}$-type phase and a rhombohedral $InMo_6S_8$-type phase appears [2.53a]. Above $x = 12$, only the latter remains and belongs to a solid solution of the $InMo_6S_{8-x}Se_x$ type.

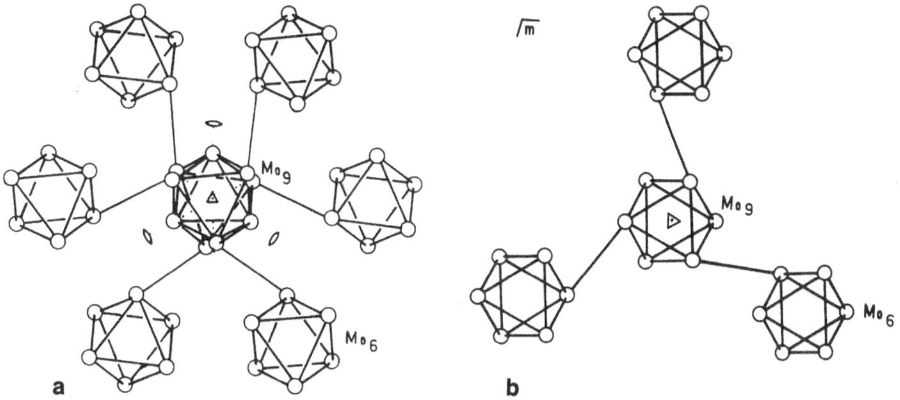

Fig.2.25a,b. Symmetry of the intercluster bonds from Mo_9 cluster: (a) symmetry 32 (D3) in $In_2Mo_{15}Se_{19}$; (b) symmetry 3/m (C3h) in $In_3Mo_{15}Se_{19}$ [2.77]

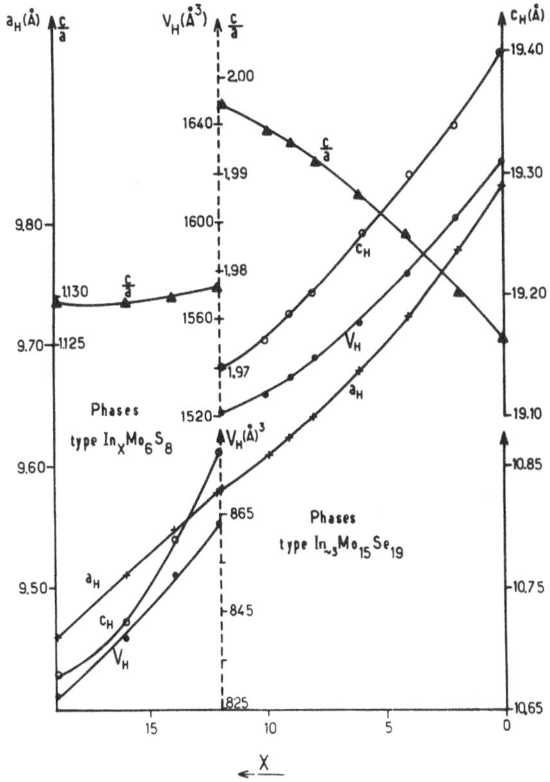

Fig.2.26. Lattice parameters for the solid solution $In_{3.33}Mo_{15}Se_{19-x}S_x$ [2.76]

These phases are metallic and are found to be superconducting (Table 2.27, see page 80) [2.83]. The phase $In_{2.9}Mo_{15}Se_{19}$ is a high field superconductor [2.83]. The initial slope $(dH_{c2}/dT)_{T_c}$ varies between 68 kG/K and 78 kG/K. The values are about 30% higher than that one found for $PbMo_6S_8$ and slightly above the value for

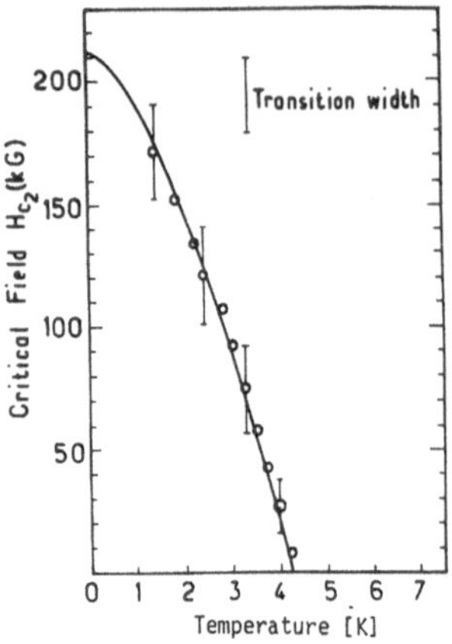

Fig.2.27. Temperature dependence of the critical field H_{c2} for $In_{2.9}Mo_{15}Se_{19}$ [2.80]

$LaMo_6Se_8$ [2.79] (Fig.2.27). For the phase $M_2Mo_{15}Se_{19}$, the initial slopes are only half of that found in $In_3Mo_{15}Se_{19}$. For example, the measurements of $(dH_{c2}/dT)_{T_c}$ for $K_2Mo_{15}Se_{19}$ and $Ba_2Mo_{15}Se_{19}$ yield 34 kG/K and 32 kG/K, respectively.

2.5.2 A New Cluster Mo_{12} in $Mo_{12}X_{14}$ Unit (X = S)

The new cluster Mo_{12} can be considered as a linear condensation of three octahedral Mo_6 clusters (Fig.2.28). This cluster has the $\bar{3}$ symmetry like one of Mo_6X_8 unit. The fourteen chalcogen atoms are of two different types. Eight chalcogen atoms bridge a triangular face of Mo atoms like the eight chalcogens of the Mo_6X_8 units, and six new chalcogen atoms bridge two triangular Mo faces sharing one edge; this edge is perpendicular to the 3-fold axis. The X_{14} polyhedron forms twelve triangular faces and six square faces of chalcogen atoms.

The synthesis of these new compounds $M_2Mo_9S_{11}$ (M = K, Tl) was carried out in the same way as the previous phases.

Single phases were obtained for the formula $M_2Mo_9S_{11}$ (M = K, Tl). These compounds crystallize with the space group $R\bar{3}$ in a rhombohedral hexagonal symmetry [2.81]. Table 2.28 (see page 80) reports their lattice parameters, and in Fig.2.29 the X-ray powder diagram of $Tl_2Mo_9S_{11}$ is shown.

Table 2.29 (see page 80) gives the rhombohedral crystallographic coordinates of the $Tl_2Mo_9S_{11}$ compound [2.82]. The $M_2Mo_9S_{11}$ phases are characterized by the Mo_6S_8 units and the new $Mo_{12}S_{14}$ units. The formula can be described as $Tl_4(Mo_6S_8)(Mo_{12}S_{14})$. The projection of the structure onto the hexagonal $(11\bar{2}0)$ plane presents a stacking sequence of the (Mo_6S_8) and $(Mo_{12}S_{14})$ units along the 3-fold axis, (Mo_6S_8)-Tl-Tl--$(Mo_{12}S_{14})$-Tl-Tl- (Fig.2.30). This sequence of chains (Mo_6S_8)-$(Mo_{12}S_{14})$ is similar to the stacking (Mo_6X_8)-(Mo_9X_{11}) found in $M_2Mo_{15}X_{19}$. Figure 2.30 shows the existence of channels running along three directions like the ones found in the MMo_6S_8- and $M_2Mo_{15}X_{19}$-type materials. These channels are occupied by two thallium atoms (Fig. 2.31). The two Tl^+ ions are situated on the ternary axis and are surrounded by

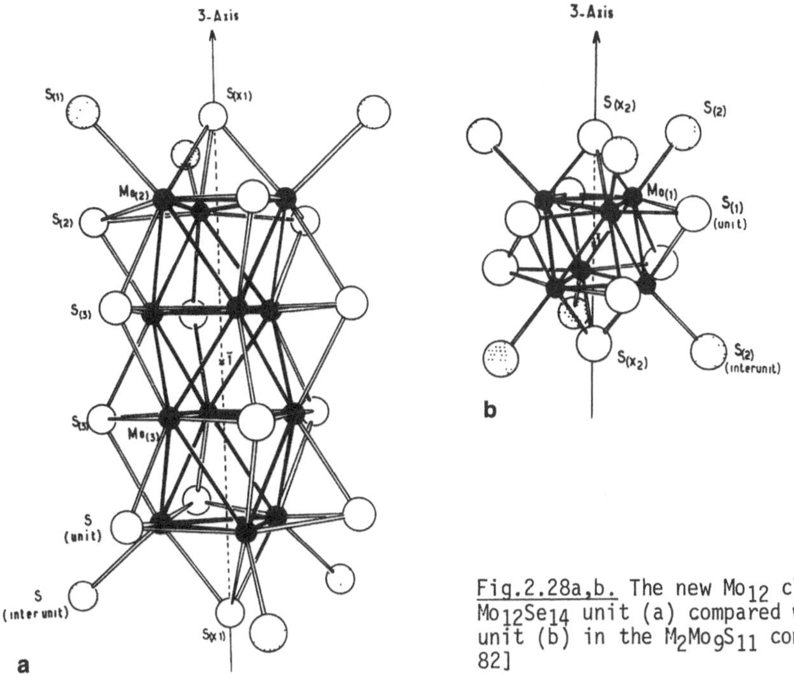

Fig.2.28a,b. The new Mo_{12} cluster in the $Mo_{12}Se_{14}$ unit (a) compared with the Mo_6S_8 unit (b) in the $M_2Mo_9S_{11}$ compound [2.81, 82]

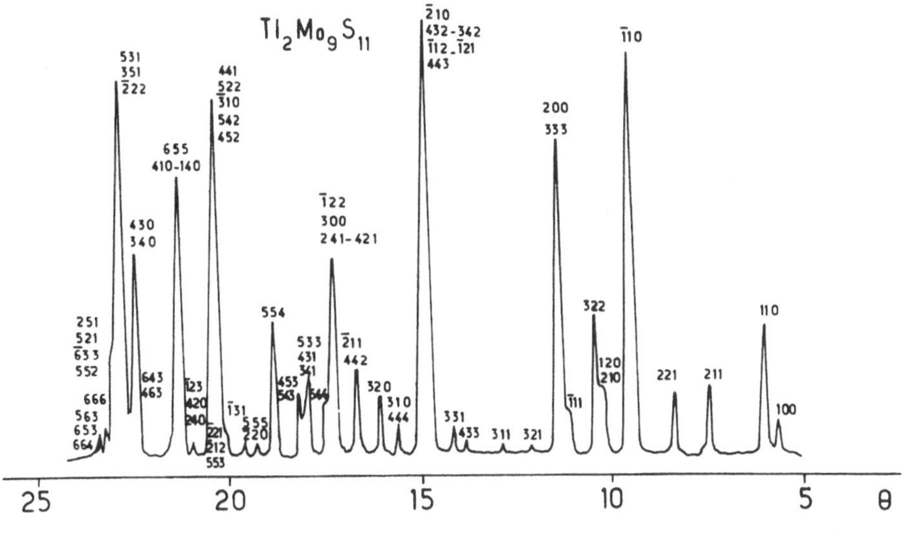

Fig.2.29. X-ray diffraction pattern of the $Tl_2Mo_9S_{11}$ compound (CuK$\bar{\alpha}$ = 1.541 Å) with rhombohedral hkl Miller indices [2.78,81]

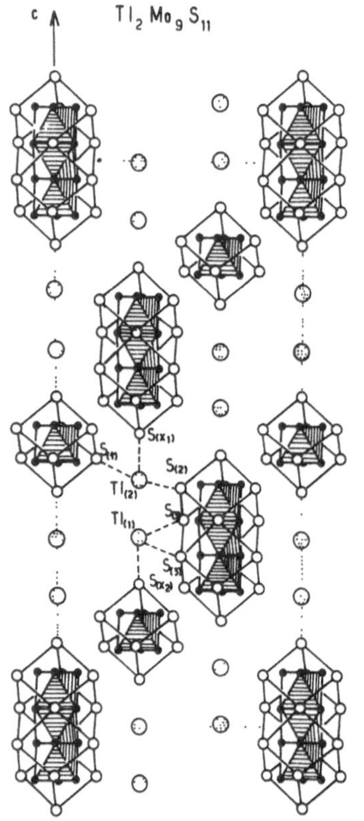

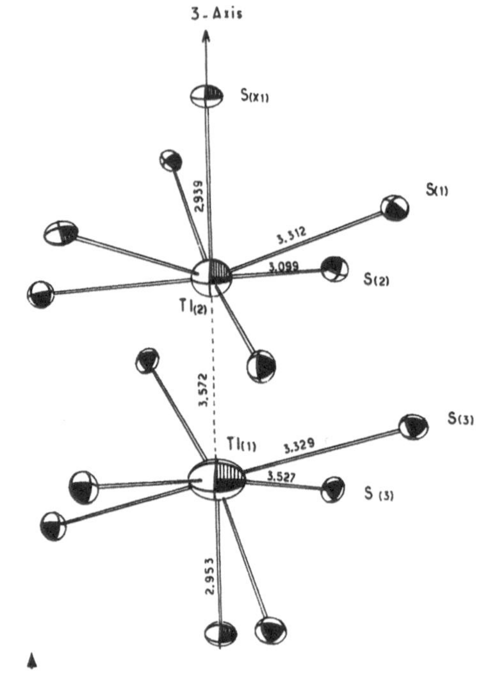

Fig.2.31. Environment of the thallium atoms in $Tl_2Mo_9S_{11}$ [2.82]

◂ Fig.2.30. Projection onto the (11$\bar{2}$0) hexagonal plane of the $Tl_2Mo_9S_{11}$ compound [2.82]

fourteen sulphur atoms belonging to three $Mo_{12}S_{14}$, three Mo_6S_8 units around the 3-fold axis, one Mo_6S_8 and one $Mo_{12}S_{14}$ unit on the 3-fold axis.

In these compounds the six molybdenum atoms located in extreme positions inside the $Mo_{12}S_{14}$ unit are each further bonded to one sulphur atom of a neighbouring Mo_6S_8 unit. The six atoms which are dotted in Fig.2.28 are the nearest sulphur atoms of the six neighbouring units, and their bonds with the extremal Mo atoms symbolize the inter-unit Mo-S bonds. The intercluster Mo-Mo bonds are realized in the same manner in the $Tl_2Mo_9S_{11}$ compound as in the MMo_6S_8-type compound (Fig.2.32).

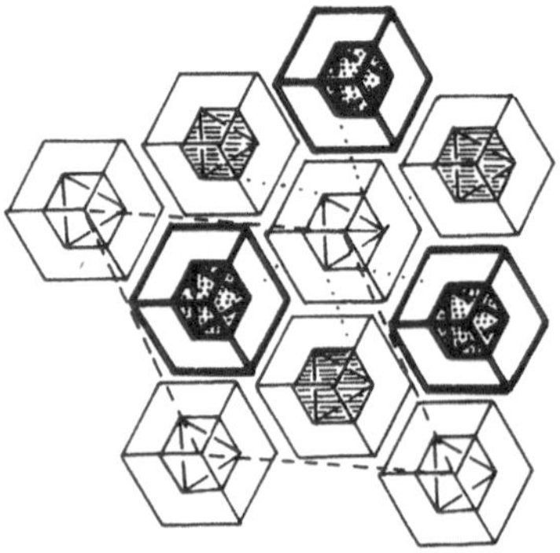

$$6\,Mo\,(Mo_6) \longrightarrow 6\,Mo_{ext}(6\,Mo_{12})$$

$$6\,Mo_{ext.}(Mo_{12}) \longrightarrow 6\,Mo\,(6\,Mo_6)$$

Fig.2.32. Intercluster bonds in the $M_2Mo_9S_{11}$ (M = Tl, K) compounds; projection onto the (001) hexagonal plane [2.78]

Each Mo atom of a Mo_6 cluster is bonded to one extreme Mo atom of a neighbouring Mo_{12} cluster and, conversely, each extreme Mo atom of a Mo_{12} cluster is bonded to one Mo atom of a neighbouring Mo_6 cluster. The Mo-Mo intercluster distance is 3.22 Å in $Tl_2Mo_9S_{11}$ (Table 2.30, see page 81).

As we have already pointed out, there is a strong similarity between the MMo_6X_8-type compounds and these two new compounds. Like the former compounds, $Tl_2Mo_9S_{11}$ and $K_2Mo_9S_{11}$ are metallic, however, they do not become superconducting down to 2.1 K [2.81]. The similarity is further corroborated by the average valence of Mo. Assuming a valence of -2 for S and +1 for Tl, we find a mean valence of 2.22 for a Mo atom. There are therefore 3.78 valence electrons per Mo. This number varies between 3.33 and 4.0 per Mo in the MMo_6X_8 compounds and is equal to 3.66 in $PbMo_6S_8$ and 3.83 in $LaMo_6S_8$. If we examine the Mo-Mo intertriangle bonds within the Mo_6S_8 unit of MMo_6S_8 compounds versus the valence electrons per Mo atom and knowing the Mo-Mo intertriangle bond observed in the Mo_6S_8 unit of $Tl_2Mo_9S_{11}$ compound, one finds in the latter compound 3.60 valence electrons per Mo atom for the Mo_6 cluster instead of 3.78 calculated valence electrons. This suggests that there is a slight charge transfer from the Mo_6 cluster to the Mo_{12} cluster. This could then be consistent with the very short interplane distance in the central octahedron of the Mo_{12} cluster.

2.5.3 New One-Dimensional Cluster $(Mo_{6/2})^1_\infty$ in the $(Mo_{6/2}X_{6/2})^1_\infty$ unit (X = S, Se, Te)

The new one-dimensional cluster $(Mo_{6/2})^1_\infty$ may be viewed as the result of a linear condensation of an infinite number of Mo_6 octahedral clusters. The $(Mo_{6/2}X_{6/2})^1_\infty$ chain is therefore the last member of the series Mo_6X_8, Mo_9X_{11}, $Mo_{12}X_{14}$ resulting from the progressive condensation of the Mo_6X_8 units.

The new compounds $M_2Mo_6Se_6$ and $M_2Mo_6Te_6$ can be made from the elements Na, In, K, Tl in the form of lumps and Mo in the form of powder, and Se and Te in form of grains [2.83]. The $M_2Mo_6S_6$ compounds were synthetized by another method. We started from M_2MoS_4 (M = alkalin) and added Mo and sulphur in correct proportions [2.78]. All manipulations were carried out in a glove box under dry argon atmosphere. The mixtures were reacted in evacuated silica tubes at 850°C for twelve hours. During this reaction, the samples were placed in Al_2O_3 crucibles in order to avoid contamination by the silica tube. A subsequent heat treatment at 1000°-1200°C for twelve hours was necessary in order to obtain simple phase samples. All black-coloured compounds are stable in air.

It is also possible to prepare these compounds by exchange reaction, for instance [2.31],

$$In_2Mo_6Se_6 + 2\ MCl \overset{t^\circ\ 350^\circ C}{\rightleftarrows} M_2Mo_6Se_6 + 2\ InCl \quad .$$

Single phases are obtained for the formula $M_2Mo_6X_6$ (X = S, Se, Te; M = Na, K, Rb, Cs, In, Tl). These compounds crystallize in a hexagonal symmetry with space

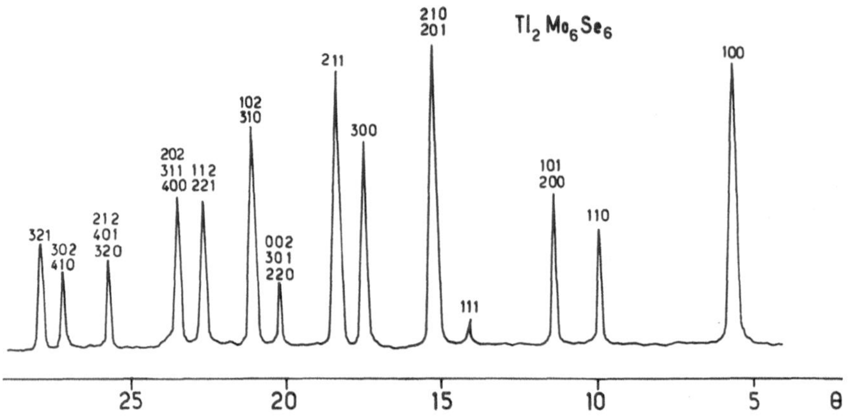

Fig.2.33. X-ray powder diffraction pattern of the hexagonal compound $Tl_2Mo_6Se_6$ ($CuK\bar{\alpha}$ = 1.541 Å) [2.78,83]

group $P6_3/m$. In Fig.2.33 the X-ray powder diffraction pattern of $Tl_2Mo_6Se_6$ is shown. The lattice parameters of these compounds are reported in Table 2.31 (see page 82).

In Table 2.32 (see page 82) the positional crystallographic coordinates of the $Tl_2Mo_6Se_6$ compound are reported. The new structure [2.84] is mainly characterized by the presence of one-dimensional $(Mo_{6/2}X_{6/2})^1_\infty$ units running along the c hexagonal axis and separated by thallium ions (Fig.2.34).

The intracluster Mo-Mo distances are very close to the intracluster bonds observed in other ternary molybdenum chalcogenides (Table 2.33). The Tl^+ ions are trapped between three $(Mo_{6/2}Se_{6/2})^1_\infty$ units in 9-prismatic sites (Fig.2.35). The large difference to the other chalcogenides comes from the fact that there is no chalcogen located on the ternary axis (6_3) in $M_2Mo_6X_6$ (Fig.2.36b). In the other structures, one unit is always connected to six other adjacent units by the Mo-Mo intercluster bonds (3.10-3.60 Å) (Fig.2.36a) whereas in this last structure the shortest Mo-Mo intercluster distance is larger than 6.30 Å (Fig.2.36b).

A very striking property of these compounds is the possibility of synthetizing new binaries, for instance for MoSe, $(Mo_{6/2}Se_{6/2})^1_\infty$. In fact, if one heats the $In_2Mo_6Se_6$ compound under a gas flow of HCl at about 350°C, one obtains the indium chloride and the new binary MoSe [2.31].

Some of these compounds show a metallic behaviour with a strongly anisotropic resistivity. So, the anisotropy ratios $\rho_\perp/\rho_{||}$ are of the order of 10^3. These results are of the same order of magnitude as the ones observed for certain organic conductors. The parallel resistivity at 293 K is about 30 μΩ cm and at 4.2 K about 2 μΩ cm. At 3.5 K, the $Tl_2Mo_6Se_6$ compound becomes superconducting and the anisotropy of the resistivity leads to a strong anisotropy of the upper critical field H_{c2}. In Fig.2.37, the anisotropy of the critical field is plotted versus the angle Θ between the applied field and the hexagonal axis of the single crystal. The ratio

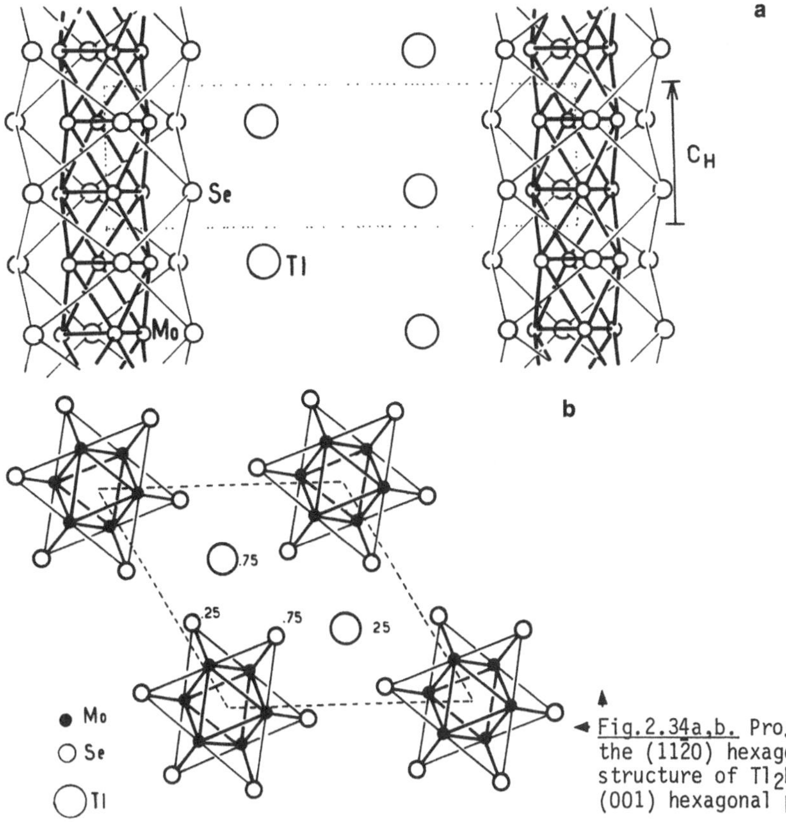

Fig.2.34a,b. Projection onto (a) the (11$\bar{2}$0) hexagonal plane of the structure of $Tl_2Mo_6Se_6$, and (b) the (001) hexagonal plane [2.78,84]

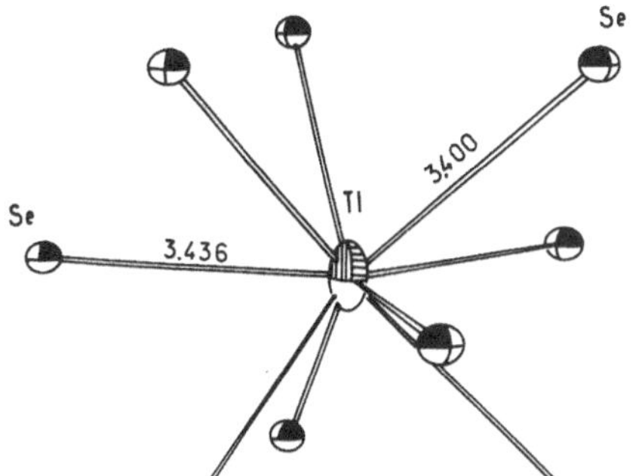

Fig.2.35. Environment of the thallium atoms in $Tl_2Mo_6Se_6$ [2.84]

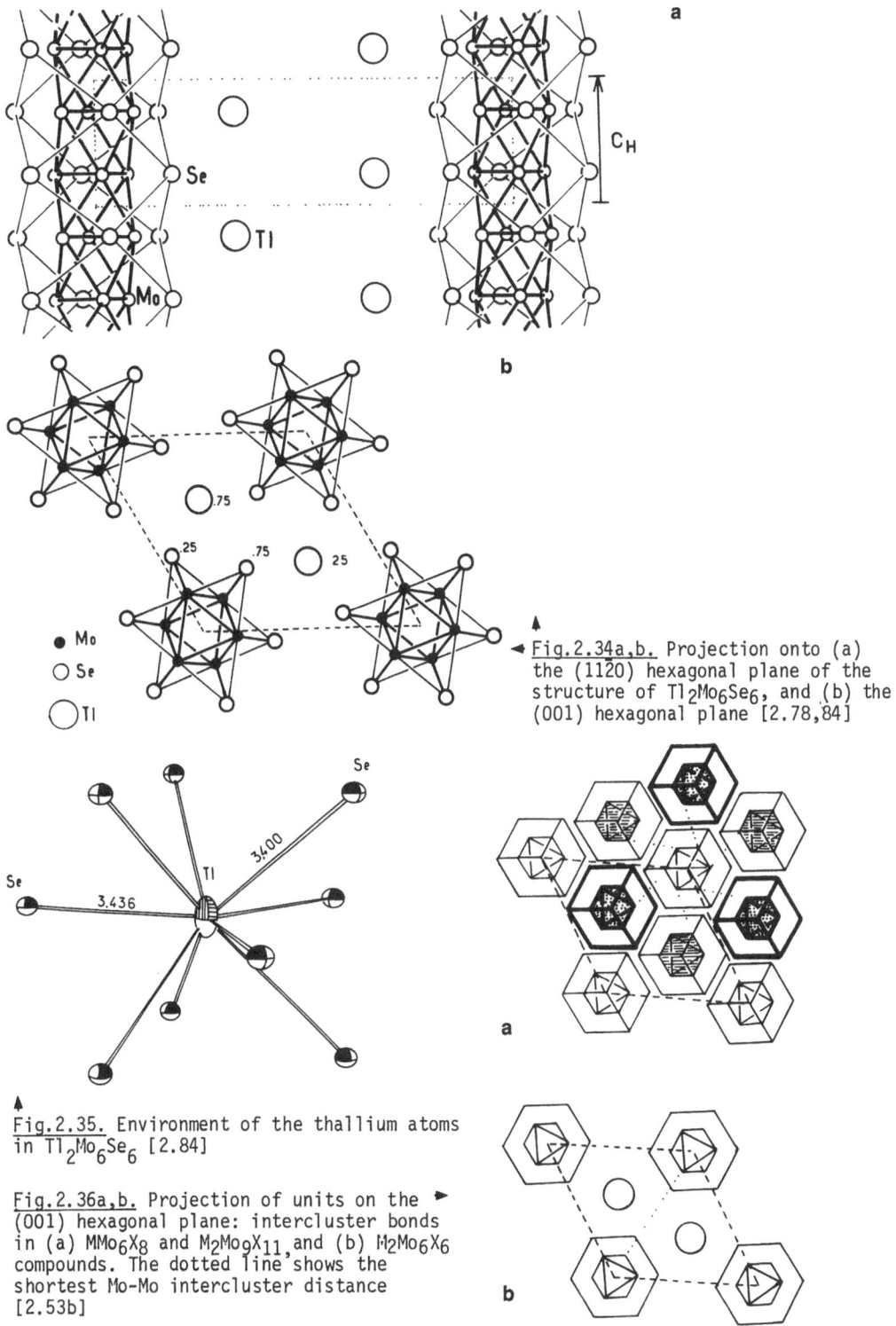

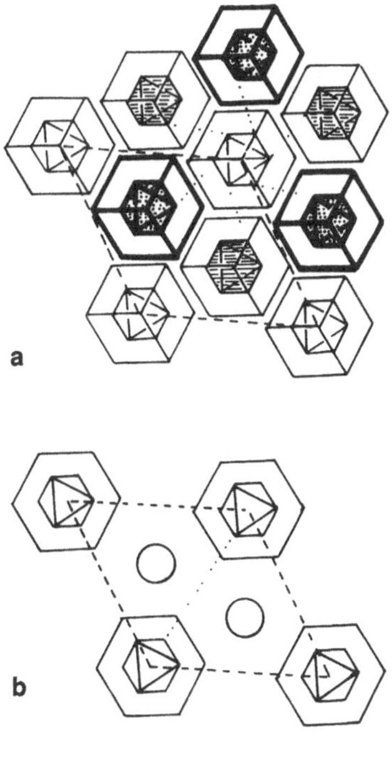

Fig.2.36a,b. Projection of units on the (001) hexagonal plane: intercluster bonds in (a) MMo_6X_8 and $M_2Mo_9X_{11}$, and (b) $M_2Mo_6X_6$ compounds. The dotted line shows the shortest Mo-Mo intercluster distance [2.53b]

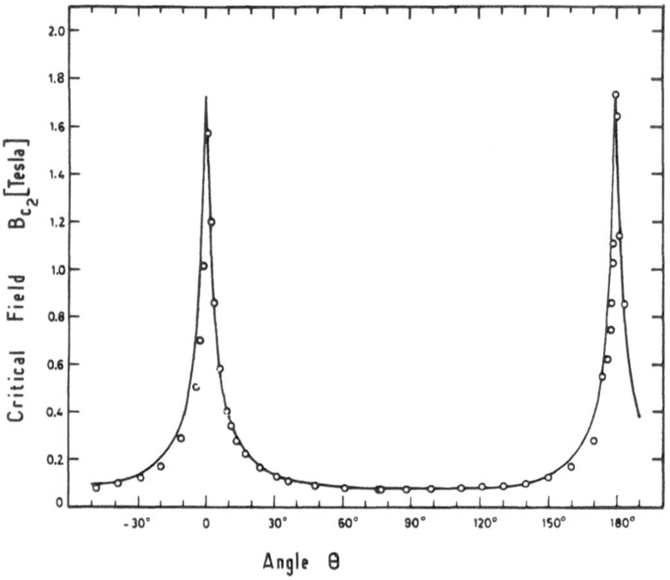

Fig.2.37. The critical field of $Tl_2Mo_6Se_6$ single crystal at T = 2.2 K as a function of the angle Θ between the applied field and the c_H axis. The solid line shows the effective mass prediction for ε = 26. Opened circles represent experimental values [2.85]

$B_{c2\parallel}/B_{c2\perp}$ is found to be larger than 26 and fits the results of the resistivity very well [2.85](Fig.2.37).

2.6 Conclusion

In this chapter we have presented the different results on the synthesis and the structure of the ternary molybdenum chalcogenides. These materials constitute a new aspect of modern solid-state chemistry which is situated between iono-covalent chemistry and metallurgical chemistry. This field of chemistry is similar to that of the organo-metallic materials due to its pseudo-molecular aspect, but is different because of the high temperatures involved. The assumption of a molecular model for the MMo_6X_8 materials has been retained and confirmed by the discovery of the new molecular Mo_9X_{11}, $Mo_{12}X_{14}$, $(Mo_{6/2}X_{6/2})^1_\infty$ units. Earlier, such units as Mo_6X_8 had been only found in the halogenides with added ligands-between the units. In the present materials, the units are made up of polyhedral metallic clusters and are weakly coupled to one another by the intercluster Mo-Mo metal bonds.

The great wealth of these materials is due to their new structure type with tri-dimensional channels constituting a host lattice for a large number of M elements. The great originality of these phases results from the large flexibility of the structure which allows all constituents of the ternary phase to be replaced by a great variety of elements differing in size, oxidation state and concentration. This is mainly due to the possibility of a variable charge transfer onto the cluster, so that one can play with the oxidation or the reduction of the cluster. Such substitutions exert a very drastic influence on the physical properties [2.53b,86].

Thus, these phases are particularly interesting materials for the study of the metal-metal bond and may even be model substances for a systematic study of the relationships between structural parameters and physical properties.

Appendix: Tables

Table 2.1. Results of the X-ray analysis of several $Pb_xMo_6S_8$ samples. The numbers in parentheses give the ratio of the intensities of the most intense line of Mo_2S_3 (d = 5.46 Å) or MoS_2 (d = 6.12 Å) and the first line of $PbMo_6S_8$ (d = 6.55 Å)

Nominal composition	Results obtained after reaction at 1100°C	
$Pb_{0.66}Mo_6S_8$	"$PbMo_6S_8$" + Mo_2S_3	(2.60)
$PbMo_6S_8$	"$PbMo_6S_8$" + MoS_2	(0.80)
$Pb_{1.14}Mo_6S_8$	"$PbMo_6S_8$" + MoS_2	(0.15)
$Pb_{1.2}Mo_6S_8$	"$PbMo_6S_8$" + MoS_2	(0.20)
$Pb_2Mo_6S_8$	"$PbMo_6S_8$" + Pb + MoS_2	(0.60)

Table 2.2. Lattice parameters for Mo_6S_8, Mo_6Se_8, and Mo_6Te_8 binaries

Compound	Rhombohedral lattice parameters		Hexagonal lattice parameters			Ref.
	a_R[Å]	α_R[°]	a_H[Å]	c_H[Å]	V_H[Å^3]	
Mo_6S_8	6.43	91.34	9.20	10.88	797,5	2.16,28
Mo_6S_8	6.66	91.58	9.54	11.21	884.2	2.22
Mo_6Te_8	7.10	92.60	10.20	11.65	1050	2.29

Table 2.3. Lattice parameters for non-stoichiometric compounds of type I and type II [2.16,18]

Compound	Rhombohedral lattice parameters		Hexagonal lattice parameters		
	a_R[Å]	α_R[°]	a_H[Å]	c_H[Å]	V_H[Å^3]
$Cu_4Mo_6S_8$	6.59	95.59	9.77	10.25	847
$Cu_{1.8}Mo_6S_8$	6.48	94.91	9.55	10.22	807
$Ni_2Mo_6S_8$	6.46	94.70	9.51	10.24	801
$Ni_{1.32}Mo_6S_8$	6.44	94.70	9.48	10.21	795
$Co_2Mo_6S_8$	6.48	95.27	9.58	10.15	807
$Co_{1.32}Mo_6S_8$	6.47	95.00	9.55	10.13	804
$Li_4Mo_6S_8$	6.62	94.53	9.73	10.53	864
$Mg_{1.14}Mo_6S_8$	6.51	93.58	9.49	10.55	822
$Zn_{1.10}Mo_6S_8$	6.49	94.68	9.54	10.28	810
$Mn_{1.04}Mo_6S_8$	6.50	93.63	9.48	10.52	819
$Cd_{1.10}Mo_6S_8$	6.52	92.82	9.44	10.72	828
$Fe_{1.32}Mo_6S_8$	6.50	94.78	9.56	10.27	813

Table 2.4. Lattice parameters for non-stoichiometric compounds of type II and type III

Compounds	Triclinic lattice parameters						
	a[Å]	b[Å]	c[Å]	α[°]	β[°]	γ[°]	Ref.
$Fe_2Mo_6S_8$	6.502	6.466	6.481	95.94	97.37	91.33	2.30
$Fe_{1.6}Mo_6S_8$	6.464	6.473	6.501	97.36	91.41	95.96	2.31
$Cr_2Mo_6S_8$	6.58	6.53	6.47	94.64	90.60	98	2.32

Table 2.5. Lattice parameters for non-stoichiometric compounds of type I

Compound	Rhombohedral lattice parameters		Hexagonal lattice parameters			Ref.
	a_R[Å]	α_R[°]	a_H[Å]	c_H[Å]	V_H[Å^3]	
$Cu_{2.8}Mo_6Se_8$	6.79	94.91	10.08	10.73	930	2.22
$Ni_{1.32}Mo_6Se_8$	6.73	91.73	9.66	11.3	913	2.22
$Co_{1.4}Mo_6Se_8$	6.71	92.45	9.69	11.1	903	2.22
$Li_{1.5}Mo_6Se_8$	6.71	92.53	9.70	11.1	905	2.31

Table 2.6. Lattice parameters for non- stoichiometric compounds of type III [2.22]

Compound	Triclinic lattice parameters a[Å]	b[Å]	c[Å]	α[°]	β[°]	γ[°]	V[Å^3]
$Fe_{1.2}Mo_6Se_8$	6.80	6.66	6.66	91.13	96.05	93.32	300
$Mn_{1.2}Mo_6Se_8$	6.82	6.67	6.74	92.35	97.02	91.95	304
$Cr_{1.2}Mo_6Se_8$	6.75	6.75	6.70	92.33	98.03	94.28	301
$V_{1.2}Mo_6Se_8$	6.73	6.75	6.69	91.93	98.20	94.48	300
$Ti_{1.2}Mo_6Se_8$	6.69	6.79	6.76	91.36	98.86	94.35	309

Table 2.7. Lattice parameters for non-stoichiometric compounds of type I

Compound	Rhombohedral lattice parameters		Hexagonal lattice parameters			Ref.
	a_R[Å]	α_R[°]	a_H[Å]	c[Å]	V[Å^3]	
$Fe_2Mo_6Te_8$	7.06	92.99	10.24	11.57	1051	2.33
$Co_{0.66}Mo_6Te_8$	7.03	92.75	10.18	11.58	1039	2.33

Table 2.8. Lattice parameters for stoichiometric compounds

Compound	Rhombohedral lattice parameters		Hexagonal lattice parameters			Ref.
	a_R[Å]	α_R[°]	a_H[Å]	c_H[Å]	V_H[Å^3]	
$NaMo_6S_8$	6.53	89.84	9.22	11.34	835	2.31
KMo_6S_8	6.57	89.14	9.24	11.57	838	2.37
$AgMo_6S_8$	6.48	91.97	9.32	10.83	815	2.18
$CaMo_6S_8$	6.50	89.88	9.19	11.29	825	2.18
$SrMo_6S_8$	6.56	89.44	9.23	11.47	846	2.18
$BaMo_6S_8$	6.64	89.00	9.31	11.70	878	2.18
$Pd_{1.6}Mo_6S_8$	6.44	92.40	9.30	10.68	800	2.31,33
$InMo_6S_8$	6.52	93.02	9.46	10.68	828	2.33
$SnMo_6S_8$	6.51	89.71	9.19	11.34	829	2.18
$PbMo_6S_8$	6.54	89.45	9.20	11.43	838	2.18
$ScMo_6S_8$		not well defined				2.24
YMo_6S_8	6.45	89.55	9.08	11.25	803	2.35
$LaMo_6S_8$	6.51	88.94	9.12	11.48	827	2.35
$PrMo_6S_8$	6.49	88.99	9.10	11.44	820	2.35
$NdMo_6S_8$	6.49	89.06	9.10	11.42	819	2.35
$SmMo_6S_8$	6.47	89.19	9.09	11.37	814	2.35
$EuMo_6S_8$	6.55	88.95	9.17	11.54	840	2.35
$EuMo_6S_8$	6.53	89.30	9.18	11.45	836	2.35
$GdMo_6S_8$	6.47	89.30	9.10	11.35	814	2.35
$TbMo_6S_8$	6.46	89.42	9.09	11.30	809	2.35
$DyMo_6S_8$	6.45	89.50	9.09	11.27	806	2.35
$HoMo_6S_8$	6.45	89.53	9.08	11.26	804	2.35
$ErMo_6S_8$	6.45	89.66	9.09	11.23	804	2.35
$TmMo_6S_8$	6.44	89.80	9.10	11.20	803	2.35
$YbMo_6S_8$	6.50	89.55	9.16	11.35	825	2.35
$YbMo_6S_8$	6.49	89.40	9.14	11.36	822	2.35
$LuMo_6S_8$	6.43	89.90	9.08	11.15	796	2.35
UMo_6S_8	6.45	89.19	9.05	11.32	803	2.23,38
$ThMo_6S_8$	6.45	89.02	9.05	11.37	806	2.23,38
$NpMo_6S_8$	6.50	88.97	9.11	11.46	823	2.39

Table 2.9. Lattice parameters for stoichiometric MMo_6Se_8 compounds

Compound	Rhombohedral lattice parameters		Hexagonal lattice parameters			Ref.
	a_R[Å]	α_R	a_H[Å]	c_H[Å]	V_H[Å^3]	
$ZnMo_6Se_8$	6.73	94.50	9.89	10.71	906	2.22
$AgMo_6Se_8$	6.72	91.36	9.65	11.30	909	2.22
$CdMo_6Se_8$	6.77	92.53	9.78	11.19	924	2.22
$SnMo_6Se_8$	6.78	89.60	9.56	11.83	936	2.22
$PbMo_6Se_8$	6.81	89.23	9.56	11.94	945	2.22
$CaMo_6Se_8$	6.76	89.36	9.50	11.83	925	2.33
$SrMo_6Se_8$	6.81	89.62	9.51	12.07	945	2.33
YMo_6Se_8	6.71	89.23	9.43	11.78	907	2.33,40
$LaMo_6Se_8$	6.80	88.96	9.52	12.00	942	2.33,40
$CeMo_6Se_8$	6.78	88.83	9.49	11.98	934	2.33,40
$PrMo_6Se_8$	6.77	88.80	9.47	11.96	929	2.33,40
$NdMo_6Se_8$	6.76	89.07	9.47	11.93	926	2.33,40
$SmMo_6Se_8$	6.73	88.93	9.43	11.87	914	2.33,40
$EuMo_6Se_8$	6.79	88.87	9.51	12.00	940	2.33,40
$GdMo_6Se_8$	6.73	89.13	9.45	11.84	916	2.33,40
$TbMo_6Se_8$	6.71	89.13	9.42	11.80	907	2.33,40
$DyMo_6Se_8$	6.70	89.27	9.42	11.76	904	2.33,40
$HoMo_6Se_8$	6.70	89.30	9.42	11.75	903	2.33,40
$ErMo_6Se_8$	6.70	89.33	9.42	11.74	902	2.33,40
$TmMo_6Se_8$	6.69	89.40	9.41	11.71	898	2.33,40
$YbMo_6Se_8$	6.78	89.37	9.54	11.88	936	2.33,40
$LuMo_6Se_8$	6.69	89.80	9.44	11.62	897	2.33,40
UMo_6Se_8	6.69	88.82	9.36	11.82	897	2.38
$NpMo_6Se_8$	6.75	88.90	9.45	11.91	921	2.41
$PuMo_6Se_8$	6.76	89.01	9.48	11.91	927	2.39
$AmMo_6Se_8$	6.75	88.89	9.45	11.91	921	2.39

Table 2.10. Structural parameters of $M_xMo_6X_8$ compounds (X = S, Se, Te) space group $R\bar{3}$, rhombohedral setting Mo and X(1) atoms in 6f, X(2) atoms in 2c. The M atoms (big cations) are in la (0, 0, 0), the M atoms (small cations) occupied one or two 6f positions as indicated in Table 2.11. The coordinates of equivalent positions are: ±[x,y,z; z,x,y; y,z,x]: 6f positions; ±[x,x,x]: 2c positions; [0, 0, 0]: la positions

				Mo			X(1)			X(2)	Ref.
	a_R[Å]	α_R[°]	V_H[Å^3]	x	y	z	x	y	z	x	
	Sulphides										
Mo_6S_8	6.432	91.56	798	0.2130	0.4055	0.5504	0.3868	0.1319	0.7367	0.2171	2.28
$Ni_2Mo_6S_8$	6.462	94.18	801	0.2173	0.4045	0.5430	0.3781	0.1330	0.7269	0.1999	2.46
$Ni_{1.4}Mo_6S_8$	6.444	94.68	795	0.2160	0.4048	0.5434	0.3766	0.1344	0.7265	0.1978	2.47
$Co_{1.6}Mo_6S_8$	6.483	95.28	806	0.2194	0.4059	0.5419	0.3743	0.1343	0.7238	0.2002	2.47
$Fe_{1.32}Mo_6S_8$	6.497	94.78	813	0.2203	0.4076	0.5441	0.3742	0.1337	0.7255	0.2053	2.47
$Cu_{1.8}Mo_6S_8$	6.503	94.93	815	0.220	0.4049	0.5421	0.3790	0.1349	0.7215	0.2015	2.49
$Cu_{2.76}Mo_6S_8$	6.560	95.51	834	0.2254	0.4058	0.5418	0.3793	0.1326	0.7172	0.2020	2.49
$Cu_{2.94}Mo_6S_8$	6.573	95.56	839	0.2269	0.4063	0.5419	0.3789	0.1327	0.7169	0.2022	2.49
$Cu_{3.66}Mo_6S_8$	6.597	95.58	848	0.2291	0.4070	0.5423	0.3797	0.1309	0.7155	0.2015	2.49
$AgMo_6S_8$	6.473	91.94	813	0.2178	0.3926	0.5674	0.3830	0.1305	0.7333	0.2223	2.53
$InMo_6S_8$	6.540	93.16	828	0.2239	0.4113	0.5512	0.3742	0.1308	0.7285	0.2220	2.33,51
$SnMo_6S_8$	6.527	89.70	834	0.2251	0.4163	0.5611	0.3794	0.1259	0.7436	0.2412	2.36
$SnMo_6S_8$	6.515	89.60	824	0.2248	0.4152	0.5605	0.3795	0.1299	0.7429	0.2424	2.52
$Pb_{0.92}Mo_6S_8$	6.544	89.48	840	0.2273	0.4162	0.5620	0.3830	0.1253	0.7421	0.2436	2.25
$Pb_{0.92}Mo_6S_8$	6.548	89.37	842	0.2273	0.4159	0.5617	0.3832	0.1256	0.7429	0.2424	2.24
$PbMo_6S_8$	6.544	89.27	840	0.2269	0.4165	0.5622	0.3823	0.1254	0.7424	0.2436	2.26
$LaMo_6S_8$	6.524	88.69	832	0.2272	0.4199	0.5636	0.3764	0.1254	0.7492	0.2448	2.26
$GdMo_6S_8$	6.461	89.18	809	0.2270	0.4197	0.5608	0.3716	0.1268	0.7529	0.2386	2.26
$HoMo_6S_8$	6.449	89.40	804	0.2231	0.4199	0.5599	0.3696	0.1267	0.75319	0.2373	2.26
$ErMo_6S_8$	6.445	89.45	803	0.2229	0.4198	0.5594	0.3683	0.1280	0.75319	0.2372	2.26

	Selenides										
Mo_6Se_8	6.658	91.58	884	0.22304	0.41298	0.54608	0.37917	0.12741	0.7447	0.2137	2.45
Mo_6Se_8	6.654	91.76	882	0.22327	0.41306	0.54647	0.3786	0.1269	0.7442	0.2136	2.26
Mo_6Se_8	6.660	91.81	885	0.2235	0.4142	0.5461	0.3791	0.1274	0.7447	0.2139	2.26
$Cu_1Mo_6Se_8$	6.708	93.79	899	0.2270	0.4124	0.5425	0.3742	0.1291	0.7344	0.2061	2.26
$Cu_{1.5}Mo_6Se_8$	6.733	94.14	908	0.2213	0.4124	0.5424	0.3738	0.1285	0.7307	0.2058	2.26
$Cu_2Mo_6Se_8$	6.784	94.72	927	0.2330	0.4117	0.5417	0.3733	0.1282	0.7251	0.2050	2.26
$AgMo_6Se_8$	6.730	91.72	913	0.2292	0.4153	0.5506	0.3792	0.1255	0.7408	0.2227	2.26
$Sn_{0.8}Mo_6Se_8$	6.754	89.44	924	0.2333	0.4195	0.5576	0.3785	0.1216	0.7496	0.2364	2.26
$PbMo_6Se_8$	6.810	89.23	947	0.2364	0.4206	0.5609	0.3819	0.1201	0.7482	0.2422	2.25
$La_{0.85}Mo_6Se_8$	6.768	88.63	929	0.2354	0.4223	0.5617	0.3783	0.1205	0.7536	0.2417	2.26
	Tellurides										
Mo_6Te_8	7.056	92.57	1051	0.2410	0.4190	0.5450	0.3770	0.1200	0.7479	0.2080	2.33

Table 2.11. Positional parameters of the two M-atom sites and their occupancy in $M_xMo_6X_8$ compounds containing small M atoms

	M in site (1)					M in site (2)					Ref.
M_x	x	y	z	occup.	B[Å^2]	x	y	z	occup.	B[Å^2]	
Sulphides											
$Ni_{2.0}$	0.0558	0.0797	0.8192	1/6	1.89	0.0144	0.3458	0.9404	1/6	1.45	2.46
$Ni_{1.4}$	0.0558	0.0831	0.8334	0.13	1.71	0.0134	0.3469	0.9418	0.10	1.76	2.47
$Co_{1.6}$	0.0536	0.0895	0.8562	0.20	1.88	0.0155	0.3479	0.9480	0.07	1.75	2.47
$Fe_{1.32}$	0.0526	0.0768	0.8706	0.17	1.89	0.0380	0.3392	0.9459	0.05	2.15	2.47
$Cu_{1.8}$	0.056	0.086	0.845	0.26		0.002	0.314	0.958	0.04		2.49
$Cu_{2.76}$	0.057	0.094	0.845	0.24		0.014	0.339	0.956	0.22		2.49
$Cu_{2.94}$	0.058	0.097	0.847	0.22		0.017	0.342	0.956	0.27		2.49
$Cu_{3.66}$	0.056	0.091	0.849	0.23		0.019	0.349	0.956	0.38		2.49
In_1	0.063	0.017	0.911	0.167	1.96						2.33,51
Selenides											
$Cu_{1.0}$	0.11	0.96	0.95	0.16							2.26
$Cu_{1.5}$	0.119	0.962	0.933	0.16		0.047	0.366	0.941	0.10		2.26
$Cu_{2.0}$	0.129	0.952	0.921	0.16		0.044	0.370	0.946	0.18		2.26

Table 2.12. Interatomic distances in Å in $M_xMo_6X_8$ compounds (X = S, Se, Te): the shortest Mo-Mo intracluster distances perpendicular to the ternary axis called $(Mo\text{-}Mo)_\Delta$ between two equivalent Mo atoms in the same triangle by 3-fold rotation and parallel to the ternary axis called $Mo_\Delta\text{-}Mo_\Delta$ distance between two Mo atoms belonging to two different triangles; the shortest Mo-Mo intercluster distance between two neighbouring Mo_6 octahedra; the shortest Mo-X intercluster distance between a Mo atom of one Mo_6X_8 unit and a chalcogen atom of a neighbouring Mo_6X_8 unit; the shortest M-Mo distance between an M atom and an Mo atom

$M_xMo_6X_8$		$(Mo\text{-}Mo)_\Delta$ intra	$Mo_\Delta\text{-}Mo_\Delta$ intra	Mo-Mo inter	Mo-X inter	M-Mo	References
Sulphides							
Mo_6S_8		2.698	2.862	3.084	2.425	-	2.28
$Ni_2Mo_6S_8$	(Ni^{2+})	2.691	2.768	3.185	2.460	2.98	2.46
$Ni_{1.4}Mo_6S_8$	(Ni^{2+})	2.698	2.766	3.160	2.455	2.96	2.47
$Co_{1.6}Mo_6S_8$	(Co^{2+})	2.687	2.733	3.217	2.494	3.02	2.47
$Fe_{1.32}Mo_6S_8$	(Fe^{2+})	2.693	2.732	3.221	2.504	3.01	2.47
$Cu_{1.8}Mo_6S_8$	(Cu^{1+})	2.683	2.752	3.237	2.494	3.03	2.49
$Cu_{2.76}Mo_6S_8$	(Cu^{1+})	2.671	2.708	3.330	2.527	3.08	2.49
$Cu_{2.94}Mo_6S_8$	(Cu^{1+})	2.665	2.697	3.353	2.541	3.08	2.49
$Cu_{3.66}Mo_6S_8$	(Cu^{1+})	2.659	2.681	3.389	2.551	3.07	2.49
$InMo_6S_8$	(In^{3+})	2.702	2.725	3.258	2.529	3.73	2.33,51
$AgMo_6S_8$	(Ag^{1+})	2.706	2.804	3.154	2.459	4.23	2.53a
$SnMo_6S_8$	(Sn^{2+})	2.688	2.737	3.232	2.550	4.20	2.36
$Pb_{0.92}Mo_6S_8$	(Pb^{2+})	2.678	2.726	3.267	2.559	4.21	2.25
$Pb_{0.92}Mo_6S_{7.5}$	(Pb^{2+})	2.672	2.737	3.271	2.566	4.21	2.24
$PbMo_6S_8$	(Pb^{2+})	2.679	2.732	3.262	2.561	4.20	2.26
$LaMo_6S_8$	(La^{3+})	2.667	2.707	3.238	2.590	4.18	2.26
$GdMo_6S_8$	(Gd^{3+})	2.660	2.714	3.163	2.568	4.16	2.26
$HoMo_6S_8$	(Ho^{3+})	2.659	2.711	3.148	2.562	4.16	2.26
$ErMo_6S_8$	(Er^{3+})	2.654	2.713	3.144	2.569	4.16	2.26

Table 2.12. (cont.)

Selenides						
Mo_6Se_8	2.684	2.836	3.266	2.598	-	2.45
Mo_6Se_8	2.688	2.820	3.269	2.599	-	2.26
Mo_6Se_8	2.691	2.827	3.273	2.596	-	2.26
$Cu_{1.0}Mo_6Se_8$	2.691	2.771	3.363	2.632	3.04	2.26
$Cu_{1.5}Mo_6Se_8$	2.685	2.752	3.412	2.643	3.07	2.26
$Cu_{2.0}Mo_6Se_8$	2.680	2.727	3.491	2.673	3.12	2.26
$AgMo_6Se_8$	2.701	2.776	3.378	2.640	4.46	2.26
$Sn_{0.8}Mo_6Se_8$	2.680	2.755	3.419	2.694	4.39	2.26
$PbMo_6Se_8$	2.697	2.734	3.490	2.722	4.42	2.25
$La_{0.8}Mo_6Se_8$	2.682	2.725	3.446	2.730	4.37	2.26
Tellurides						
Mo_6Te_8	2.700	2.772	3.674	2.837	-	2.33

Table 2.13. Angles (Θ^o) of the MS_8 cube (site 1) in $PbMo_6S_8$, $SnMo_6S_8$, $InMo_6S_8$, and $Cu_{1.8}Mo_6S_8$

	S_1-S_2-S_1	S_1-S_1-S_1	S_1-S_1-S_2	Ref.
PbS_8	94.98	93.53	85.74	2.25
SnS_8	95.34	93.68	85.48	2.36
InS_8	101.69	99.46	79.41	2.33
CuS_8	107.26	101.24	75.59	2.33

Table 2.14. Root-mean-square displacements of tin and lead atoms in the $SnMo_6S_8$ and $PbMo_6S_8$ compounds

Compounds	X[Å]⊥	X[Å]∥	References
$SnMo_6S_8$	0.2208	0.1310	2.36
$PbMo_6S_8$	0.151	0.169	2.24

Table 2.15. Interatomic distances in Å of the origin site in $M_xMo_6X_8$ (X = S, Se, Te) compounds: the shortest distances M_1-X_1, M_1-X_2, origin-M_1 between the M_1 atom, the peripherical atom X_1 of the origin cubic site, and the axial chalcogen atom X_2 situated on the ternary axis; the occupancy factors of the M_1 atom in the site 1 and the M_2 atom in the site 2

$M_xMo_6X_8$	Envir. M_1	M_1-X_1	M_1-X_2	orig.-M_1	τ^{M_1}	τ^{M_2}	orig.-X_1	orig.-X_2	X_2-X_1	X_1-X_1	Ref.
Sulphides											
Mo_6S_8							3.153	2.362	3.310	3.550	2.28
$Ni_2Mo_6S_8$	4	2.224,2.230,3.410, 3.606,4.433,4.436	2.341,2.583	1.368	1	1	3.229	2.047	3.362	3.527	2.46
$Ni_{1.4}Mo_6S_8$	4				0.78	0.6	3.218	2.019	3.341	3.509	2.47
$Co_{1.6}Mo_6S_8$	4	2.335,2.325,3.456, 3.472,4.312,4.307	2.359,2.363	1.204	1.20	0.4	3.249	2.032	3.389	3.522	2.47
$Fe_{1.32}Mo_6S_8$	4	2.434,2.377,3.371, 3.451,4.198,4.165	2.368,2.368	1.078	1	0.33	3.237	2.109	3.399	3.520	2.47
$Cu_{1.8}Mo_6S_8$	4	2.321,2.343,3.473, 3.562,4.389,4.401	2.382,2.456	1.259	1.56	0.24	3.285	2.066	3.425	3.568	2.49
$Cu_{2.76}Mo_6S_8$	4	2.347,2.354,3.573, 3.603,3.613	2.427,2.455	1.305	1.44	1.32	3.342	2.063	3.494	3.605	2.49
$Cu_{2.94}Mo_6S_8$	4	2.349,2.357,3.598, 3.595,3.596	2.456,2.439	1.311	1.32	1.62	3.349	2.067	3.503	3.610	2.49
$Cu_{3.66}Mo_6S_8$	4	2.390,2.399,3.590, 3.610	2.421,2.437	1.276	1.38	2.28	3.369	2.066	3.525	3.625	2.49
$InMo_6S_8$	5	2.518,2.853,2.977, 3.580,3.679,3.917	2.452,2.518	0.739	1	0	3.208	2.372	3.460	3.517	2.33, 51
$AgMo_6S_8$	6+2	6 × 3.179	2 × 2.407	0	1	0	3.179	2.407			2.53
$SnMo_6S_8$	6+2	6 × 3.094	2 × 2.741	0	1	0	3.094	2.741	3.452	3.499	2.36
$Pb_{0.92}Mo_6S_8$	6+2	6 × 3.121	2 × 2.786	0	0.92	0	3.121	2.786	3.490	3.531	2.25
$Pb_{0.92}Mo_6S_{7.5}$	6+2	6 × 3.117	2 × 2.786	0	0.92	0	3.117	2.786	3.428	3.488	2.24
$PbMo_6S_8$	6+2	6 × 3.110	2 × 2.796	0	1	0	3.110	2.796			2.26
$LaMo_6S_8$	6+2	6 × 3.037	2 × 2.827	0	1	0	3.037	2.827			2.26

$GdMo_6S_8$	6+2	6 × 2.982	2 × 2.708	0	1	0	2.982	2.708			2.26
$HoMo_6S_8$	6+2	6 × 2.969	2 × 2.678	0	1	0	2.969	2.678			2.26
$ErMo_6S_8$	6+2	6 × 2.970	2 × 2.623	0	1	0	2.970	2.623			2.26
Selenides											
Mo_6Se_8							3.190	2.385	3.380	3.581	2.45
Mo_6Se_8							3.191	2.396			2.26
Mo_6Se_8							3.191	2.388			2.26
$Cu_{1.0}Mo_6Se_8$					0.96	0	3.276	2.231			2.26
$Cu_{1.5}Mo_6Se_8$					0.96	0.60	3.310	2.220			2.26
$Cu_2Mo_6Se_8$					0.96	1.08	3.300	2.202			2.26
$AgMo_6Se_8$					1	0	3.240	2.517			2.26
$Sn_{0.8}Mo_6Se_8$					0.8	0	3.163	2.792			2.26
$PbMo_6Se_8$					1	0	3.205	2.895	3.589	3.641	2.25
$La_{0.8}Mo_6Se_8$					0.8	0	3.136	2.899			2.26
Tellurides											
Mo_6Te_8							3.376	2.425			2.33

Table 2.16. Positional parameters (x 104) for the triclinic modification of $M_xMo_6X_8$ (space group $P\bar{1}$)

	Mo(1)			Mo(2)			Mo(3)			S(1)			occupancy	Ref.
	x	y	z	x	y	z	x	y	z	x	y	z		
$Cu_{1.8}Mo_6S_8$	2204	4064	5433	4057	5371	2208	5442	2194	4093	3734	1405	7155	Cu = 0.9	2.50
$Fe_2Mo_6S_8$	2241	4155	5373	4019	5429	2200	5491	2225	4087	3885	1430	7227	Fe = 1	2.30
$Fe_{1.68}Mo_6S_8$	2241	4152	5373	4019	5492	2199	5492	2224	4089	3886	1424	7223	Fe = 0.84	2.31
$Ni_{0.66}Mo_6Se_8$	2291	4216	5469	4132	5491	2285	5456	2236	4158	3872	1269	7459	Ni = 0.33	2.44

	S(2)			S(3)			S(4)			M			occupancy	Ref.
	x	y	z	x	y	z	x	y	z	x	y	z		
$Cu_{1.8}Mo_6S_8$	1373	7174	3814	7282	3828	1261	2044	1958	1995	1593	-573	9088	Cu = 0.9	2.50
$Fe_2Mo_6S_8$	1191	7330	3634	7280	3609	1381	2137	2058	1877	951	- 3	8431	Fe = 1	2.30
$Fe_{1.68}Mo_6S_8$	1194	7340	3632	7285	3612	1385	2144	2054	1878	949	0	843	Fe = 0.84	2.31
$Ni_{0.66}Mo_6Se_8$	1172	7430	3730	7385	3702	1301	2141	2224	2067	8417	9368	5072	Ni = 0.33	2.44

Table 2.17. Mean distances (in Å) of triclinic $M_xMo_6X_8$ compounds

Compounds	$\bar{d}$ Mo-Mo intracluster	$\bar{d}$ Mo-Mo intercluster	$\bar{d}_{Mo-X}$	$\bar{d}_{M-Mo}$	Ref.
$Cu_{1.8}Mo_6S_8$ at 250 K	2.73	3.26	2.46	3.05	2.50
$Fe_2Mo_6S_8$	2.68	3.20	2.47	3.58	2.30
$Fe_{1.68}Mo_6S_8$	2.68	3.20	2.47	3.58	2.31
$Ni_{0.66}Mo_6Se_8$	2.73	3.31	2.57	2.73	2.44

Table 2.18. Lattice parameters for pseudo-binary compounds of the type $Mo_6X_{8-x}Y_x$ (X = S, Se, Te; Y = Cl, Br, I) [2.66]

Compounds	a_R[Å]	α_R[°]	a_H[Å]	c_H[Å]	V[Å^3]	T_c[K]
Mo_6S_8	6.43	91.34	9.20	10.88	797	1.8
$Mo_6S_6Br_2$	6.50	94.43	9.55	10.36	817	13.8
$Mo_6S_6I_2$	6.56	94.50	9.64	10.44	841	14.0
Mo_6Se_8	6.66	91.58	9.54	11.21	797	6.2
$Mo_6Se_5Cl_3$	6.67	92.40	9.64	11.05	889	5.7
Mo_6Se_7Br	6.67	92.60	9.65	11.02	889	7.1
$Mo_6Se_5Br_3$	6.72	93.86	9.82	10.83	904	7.1
Mo_6Se_7I	6.72	93.62	9.80	10.88	905	7.6
$Mo_6Se_5I_3$	6.80	94.89	10.02	10.73	933	4.2
Mo_6Te_8	7.10	92.60	10.20	11.65	1050	1.7
$Mo_6Te_6Cl_2$	7.05	92.47	10.19	11.68	1050	1.7
$Mo_6Te_5Cl_3$	7.04	92.60	10.20	11.66	1050	1.7
$Mo_6Te_6I_2$	7.09	93.43	10.32	11.52	1062	2.6
$Mo_6Te_5I_3$	7.09	93.46	10.32	11.51	1061	2.4

Table 2.19. Positional parameters and thermal vibrations of the $Mo_6S_6Br_2$ compound

	Mo(6f)	S(6f)	Br(2c)
x	0.22024	0.37551	0.18894
y	0.40792	0.13347	0.18894
z	0.54803	0.71990	0.18894
Beq [Å^2]	0.33	0.41	0.66

Table 2.20. Interatomic distances in Å in $Mo_6S_6Br_2$ compound

$(Mo-Mo)_\Delta$	2.719	origin site:	
$Mo_\Delta-Mo_\Delta$	2.732	S - Br	3.737
Mo-Mo interunit	3.225	S - S	3.537
Mo-S	2.401	Br-Br via origin	3.914
	2.475		
	2.422		
Mo-S interunit	2.485		

Table 2.21. Lattice parameters of the $(Mo_{6-x}M_x)X_8$ mixed cluster compounds (M = Nb, Ta, Re, Ru, Rh)

Compounds	a_H[Å]	c_H[Å]	a_R[Å]	α_R[°]	V[Å^3]	T_c	Ref.
$Mo_2Re_4S_8$	9.34	10.42	6.41	93.43	787	semiconductor	2.68
$Mo_2Re_4Se_8$	9.67	10.74	6.63	93.61	870	semiconductor	2.68
$Mo_4Ru_2Se_8$	9.69	10.82	6.65	93.41	879	semiconductor	2.66
$Mo_{5.25}Nb_{0.75}Se_8$	9.60	11.16	6.68	91.96	892	6.2	2.66
$Mo_4Re_2Te_8$	10.22	11.51	7.04	93.08	1041	3.5	2.68
$Mo_4Ru_2Te_8$	10.28	11.35	7.03	93.82	1037	1.7	2.66
$Mo_{5.25}Nb_{0.75}Te_8$	10.21	11.70	7.07	92.49	1057	1.7	2.66
$Mo_{5.25}Ta_{0.75}Te_8$	10.21	11.70	7.07	92.48	1056	1.7	2.66
$Mo_{4.66}Rh_{1.33}Te_8$	10.25	11.53	7.06	93.15	10.49	1.7	2.66

Table 2.22. Lattice parameters for $M_xMo_6S_6O_2$ (M = Cu, Ni, Co, Pb) and $Mo_6S_6O_2$ [2.72]

Compounds	a_H[Å]	c_H[Å]	a_R[Å]	α_R[°]	T_c[K]
$Mo_6S_6O_2$	9.14	10.82	6.39	91.28	
$Cu_2Mo_6S_6O_2$	9.69	10.19	6.54	95.51	9
$Ni_2Mo_6S_6O_2$	9.47	10.19	6.44	94.72	
$Co_2Mo_6S_6O_2$	9.49	10.11	6.43	95.06	
$PbMo_6S_6O_2$	9.14	11.50	6.52	88.98	11.7

Table 2.23. Lattice parameters for (M, M')Mo_6S_8 compounds

Compounds	a_H[Å]	c_H[Å]	a_R[Å]	α_R[°]	References
$PbMo_6S_8$	9.20	11.43	6.54	89.47	2.18
$Cu_{0.1}PbMo_6S_8$	9.25	11.48	6.57	89.50	2.23
$Cu_{0.2}PbMo_6S_8$	9.267	11.49	6.58	89.53	2.23
$Cu_{0.3}PbMo_6S_8$	9.27	11.47	6.58	89.60	2.23
$SnMo_6S_8$	9.19	11.34	6.52	89.73	2.18,36
$Zn_{0.4}SnMo_6S_8$	9.18	11.37	6.51	89.55	2.73
$Fe_{0.4}SnMo_6S_8$	9.22	11.30	6.52	89.99	2.73
$Fe_{0.4}SnMo_6S_8$	9.23	11.34	6.53	89.85	2.74
$FeZnMo_6S_8$	9.59	10.28	6.51	94.85	2.73

Table 2.24. Lattice parameters for $In_xMo_{15}Se_{19}$ (x = 2.9 and 3.3) (space group $P6_3/m$) Z = 2 and for $M_2Mo_{15}Se_{19}$ (M = In, K, Ba, Tl) and $K_2Mo_{15}S_{19}$ (space group $R\bar{3}c$) Z_R = 2

Compounds	a_H[Å]	c_H[Å]	V_H[Å^3]	a_R[Å]	α_R[°]	References
$In_{3.3}Mo_{15}Se_{19}$	9.83	19.40	1624			2.76
$In_{2.9}Mo_{15}Se_{19}$	9.80	19.49	1622			2.76
$In_2Mo_{15}Se_{19}$	9.69	58.10	4725	20.16	27.81	2.77
$K_2Mo_{15}Se_{19}$	9.74	58.16	4779	20.19	27.92	2.77
$Ba_2Mo_{15}Se_{19}$	9.88	57.60	4869	20.03	28.56	2.77
$Tl_2Mo_{15}Se_{19}$	9.80	58.23	4843	20.22	28.05	2.77
$K_2Mo_{15}S_{19}$	9.36	56.22	4266	19.50	27.77	2.77

Table 2.25. Fractional atomic coordinate ($\times 10^4$) isotropic thermal parameters [$Å^2$] for $In_{2.9}Mo_{15}Se_{19}$, $In_{3.3}Mo_{15}Se_{19}$, and $In_2Mo_{15}Se_{19}$, and occupancy factors for $In_{2.9}Mo_{15}Se_{19}$ and $In_{3.3}Mo_{15}Se_{19}$

		$In_{2.9}Mo_{15}Se_{19}$ ($P6_3/m$)				$In_{3.3}Mo_{15}Se_{19}$ ($P6_3/m$)			
Position		x	y	z	B[$Å^2$]	x	y	z	B[$Å^2$]
Mo(1)	12i	142	1647	571	1.2	145	1646	564	0.36
Mo(2)	12i	3185	5035	1334	0.8	3186	5039	1336	0.19
Mo(3)	6h	1692	5127	2500	0.5	1661	5092	2500	0.28
Se(1)	12i	7142	340	509	1.2	7133	342	522	0.04
Se(2)	12i	91	3794	1393	1.2	113	3812	1376	0.22
Se(3)	6h	3136	3543	2500	1.0	3178	3544	2500	0.10
Se(4)	4e	0	0	1617	2.5	0	0	1605	0.37
Se(5)	4f	3333	6667	297	1.0	3333	6667	307	0.30
In(1)	4f	6667	3333	1288	4.5	6667	3333	1259	2.96
In(2)	6h	2127	458	2500		2254	472	2500	1.23
(occup. 0.29)						(occup. 0.44)			

		$In_2Mo_{15}Se_{19}$ ($R\bar{3}c$) (rhombohedral setting)			
Position		x	y	z	B[$Å^2$]
Mo(1)	12f	4999	6889	3706	0.57
Mo(2)	12f	2225	457	3637	0.49
Mo(3)	6e	4097	903	2500	0.53
Se(1)	12f	1898	8034	5567	0.55
Se(2)	12f	5381	9201	1847	0.78
Se(3)	6e	598	4402	7500	0.59
Se(4)	4c	4454	4454	4454	0.80
Se(5)	4c	1761	1761	1761	0.80
In	4c	1143	1143	1143	5.18

Table 2.26. Interatomic distances [Å] in $In_2Mo_{15}Se_{19}$, $In_{2.9}Mo_{15}Se_{19}$ and $In_{3.33}Mo_{15}Se_{19}$ compounds

Valence electrons per Mo atom	$In_2Mo_{15}Se_{19}$ 3.60 e^- with In^{1+}	$In_{2.9}Mo_{15}Se_{19}$ 3.78 e^- with $In^{3+}{}_{0.9}In^{1+}{}_2$	$In_{3.33}Mo_{15}Se_{19}$ 3.86 e^- with $In^{3+}{}_{1.33}In^{1+}{}_2$
Distances in Mo_6Se_8			
$\|Mo(1)-Mo(1)\|_\Delta$	2 × 2.686	2 × 2.684	2 × 2.68
$Mo(1)_\Delta-Mo(1)_\Delta$	2 × 2.772	2 × 2.713	2 × 2.683
Mo(1)-Mo(1) interplane	2.301		2.188
Mo(1)-Se(4)	2.548	2.561	2.547
Mo(1)-Se(1)	2.559	2.547	2.557
	2.590	2.557	2.573
	2.605	2.616	2.628
Mo(1)-Se(2) interunit	2.589	2.666	2.661
Distances in Mo_9Se_{11}			
$[Mo(2)-Mo(2)]_\Delta$ ext.	2 × 2.673	2 × 2.654	2 × 2.655
$[Mo(3)-Mo(3)]_\Delta$ med.	2 × 2.680	2 × 2.705	2 × 2.768
$Mo(2)_\Delta-Mo(3)_\Delta$	2 × 2.712	2 × 2.729	2 × 2.725
	2 × 2.808	2 × 2.770	2 × 2.771
Mo(2)-Mo(3) interplane	2.28		2.258
Mo(2)-Se(5)	2.533	2.537	2.517
Mo(2)-Se(2)	2.582	2.579	2.568
Mo(2)-Se(2)	2.638	2.647	2.635
Mo(2)-Se(3)	2.715	2.690	2.692
Mo(2)-Se(1) interunit	2.641	2.677	2.665
Mo(3)-Se(3)	2 × 2.600	2.575	2.584
		2.572	2.609
Mo(3)-Se(2)	2 × 2.542	2 × 2.603	2 × 2.596
Intercluster distances			
Mo(1)-Mo(2)	3.389	3.498	3.512

Table 2.27. Critical temperatures of $In_{x\sim3}Mo_{15}Se_{19}$ and $M_2Mo_{15}X_{19}$ (X = S, Se) compounds

Compounds	T_c[K]	Compounds	T_c[K]
$In_{2.9}Mo_{15}Se_{19}$	4.24	$K_2Mo_{15}S_{19}$	3.32
$In_{3.2}Mo_{15}Se_{19}$	3.74	$K_2Mo_{15}Se_{19}$	2.45
$In_{3.3}Mo_{15}Se_{19}$	1.88	$Ba_2Mo_{15}Se_{19}$	2.75
$In_{3.4}Mo_{15}Se_{19}$	1.8	$In_2Mo_{15}Se_{19}$	1.50
		$Tl_2Mo_{15}Se_{19}$	1.65

Table 2.28. Lattice parameters of $M_2Mo_9S_{11}$ (M = K, Tl) compounds

$M_2Mo_9S_{11}$	Hexagonal-rhombohedral symmetry		Space group $R\bar{3}$ Laüe group $\bar{3}$ Z_R = 2
$K_2Mo_9S_{11}$	a_R = 13.13 Å a_H = 9.27 Å	α_R = 41.34° c_H = 35.97 Å	V_R = 892.4 Å^3
$Tl_2Mo_9S_{11}$	a_R = 12.95 Å a_H = 9.30 Å	α_R = 42.09° c_H = 35.37 Å	V_R = 883.3 Å^3

Table 2.29. Fractional atomic coordinates ($\times 10^4$) and isotropic thermal parameters [Å^2] for the $Tl_2Mo_9S_{11}$ compound in a rhombohedral setting ($R\bar{3}$)

	Position	x	y	z	Beq [Å^2]
Mo(1)	6f	3082	6415	4525	0.37
Mo(2)	6f	2527	9237	1094	0.36
Mo(3)	6f	8694	2025	201	0.35
S(1)	6f	2394	8461	4941	0.55
S(2)	6f	607	7990	4073	0.69
S(3)	6f	514	3302	7154	0.67
S(×1)	2c	1474	1474	1474	0.69
S(×2)	2c	4150	4150	4150	0.85
Tl(1)	2c	3315	3315	3315	2.54
Tl(2)	2c	2305	2305	2305	1.40

Table 2.30. Interatomic distances [Å] in $Tl_2Mo_9S_{11}$

Distances in Mo_6S_8	
$\|Mo(1)-Mo(1)\|_{\Delta}$	2 × 2.693
$Mo(1)_{\Delta}-Mo(1)_{\Delta}$	2 × 2.780
$\Delta(1)-\Delta(1)$ interplane	2.30
Mo(1)-S(1)	2.445
	2.451
	2.491
Mo(1)-S(×2)	2.419
Mo(1)-S(2)	2.469
Distances in $Mo_{12}S_{14}$	
$\|Mo(2)-Mo(2)\|_{\Delta}$	2 × 2.658
$\|Mo(3)-Mo(3)\|_{\Delta}$	2 × 2.688
$Mo(2)_{\Delta}-Mo(3)_{\Delta}$	2.743
	2.771
$Mo(3)_{\Delta}-Mo(3)_{\Delta}$	2 × 2.669
$\Delta(2)-\Delta(3)$ interplane	2.28
$\Delta(3)-\Delta(3)$ interplane	2.17
Mo(2)-S(2)	2.461
	2.481
Mo(2)-S(3)	2.593
Mo(2)-S(×1)	2.399
Mo(2)-S(1)	2.504
Mo(3)-S(3)	2.483
	2.486
Mo(3)-S(2)	2.431
Mo(3)-S(3)	2.587
Intercluster distance	
Mo(1)-Mo(2)	3.217

Table 2.31. Lattice parameters of $M_2Mo_6X_6$ compounds

$M_2Mo_6X_6$	Hexagonal symmetry		Space group $P\bar{6}_3/m$ Laüe group 6/m Z = 1	
	a_H[Å]	c_H[Å]	V_H[Å^3]	Ref.
$K_2Mo_6S_6$	8.76	4.42	243.7	2.83
$Rb_2Mo_6S_6$	8.82	4.44	299.1	2.83
$Cs_2Mo_6S_6$	8.96	4.46	310.1	2.83
$Na_2Mo_6Se_6$	8.74	4.42	296.4	2.83
$In_2Mo_6Se_6$	8.88	4.49	306.6	2.83
$K_2Mo_6Se_6$	9.01	4.50	316.4	2.83
$Rb_2Mo_6Se_6$	9.27	4.48	333.4	2.31
$Cs_2Mo_6Se_6$	9.51	4.48	350.9	2.31
$Tl_2Mo_6Se_6$	8.9398	4.4997	311.4	2.84
$Ba_{2-\varepsilon}Mo_6Se_6$	9.07	4.31	307.0	2.31
$Na_2Mo_6Te_6$	9.23	4.60	339.4	2.83
$In_2Mo_6Te_6$	9.38	4.58	349.0	2.83
$K_2Mo_6Te_6$	9.60	4.60	367.1	2.83
$Rb_2Mo_6Te_6$	9.76	4.60	379.5	2.31
$Cs_2Mo_6Te_6$	10.08	4.60	404.8	2.31
$Tl_2Mo_6Te_6$	9.44	4.59	354.2	2.83
$Ba_2Mo_6Te_6$	9.41	4.57	350.4	2.31
MoSe	8.35	4.44	268.1	2.31

Table 2.32. Fractional atomic coordinates ($\times 10^4$) and equivalent thermal parameters of the $Tl_2Mo_6Se_6$ compound

Position	x	y	z	B[Å^2]
Mo 6h	1861	1530	1/4	0.35
Se 6h	663	3661	1/4	0.71
Tl 2d	2/3	1/3	1/4	1.60

Table 2.33. Interatomic distances [Å] in the $Tl_2Mo_6Se_6$ compound

Intra-unit distances $(Mo_{6/2}Se_{6/2})^1_\infty$			
Mo-Mo	2 × 2.663	Mo-Se	2 × 2.695
	4 × 2.725		2.611
	2 × 3.810		2.662
		Se-Se	4 × 3.767
Inter-unit distances $(Mo_{6/2}Se_{6/2})^1_\infty$			
Mo-Mo	2 × 6.343	Se-Se	2 × 3.877
	6.511		
	7.094		
Other distances			
Tl-Se	6 × 3.400	Tl-Mo	3 × 3.759
	3 × 3.436		6 × 4.673
Tl-Tl	2 × 4.499		
Mo-Mo interplane	2.250		

Acknowledgements. It is a great pleasure for the authors to thank Professor J. Prigent for his encouragement in this research on cluster chemistry. We are grateful to Professor Ø. Fischer and his co-workers for the enthusiastic collaboration and the many fruitful discussions. The authors thank Dr. Potel for having communicated to them several results prior to publication, and Drs. C. and A. Perrin and for many stimulating discussions.

We gratefully acknowledge the invaluable help of J.C. Jegaden and Mrs. P. Poulain for the drawings and the preparation of this manuscript.

References

2.1 F.A. Cotton: Q. Rev. Chem. Soc. *20*, 389 (1966)
2.2 M.C. Baird: Prog. Inorg. Chem. *9*, 1 (1968)
2.3 D.L. Kepert: *The Early Transition Metals* (Academic Press, New York 1972)
2.4 R.B. King: Prog. Inorg. Chem. *15*, 287 (1972)
2.5 J.B. Brandt, A.C. Skapski: Acta Chem. Scand. *21*, 66 (1967); B. Morosin, A. Narath: J. Chem. Phys. *40*, 1958 (1964)
2.6 G.B. Ansel, L. Katz: Acta Crystallogr. *21*, 482 (1966)
2.7 H.G. Von Schnering, H. Wöhrle: Naturwissenschaften *50*, 91 (1963)
2.8 J. Guillevic, J.Y. Le Marouille, D. Grandjean: Acta Crystallogr. B*30*, 111 (1974)
R. Chevrel, M. Sergent, J.L. Meury, Dang Tran Quan, Y. Colin: J. Solid State Chem. *10*, 260-269 (1974)
2.9 C. Perrin, R. Chevrel, M. Sergent: C.R. Acad. Sci. C*280*, 949 (1975)
2.10 C. Perrin, R. Chevrel, M. Sergent: C.R. Acad. Sci. C*281*, 23 (1975)
2.11 C.C. Torardi, R.E. MacCarley: J. Am. Chem. Soc. *101*, 3963 (1979)

2.12 H. Schäfer, H.G. Von Schnering, J. Tillack, F. Kuhnen, H. Wöhrle, H. Baumann: Z. Anorg. Allgm. Chem. *353*, 287 (1967)
2.13 C. Perrin, M. Sergent, F. Le Traon, A. Le Traon: J. Solid State Chem. *25*, 197 (1978)
2.14 A. Morette: Ann. Chim. Paris *19*, 130 (1944)
2.15 A.A. Opalovski, V.E. Fedorov: Izv. Akad. Nauk. SSSR Neorg. Mater. *2*, 443 (1966); V.V. Bakakine, Y.I. Mironov, A.A. Opalovski, V.E. Fedorov: Izv. Sib. Otd. Akad. Nauk. SSSR *3*, 98 (1966); M. Spiesser, J. Rouxel, M. Kerriou, G. Goureaux: Bull. Soc. Chim. *5*, 1427 (1969) M. Spiesser, C. Marchal, J. Rouxel: C.R. Acad. Sci. C*266*, 1583 (1968) M. Spiesser, J. Rouxel: C.R. Acad. Sci. C*265*, 92 (1967)
2.16 R. Chevrel: Thesis No. B112, University of Rennes (1974)
2.17 J. Hauck: Mater. Res. Bull. *12*, 1015 (1977)
2.18 R. Chevrel, M. Sergent, J. Prigent: J. Solid State Chem. *3*, 515 (1971)
2.19 R. Flükiger, R. Baillif, E. Walker: Mater. Res. Bull. *13*, 743 (1978)
2.20 C.K. Banks, L. Kammerdiner, H.L. Luo: J. Solid State Chem. *15*, 271 (1975); K.C. Chi, R.O. Dillon, R.F. Bunshah, S. Alterovitz, D.C. Martin, J.A. Woollam: Thin Solid Films *47*, L9 (1977); K.C. Chi, R.O. Dillon, R.F. Bunshah, S. Alterovitz, J.A. Woollam: Thin Solid Films *54*, 259 (1978); P. Przyslupski, R. Horyn, J. Szymaszek, B. Gren: Solid State Commun. *28*, 869 (1978); T. Luhmann, D. Dew-Hughes: J. Appl. Phys. *49*, 936 (1978); C. Rossel, B. Seeber, Ø. Fischer: *Proc. Conf. Ternary Superconductors*, Lake Geneva, WI, Sept. 1980, ed. by G.K. Shenoy, B.D. Dunlap, F.Y. Fradin (North-Holland, Amsterdam 1981); M. Decroux, Ø. Fischer, R. Chevrel: Cryogenics *17*, 291 (1977)
2.21 Ø. Fischer, M.B. Maple (eds.): *Superconductivity in Ternary Compounds II*, Topics in Current Physics (Springer, Berlin, Heidelberg, New York) forthcoming
2.22 M. Sergent, R. Chevrel: J. Solid State Chem. *6*, 433 (1973)
2.23 M. Sergent, R. Chevrel, C. Rossel, Ø. Fischer: J. Less Common Met. *58*, 179 (1978)
2.24 M. Marezio, P.D. Dernier, J.P. Remeika, E. Corenzwit, B.T. Matthias: Mater. Res. Bull. *8*, 657 (1973)
2.25 J. Guillevic, H. LeStrat, D. Grandjean: Acta Crystallogr. B*32*, 1342 (1976)
2.26 K. Yvon: In *Current Topics in Materials Science*, Vol.3, ed. by E. Kaldis (Elsevier, Amsterdam 1979) p.53
2.27 R. Baillif, K. Yvon, R. Flukiger, J. Muller: J. Low Temp. Phys. *37*, 231 (1979)
2.28 R. Chevrel, M. Sergent, J. Prigent: Mater. Res. Bull. *9*, 1487 (1974)
2.29 O. Bars, D. Grandjean, A. Meerschaut, M. Spiesser: Bull. Soc. Fr. Mineral. Cristallogr. *93*, 498 (1970)
2.30 K. Yvon, R. Chevrel, M. Sergent: Acta Crystallogr. B*36*, 685 (1980)
2.31 M. Potel: Private communication
2.32 R. Chevrel, J. Guillevic, M. Sergent: C.R. Acad. Sci. *271*, 1240 (1970)
2.33 R. Chevrel, M. Sergent: Unpublished results
2.34 A. Espelund: Acta Chem. Scand. *21*, No.3 (1967)
2.35 Ø. Fischer, A. Treyvaud, R. Chevrel, M. Sergent: Solid State Commun. *17*, 721 (1975)
2.36a R. Chevrel, C. Rossel, M. Sergent: J. Less Common. Met. *72*, 31 (1980)
2.36b N.E. Alekseevski, N.M. Dobrovolskii, V.I. Tzebro: JETP Lett. *23*, 639 (1976)
2.37 M. Potel, R. Chevrel, M. Sergent, M. Decroux, Ø. Fischer: C.R. Acad. Sci. C*288*, 429-432 (1979)
2.38 H. Noël, R. Chevrel, M. Sergent: Actinides 1981, Los Angeles, CA, 10-15 sept. 1981
2.39 C.H. De Novion, D. Damien, H. Hubert: J. Solid State Chem. *39*, 360 (1981)
2.40 R.N. Shelton, A.C. Lawson, D.C. Johnston: Mater. Res. Bull. *10*, 297 (1975)
2.41 D. Damien, C.H. De Novion, J. Gal: Solid State Commun. *38*, 437 (1981)
2.42 R. Flükiger, A. Junod, R. Baillif, P. Spitzli, A. Treyvaud, A. Paoli, H. Devantay, J. Muller: Solid State Commun. *23*, 699 (1977)

2.43 D.C. Johnston, R.N. Shelton, J.J. Bugaj: Solid State Commun. *21*, 949 (1977)
2.44 O. Bars, J. Guillevic, D. Grandjean: J. Solid State Chem. *6*, 48 (1973)
2.45 O. Bars, J. Guillevic, D. Grandjean: J. Solid State Chem. *6*, 335 (1973)
2.46 J. Guillevic, O. Bars, D. Grandjean: J. Solid State Chem. *7*, 158 (1973)
2.47 J. Guillevic, O. Bars, D. Grandjean: Acta Crystallogr. B*32*, 1338 (1976)
2.48 C. Perrin, R. Chevrel, M. Sergent, Ø. Fischer: Mater. Res. Bull. *14*, 1505 (1979)
2.49 K. Yvon, A. Paoli, R. Flükiger, R. Chevrel: Acta Crystallogr. B*33*, 3066 (1977)
2.50 K. Yvon, R. Baillif, R. Flükiger: Acta Chrystallogr. B*35*, 2859 (1979)
2.51 R. Chevrel, J. Prigent, M. Sergent: Rev. Gen. Electr. *88*, 114 (1979)
2.52 N.E. Alekseevski, N.M. Dobrovolskii: Dokl. Akad. Nauk. SSSR *242*, 87 (1978)
2.53a R. Chevrel, M. Sergent, K. Yvon: Unpublished results
2.53b R. Chevrel: In *Superconducting Materials Science: Metallurgy, Fabrication and Applications*, ed. by S. Foner, B.B. Schwartz, Proc. NATO ASI 1980 (Plenum, New York 1981) Chap.10
2.53c A.C. Lawson: Mater. Res. Bull. *7*, 773 (1972)
2.53d A.C. Lawson, R.N. Shelton: Mater. Res. Bull. *12*, 375 (1977)
2.53e Ø. Fischer, M. Decroux, R. Chevrel, M. Sergent: *Superconductivity in d- and f-band metals*, Proc. 2nd Rochester Conf., ed. by D.H. Douglas (Plenum, New York 1977) pp.175-187
2.53f R. Flükiger, H. Devantay, J.L. Jorda, J. Muller: IEEE Trans. M*13*, 818 (1977)
2.54 M. Sergent, Ø. Fischer, M. Decroux, C. Perrin, R. Chevrel: J. Solid State Chem. *22*, 87 (1977)
2.55 J. Bolz, J. Hauck, F. Pobell: Z. Phys. B*25*, 351 (1976)
2.56 C.W. Kimball, L. Weber, G. Van Landuyt, F.Y. Fradin, B.D. Dunlap, G.K. Shenoy: Phys. Rev. Lett. *36*, 412 (1976)
2.57 K. Yvon: Solid State Commun. *25*, 327 (1978)
2.58 O.K. Andersen, W. Klose, H. Nohl: Phys. Rev. B*17*, 1209 (1978)
2.59 L.F. Mattheiss, C.Y. Fong: Phys. Rev. B*15*, 1760 (1977)
2.60 D.W. Bullett: Phys. Rev. Lett. *39*, 664 (1977)
2.61 T. Jarlborg, A.J. Freeman: Phys. Rev. Lett. *44*, 178 (1980)
2.62 R. Schollhorn, M. Kümpers, J.O. Besenhard: Mater. Res. Bull. *12*, 781 (1977)
2.63 R. Schollhorn, M. Kümpers, D. Plorin: J. Less Common Met. *58*, 55 (1978)
2.64 G.J. Dudley, K.Y. Cheung, B.C.H. Steele: Proc. 2nd Int. Conf. on Solid Electrolytes, St. Andrews, Scotland, Sept. 20-22, 1978
2.65 A.M. Umarji, G.V. Subba Rao, M.P. Janawadkar, T.S. Radhakrishna: J. Phys. Chem. Solids *41*, 421 (1980)
2.66 A. Perrin, R. Chevrel, M. Sergent, Ø. Fischer: J. Solid State Chem. *33*, 43 (1980)
2.67 F.J. Culetto, F. Pobell: Mater. Res. Bull. *14*, 473 (1979)
2.68 A. Perrin, M. Sergent, Ø. Fischer: Mater. Res. Bull. *13*, 259 (1978)
2.69 R. Chevrel, M. Sergent, Ø. Fischer: Mater. Res. Bull. *10*, 1169 (1975)
2.70 F.S. Delk, M.J. Sienko: Inorg. Chem. *19*, 1352 (1980)
2.71 F.Y. Fradin, J.W. Downey, T.E. Klippert: Mater. Res. Bull. *11*, 993 (1976)
2.72 A.M. Umarji, G.V. Subba Rao, V. Sankaranarayanan, G. Rangarajan, R. Srinivasan: Mater. Res. Bull. *15*, 1025 (1980)
2.73 F.Y. Fradin, J.W. Downey: Mater. Res. Bull. *14*, 1525 (1979)
2.74 J.D. Jorgensen, D.G. Hinks, F.J. Rotella: Proc. Conf. Ternary Superconductors, Lake Geneva, WI, Sept. 1980, ed. by G.K. Shenoy, B.D. Dunlap, F.Y. Fradin (North-Holland, Amsterdam 1981)
2.75 A. Grüttner, K. Yvon, R. Chevrel, M. Potel, M. Sergent, B. Seeber: Acta Crystallogr. B*35*, 285 (1979)
2.76 R. Chevrel, M. Sergent, B. Seeber, Ø. Fischer, A. Grüttner, K. Yvon: Mater. Res. Bull. *14*, 567 (1979)
2.77 R. Chevrel, M. Potel, M. Sergent, M. Decroux, Ø. Fischer: Mater. Res. Bull. *15*, 867 (1980)
2.78 R. Chevrel, M. Potel, M. Sergent: Proc. Conf. Ternary Superconductors, Lake Geneva, WI, Sept. 1980, ed. by G.K. Shenoy, B.D. Dunlap, F.Y. Fradin (North-Holland, Amsterdam 1981)
2.79 M. Potel, R. Chevrel, M. Sergent: Acta Crystallogr. B*37*, 1007 (1981)
Ø. Fischer, B. Seeber, M. Decroux, R. Chevrel, M. Potel, M. Sergent: Proc. Conf. on Superconductivity in d- and f-Band Metals, La Jolla, CA, 1979

2.80 B. Seeber, M. Decroux, Ø. Fischer, R. Chevrel, M. Sergent, A. Grüttner: Solid State Commun. *29*, 419 (1979)
2.81 R. Chevrel, M. Potel, M. Sergent, M. Decroux, Ø. Fischer: J. Solid State Chem. *34*, 247 (1980)
2.82 M. Potel, R. Chevrel, M. Sergent: Acta Crystallogr. B*36*, 1319 (1980)
2.83 M. Potel, R. Chevrel, M. Sergent, J.C. Armici, M. Decroux, Ø. Fischer: J. Solid State Chem. *35*, 286 (1980)
2.84 M. Potel, R. Chevrel, M. Sergent: Acta Crystallogr. B*36*, 1545 (1980)
2.85 J.C. Armici, M. Decroux, Ø. Fischer, M. Potel, R. Chevrel, M. Sergent: Solid State Commun. *33*, 607 (1980)
2.86 Ø. Fischer: Appl. Phys. *16*, 1 (1978)

3. Structure and Bonding of Ternary Superconductors

K. Yvon

With 7 Figures

Relationships between structural parameters and superconducting properties of ternary compounds have been discussed recently [3.1]. In this chapter the emphasis is placed on their crystal chemistry. Of particular interest are the factors governing structural stability, such as the valence, atomic size, and electronegativity of the constituents, and the formation of chemical bonds. Since these factors limit the occurrence of stable compounds within a given structural series, their study is useful for the search of new superconducting materials.

3.1 Introductory Comments

In this chapter the two largest families of ternary superconductors presently known, i.e., the $PbMo_6S_8$-type chalcogenides and the $CeCo_4B_4$-type borides, are reviewed and their bonding is investigated. For comparison, a few closely related compounds with similar bonding are included in the discussion. Special attention is given to the environment of the transition metal (T) atom and, in particular, to the formation of T-X bonds (X: nonmetal) and T-T bonds. While the former bonds lead to extended structural frameworks which are essential for the overall stability and the lattice properties of the crystals, the latter bonds stabilize more or less isolated T-atom clusters which presumably contribute less to cohesion but play an important role for the electronic properties.

A central question concerning the formation of the T atom clusters is their valence electron requirement. In general it is unknown and cannot be easily determined without detailed theoretical calculations. In this chapter an attempt will be made to estimate its value for a few compounds from interatomic distances. According to valence bond theory [3.2], a metal-metal bond can be considered as an electron-deficient, covalent, i.e., two-center two-electron bond, the strength of which increases as the separation of the metal atoms is decreased. Its electron requirement, 2n, is given by Pauling's [3.2] empirical relation

$$d(n) = d(1) - 0.6 \log n \quad , \tag{3.1}$$

where n represents the *bond order* of a bond having the length d(n) and d(1) is twice the tabulated [3.2] single-bond (n = 1) radius [Å] for a particular element.

Although the use of a unique set of single-bond radii for different structure types is rarely justified and may even become problematic even if applied to a restricted series of isostructural compounds, valence-bond theory often provides a good starting point for a qualitative understanding of the bonding. Compared to more refined bond descriptions such as molecular orbital (MO) theory,[1] its main virtues are conceptional ease and the possibility to discuss numerical quantities which are easy to measure, such as bond lengths and bond angles.

3.2 Ternary Molybdenum Chalcogenides MMo_6X_8 (M: Metal, X: Chalcogen)

These compounds form the largest class of ternary superconductors presently known. Their physical properties have been reviewed recently [3.3], and a study of relationships between structure and properties has been reported [3.4].

3.2.1 Structure

So far, accurate structural parameters for more than 30 MMo_6X_8 representatives are available (Chap.2 and [3.4-9]). Their prototype is $PbMo_6S_8$ that has a filled Mo_3Se_4-type structure [3.9] and derives from a rhombohedral stacking of quasi-cubic Mo_6S_8 building blocks. Each unit consists of a slightly elongated (along the ternary axis) Mo_6 octahedron, the faces of which are bridged by S atoms (Fig.3.1).

Building blocks Mo_6X_8 (X: halogen) of the same geometry also occur in transition metal halides, such as $MoCl_2$ [3.11] and Nb_6I_{11} [3.12]. In these compounds they represent quasi-molecular units which are connected to relatively open and unstable structures via additional halogen ligands. In the chalcogenides, however, the Mo_6S_8 units are directly interconnected via short Mo-S *inter*cluster bonds (see d^{inter}_{Mo-S} in Fig.3.2a) and form a compact structure in which each unit is covalently bonded to six surrounding units. This results in high structural stability as can be seen, for instance, from the melting points, typically around 1700°C (Chap.4 and [3.13]). Evidence for strong Mo-S *inter*cluster bonds comes also from a study of the principal directions of atomic thermal vibrations. Both the Mo and S atoms vibrate preferentially in directions away from the Mo-S intercluster bond axis [3.4]. Thus, the Mo_6X_8 building blocks in the chalcogenides can hardly be considered as quasi-molecular units from either a bonding or a lattice dynamics point of view.

An analysis of the interatomic distances shows that the Mo atoms have nine ligands (four Mo and four S atoms belonging to the same $[Mo_6S_8]$ unit, and one S atom to an adjacent one). Whereas the average Mo-S distances ($\bar{d}$[Mo-S] ~ 2.5 Å) are

---

1 In MO theory the bond order corresponds to the difference between the number of electron pairs which are in bonding states and those which are in antibonding states [3.10].

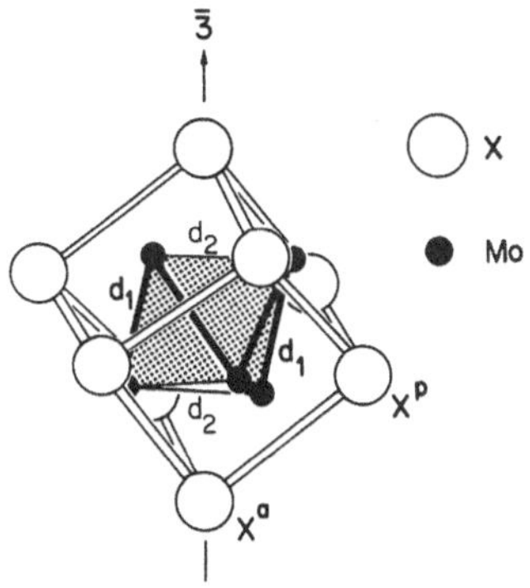

Fig.3.1. The Mo_6X_8 building block (X = S,Se,Te) in $PbMo_6S_8$-type compounds). X^a: axial, X^p: peripheral chalcogen atom. The Mo-Mo bonds perpendicular to the ternary axis ($\underline{d}_2$, thin lines) are shorter than those which are not perpendicular to it ($\underline{d}_1$, thick lines)

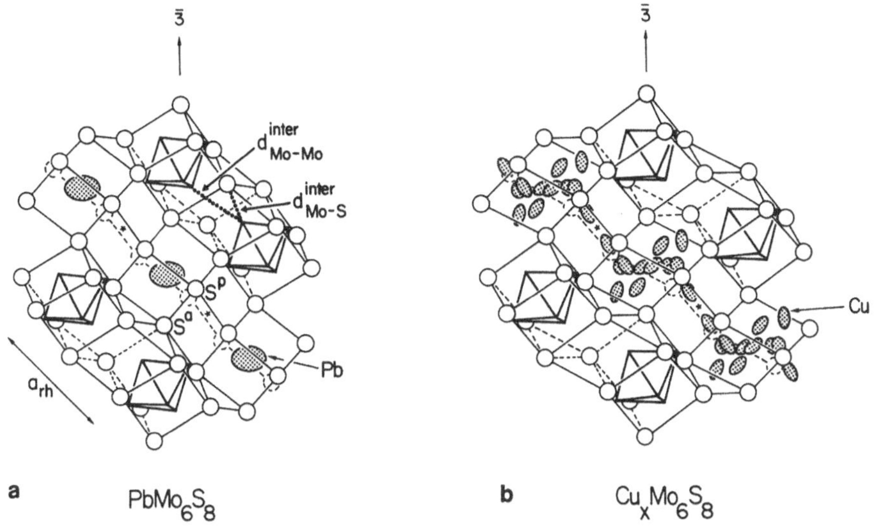

Fig.3.2a,b. The arrangement of the Mo_6S_8 units and the distribution of the M atoms in a) $PbMo_6S_8$ and b) $Cu_xMo_6S_8$

typical for covalent Mo-S bonds, the average Mo-Mo distances ($\bar{d}$[Mo-Mo] ~ 2.70 Å) are longer than the Mo-Mo single-bond distance given by Pauling (Mo : d(1) = 2.592 Å [3.2]) but slightly shorter than the Mo-Mo contact distance in the *bcc* metal (d[Mo-Mo] = 2.73 Å). In contrast to the Mo_6 clusters in the nonmetallic halides, which presumably do not interact with each other because of their wide separation in the structure, the Mo_6 clusters in the metallic chalcogenides do interact because they are only 3.1-3.2 Å (sulfides) apart (see d^{inter}_{Mo-Mo} in Fig.3.2a). As shown by theoretical band-structure calculations (Chap.6) these interactions are relatively weak; i.e., they do not contribute much to the cohesion of the structure. They do, however, contribute to the stability and the size of the Mo_6 atom clusters (see below). They are also very important for the electronic properties because they provide a three-dimensional network of possible electronic conduction paths in the crystal.

The Pb atoms are situated between the Mo_6S_8 units and have an unusual atomic environment. They occupy the center of a compressed S atom cube of $\bar{3}$ symmetry and form strong 180° bonds with the two S^a atoms on the ternary axis ($d[Pb\text{-}S^a] = 2.80$ Å), and weak bonds with the six S^p atoms away from it ($d[Pb\text{-}S^p] = 3.11$ Å) (Fig.3.2a). Consequently, their thermal vibrations are strongly anisotropic and anharmonic, and appear mainly in the low-frequency part of the phonon spectrum (Chap.7). The Pb-S distances suggest that Pb is in a divalent (Pb^{2+}) state.

The axial S^a and peripheral S^p atoms are crystallographically independent but have a similar metal atom environment. Both are covalently bonded to four metal atoms, the peripheral one to three Mo atoms of the same and one Mo atom of the different Mo_6 cluster, and the axial one to three Mo atoms of the same Mo_6 cluster and to one Pb atom. The distances between nearest S atoms ($\bar{d}[S\text{-}S] = 3.56$ Å) are all nonbonding and typical for sulfur in a divalent (S^{2-}) state. The crystal chemical formula [3.14] of $PbMo_6S_8$ can be written as $Pb^{[2S]}$⊗ $[Mo_6^{[4Mo+5S]}]S_6^{[4Mo]}S_2^{[3Mo+1Pb]}$.

3.2.2 Occurrence

So far more than a hundred compounds with this structure are known. Most have been obtained by replacing Pb by other nontransition metals, such as alkali and alkaline earth metals – Al, In, Tl, Sn [3.3], Ga, and Hg [3.15] – by lanthanides and actinides, and by transition metals of the 3d group – Sc, Cr, Fe, Cu, Ni, and Zn [3.3]. Depending on their ionic size, electronic configuration, and concentration, the metal atoms do not occupy the exact center of the chalcogen atom cube but are more or less delocalized [3.4]. Small ions (Cu^+, Ni^{2+}, In^{3+}) are usually more delocalized than big ions (Ag^+, La^{3+}). As the delocalization increases, the structure contracts along the ternary axis, (i.e., its rhombohedral cell angle increases) and a gradual transition occurs from a quasi-linear (big M atoms) to a tetrahedral (small M atoms) chalcogen atom coordination of the metal atoms. In compounds containing strongly delocalized M atoms, such as $Cu_xMo_6S_8$ ($1 \leq x \leq 4$), more than one M atom per formula unit can be inserted in the structure. At room temperature they are disordered over 12 interstices (Fig.3.2b) and some (mainly Cu^+) show high mobility [3.16]. At low temperatures they may order by condensing into pairs, such as in $Cu_{1.8}Mo_6S_8$ [3.17] and $Fe_2Mo_6S_8$ [3.7], both of which are characterized by a triclinic distortion of the lattice and a triclinic deformation of the Mo_6 octahedron.

Attempts to substitute Mo by transition metals of the 5d or 6d group, such as T = Ru, Rh or Re, lead to the discovery of compounds containing "mixed" octahedral $(Mo,T)_6$ clusters [3.18]. Complete substitution of Mo, however, has never been achieved. As can be seen from the limiting compositions, $Mo_2Re_4Se_8$, $Mo_4Ru_2Se_8$ and $Mo_{4.66}Rh_{1.33}Te_8$, the range of substitution decreases as the group number of the T element is increased. Sometimes unexpected miscibility gaps exist, such as that in the Ru substituted selenide $Mo_{6-x}Ru_xSe_8$ ($0 \leq x \leq 1.3$, $x \sim 2$ [3.18]. It is also surprising that so far, tungsten can not be substituted for molybdenum in any significant amount. This contrasts with the transition metal halides among which tungsten-based compounds containing W_6 octahedra, such as WX_2 (X = Cl, Br, I) [3.11], are not uncommon.

In most MMo_6S_8 compounds, sulfur can be replaced by other chalcogens (X = O, Se, Te). However, complete substitution is only possible for Se and Te. Generally, the number of compounds and the homogeneity range decreases as one goes from the sulfides to the selenides and tellurides. For instance, no stable tellurides containing big divalent or trivalent metal ions such as Pb^{2+}, Sn^{2+}, or Ln^{3+} (Ln: lanthanide) have been reported as yet. The homogeneity range, x, of the selenide $Cu_xMo_6Se_8$ ($0 \leq x \leq \sim 3$) is reduced with respect to that of the corresponding sulfide [3.19]. Substitution of sulfur by oxygen is only possible to concentrations up to 25% [3.20]. Halogen-substituted compounds also exist, such as $Mo_6S_6Br_2$ [3.21], which shows a relatively high superconducting transition temperature ($T_c \sim 13.8$ K). For comparison, the nonsubstituted binary sulfide Mo_6S_8 is superconducting only below 1.8 K [3.19].

3.2.3 Bonding

The preceeding survey shows that $PbMo_6S_8$ has a very flexible structure, in particular with respect to substitution by elements of different atomic size. Thus geometrical factors are presumably not very critical for its stability. By contrast, electronic factors are rather critical because the $PbMo_6S_8$-type compounds occur only within a narrow valence-electron range. The interval extends over four electrons, the most electron-poor compounds being the binaries Mo_6X_8 (X = S, Se, Te) having 84 valence electrons, and the most electron-rich ones $Cu_4Mo_6S_8$, $Mo_4Ru_2S_8$, $Mo_2Re_4S_8$, etc., having 88 valence electrons per formula unit.

At the electron-rich limit the bonding can be rationalized in terms of covalent two-center two-electron bonds. Each chalcogen atom forms four such bonds to its metal ligands, thereby using 64 valence electrons per formula unit. This leaves 88 - 64 = 24 electrons available for metal-metal bonding on the cluster. By assuming formally two center bonds[2] this number allows for 12 covalent bonds along the edges of the metal atom octahedron. As one might expect from such a description, compounds with 24 valence electrons per T_6 cluster such as $Mo_2Re_4Se_8$ and $Mo_4Ru_2Se_8$ are semiconducting [3.18] because the metal-metal bonds are formally saturated. Compounds with less than 24 valence electrons are metallic [3.3] because the metal-metal bonds are formally electron deficient.

An alternative way to rationalize the bonding is that usually followed for ionic crystals. As suggested from the interatomic distances and cell volumes [3.4], the

2 Octahedral metal atom clusters which can be described formally by three-center (two-electron) bonds occur in halides such as T_6X_{14} (T = Nb, Ta; X = Cl, Br, I [3.11]) which contain $[T_6X_{12}]$ building blocks and have 16 electrons per T_6 cluster.

formal electron charges on the atoms are -2 for the chalcogens, -1 for the halogens, +1 for Cu, Ag, and alkali metals, +2 for Pb, Sn and alkaline earths, +3 for Al, Ga, In(?), and most of the lanthanides, and +4 for Th and U. Consequently, the formal oxidation state of the metal component forming the T_6 cluster varies between +2 and +2.5, which corresponds to an average number of 3.5 to 4 valence electrons per cluster atom (also called *cluster VEC* [3.4]). These numbers agree with those found above by using a purely covalent bond description.

The true bonding situation, of course, is intermediate between those of the ionic and covalent limits. Assuming, for instance, an average 60% ionic character for all Mo-S bonds, the net atomic charges on both Mo and S are approximately zero. Clearly, these values correspond better to reality than those indicated by either the ionic or the covalent limit.

Molecular orbital and theoretical band-structure calculations (Chap.6) confirm the main features of the above bonding models. In particular, they predict the "magic" number of 24 electrons per T_6 cluster and show that the *cluster-VEC* parameter plays an important role for the stability and the electronic properties of these substances. In the halide complex $[Mo_6^{2+}X_8^{1-}]^{4+}$, for instance, there are 24 orbitals available for the formation of Mo-Mo bonds [3.10], i.e., this unit is particularly stable if the *cluster VEC* is four electrons per Mo atom. In the chalcogenides the situation is very similar. The nonmetal p states are all occupied, and the T metal d states are divided by a forbidden energy gap into 12 bonding and 18 antibonding states. The Fermi level E_F is situated just below this gap and falls into narrow bands of mainly 4d character. Thus, assuming rigid bands, an increase of the *cluster VEC* corresponds to an upward shift of E_F and a gradual filling of the conduction bands. Semiconductivity occurs if the *cluster VEC* is four, i.e., if the bands are filled, and high superconducting transition temperatures are found if the *cluster VEC* is close to 3.7, i.e., if there are about two holes in the conduction bands [3.4].

The electronic states of the conduction bands were first thought to be *non*bonding (Chap.6, and references therein). However, subsequent X-ray studies of the Mo-Mo *intracluster* distances as a function of the band filling suggested that the states at E_F have significant bonding character [3.22].As can be seen in Fig. 3.3, the Mo_6 octahedra contract as the number of valence electrons on the cluster is increased. This trend is difficult to predict from MO and band-structure calculations, but can be expected from valence bond theory. Interestingly, the contraction is anisotropic, such that the electron-rich compounds have generally the most regular Mo_6 octahedra. This behavior is consistent with theoretical calculations and has been explained by an energy-band effect. It involves the presence of at least two Mo-d energy bands at E_F, the overlap of which changes as a function of the cluster contraction and the *cluster VEC* [3.4]. However, the importance of this effect is difficult to evaluate

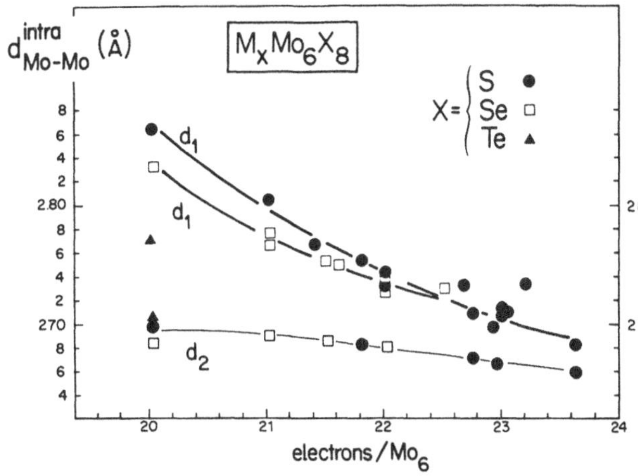

Fig.3.3. The contraction of the Mo-Mo intracluster bonds $\underline{d}_1$ and $\underline{d}_2$ (Fig.3.1) as a function of the number of valence electrons on the Mo_6 cluster in MMo_6X_8 chalcogenides

quantitatively because the elongation of the Mo_6 octahedra may also depend on other factors, such as the nature and the packing of the anions. As can be seen in Fig.3.3, the octahedra tend to become more regular as the size of the anions at a particular *cluster VEC* is increased $[r(S^{2-}) < r(Se^{2-}) < r(Te^{2-})]$. On the other hand, if one considers only compounds belonging to a given chalcogen series, the influence of the anion packing on the elongation of the Mo_6 octahedra appears to be small. In the sulfide series, for instance, the Mo-Mo *intracluster* distances $\underline{d}_1$ and $\underline{d}_2$ within the isoelectronic groups of compounds MMo_6S_8 (M = $Cu_{\sim 2}$, Sn, Pb and M = Er, La) change very little although the rhombohedral cell angle or cell volume and thus the chalcogen packing are rather different (Table 3.1). It is also remarkable that the extrapolated ratio $\underline{d}_1/\underline{d}_2$ of the electron-rich (i.e., 24-electron) compounds based on sulfur is close to unity (Fig.3.3).

With regard to the stability of the MMo_6X_8 compounds one can assume that the nonpolar metal-metal bonds contribute much less to the cohesion than the polar metal-nonmetal bonds. Yet the metal-metal bonds clearly have a stabilizing influence on the structure. This can be seen, for instance, from binary Mo_6S_8. It is one of the most electron-deficient compounds and has the biggest and most elongated Mo_6 octahedron. Contrary to other, more electron-rich compounds, which have a smaller and more regular octahedron, it is relatively unstable [3.23]. This is presumably due to its weaker Mo-Mo *intracluster* bonds, which tend to destabilize the structure in favor of the more stable binary sulfide Mo_2S_3. On the other hand, the halogen-substituted compound $Mo_6S_6Br_2$, which has two valence electrons more and a smaller Mo octahedron, is relatively stable [3.21], which suggests that it is stabilized by reinforced Mo-Mo bonds rather than by Br-Br bonds, as has been claimed.

We now try to estimate the electron requirement of the Mo_6 clusters by calculating bond orders from the logarithmic relation (3.1). Using Pauling's single-bond radius of Mo{r(1) = 1.296 Å [3.2]} and the experimentally measured Mo-Mo bond

Table 3.1. Mo-Mo bond lengths [Å], rhombohedral cell angle, hexagonal cell volume [Å^3], sum of bond orders[a] and *cluster VEC* [e^-/Mo_6] of MMo_6X_8 compounds

	d_1	d_2	d^{inter}_{Mo-Mo}	α	V	$2\Sigma n$	*cluster VEC*
Mo_6S_8	2.862	2.698	3.084	91.6	798	17.7	20
Mo_6Se_8	2.836	2.684	3.266	91.6	884	18.3	20
Mo_6Te_8	2.772	2.700	3.674	92.6	1051	19.3	20
$AgMo_6S_8$	2.804	2.706	3.154	91.6	813	18.6	21
$AgMo_6Se_8$	2.776	2.701	3.378	91.7	913	19.0	21
$PbMo_6S_8$	2.732	2.679	3.262	89.3	840	21.7	22
$PbMo_6Se_8$	2.734	2.697	3.490	89.2	948	20.5	22
$SnMo_6S_8$	2.742	2.681	3.228	89.5	829	21.3	22
$LaMo_6S_8$	2.707	2.667	3.238	88.7	832	23.2	23
$ErMo_6S_8$	2.713	2.654	3.144	89.5	803	23.9	23

[a]The sums of bond orders, Σn, are calculated for $6 \times d_1$, $6 \times d_2$, and $6/2 = 3 \times d^{inter}_{Mo-Mo}$ distances, using a modified Mo-Mo single-bond distance of $\bar{d}'(1) = 2.67$ Å. Data are taken from [3.4]

lengths of, say, Mo_3S_4 as listed in Table 3.1, one obtains for the sum of the bond orders over the 12 Mo-Mo *intracluster* bonds and the 6 Mo-Mo *intercluster* bonds (counted half), $\Sigma n = \sim 6.5$. This corresponds to approximately 13 valence electrons per Mo_6 cluster, which is far below the expected number of 20 electrons. Thus the single-bond radius of Mo in (3.1) needs to be increased for this class of compounds. A value which appears reasonable is d'(1) = 2.67 Å because it corresponds to the extrapolated average Mo-Mo *intracluster* distance for 24 electron compounds as shown in Fig.3.3. The valence electron counts per Mo_6 cluster calculated with this modified single-bond distance are given in Table 3.1. As expected they agree well with those calculated from the *cluster-VEC* parameter for electron-rich compounds such as $LnMo_6S_8$ (Ln = La, Er), and they decrease as the number of electrons on the Mo_6 cluster is decreased. However, they agree poorly for electron-deficient compounds such as the binaries Mo_3X_4. Moreover, systematic differences occur between the sum of bond orders in the sulfides, selenides, and tellurides. As pointed out earlier in the text, the Mo_6 octahedra in the selenides are systematically smaller than those in the sulfides, and that in the telluride Mo_3Te_4 is smaller than that in the selenide Mo_3Se_4 (Fig.3.3). Interestingly, this trend is opposite to that found in the Mo-cluster halides in which the Mo_6 octahedra expand as the light halogens are replaced by their heavier congeners [3.11]. The expansion in the latter compounds has been interpreted by a matrix effect [3.11] resulting from the different size of the repelling anions. An analogous effect presumably also exists in the chalcogenides (see below) but is overcompensated by at least two other effects. One is related to the different electronegativities of the chalcogens. As one goes from the sulfides to the selenides and tellurides, the ionicity of the Mo-X bonds decreases, which could lead to an increase of the formal electric charge on the Mo_6 cluster and thus to a shortening of the Mo-Mo *intracluster* bonds. An observation which supports this hypothesis [3.4] is the apparent absence of Mo_6-cluster selenides having more than about 23 cluster electrons (assuming Se^{2-}) and the apparent absence of Mo_6-cluster tellurides having more than about 22 cluster electrons (assuming Te^{2-}) per formula unit. The only selenides and tellurides reported so far which have higher formal cluster electron counts are those containing mixed $(Mo,T)_6$ clusters, such as $Mo_2Re_4Se_8$ (24 cluster electrons). Their structure, however, has not yet been sufficiently characterized in order to allow a discussion of its bonding.

Another effect which presumably also leads to a strengthening of the Mo-Mo *intracluster* bonds is the weakening of the Mo-Mo *intercluster* bonds. This can be seen from the metal-metal distances listed in Table 3.1 which show that a stretching of the Mo-Mo *intercluster* bonds (d^{inter}_{Mo-Mo}) in isoelectronic compounds correlates with a shortening of the Mo-Mo *intracluster* bonds $\underline{d}_1$. Such a situation occurs, for instance, if small chalcogen atoms (S) are replaced by big chalcogen atoms (Se, Te), or if small M atoms (Sn, Er) are replaced by their bigger congeners (Pb, La). This trend is expected from valence bond theory and can be illustrated qualitatively by using a model in which the Mo-Mo *intra*- and *intercluster* bonds in the structure are replaced by mechanical springs. As the Mo_6 clusters are separated, their wave functions presumably overlap less; i.e., the force constants of the Mo-Mo *intercluster* bonds decrease. Since the latter bonds act against the Mo-Mo *intracluster* bonds, a contraction of the Mo_6 octahedra should result. However, it appears to be difficult at present to assess the relative importance of this effect because the relative strengths of the different Mo-Mo bonds (i.e., their force constants) are unknown. In particular, it is not clear how the rather long Mo-Mo *intercluster* distances [d^{inter}_{Mo-Mo} $\sim$3.2 Å (sulfide), $\sim$3.4 Å (selenides), $\sim$3.7 Å (tellurides)] should be properly included in a bond-order calculation because on the one hand the amplitude of the Mo wave functions at such large distances from the nuclei are poorly defined, and on the other hand their spatial extent and direction presumably varies as a function of the formal charge on the Mo_6 cluster.

The influence of the matrix effect mentioned above, i.e., the stretching of the Mo-Mo *intracluster* distances by the repelling X anions which cap the triangular faces of the Mo_6 octahedron, can be seen from the pseudoternary compound $Mo_6S_6Br_2$. Due to the substitution of the axial S^{2-} ions by the bigger Br^- ions, the Mo triangles perpendicular to the ternary axis (i.e., those capped by the Br^- ions) expand (d_2 = 2.73 Å) with respect to those in nonsubstituted binary Mo_6S_8 (d_2 = 2.70 Å) or isoelectronic $PbMo_6S_8$ (d_2 = 2.68 Å). This behavior is striking because in all $PbMo_6S_8$ type chalcogenides, the Mo_6 cluster dimensions perpendicular to this axis are remarkably constant, i.e., they change only little as a function of the *cluster VEC* or the size of the chalcogen atoms (Fig.3.3).

Altogether, the above discussion shows that at least for this class of compounds one cannot give a unique Mo-Mo "single-bond" distance which describes the average Mo-Mo distances of all representatives in a satisfactory manner. This difficulty is well known from the study of other metal-metal bonded compounds [3.10] and is mainly due to the influence of the nonmetal ligands, as will appear more clearly in the following class of compounds.

3.3 Ternary Molybdenum Chalcogenides Built Up by Condensed $Mo_{3n}X_{3n+2}$ Units (X = S, Se, Te)

3.3.1 Structure

There exists a series of Mo-cluster chalcogenides that are closely related to $PbMo_6S_8$ from both a structural and a bonding point of view. Their structures can be described by a stacking of $[Mo_6X_8]$, $[Mo_9X_{11}]$, and $[Mo_{12}X_{14}]$ building blocks which contain, respectively, Mo_6 octahedra and clusters of face-sharing Mo_9 bioctahedra and Mo_{12} trioctahedra. As shown in Fig.3.4 these units belong to a structural series of composition $Mo_{3n}X_{3n+2}$ (n = 2,3,4,∞) which derives from a progressive condensation of $[Mo_6X_8]$ units along a common ternary axis. The end member of this series represents a fiber of composition $^1_\infty[Mo_3X_3]$ which contains quasi-infinite columns of face-sharing Mo_6 octahedra. An alternative way to derive the $[Mo_{3n}X_{3n+2}]$ units is by stacking triangular shaped, almost planar $[Mo_3X_3]$ subunits in staggered positions and by capping the Mo_3 triangles of the terminal $[Mo_3X_3]$ subunits with additional X atoms (Fig.3.4). (For a detailed description of these materials, see also Chap.2).

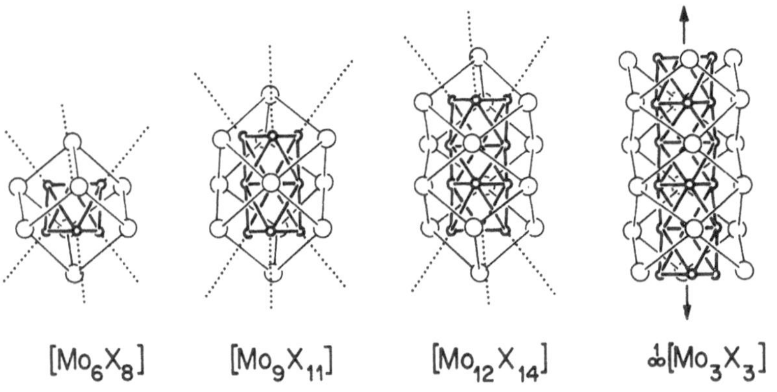

Fig.3.4. Cluster condensation in Mo-chalcogenides. The dotted lines present the approximate directions of the Mo-Mo and Mo-X *intercluster* bonds

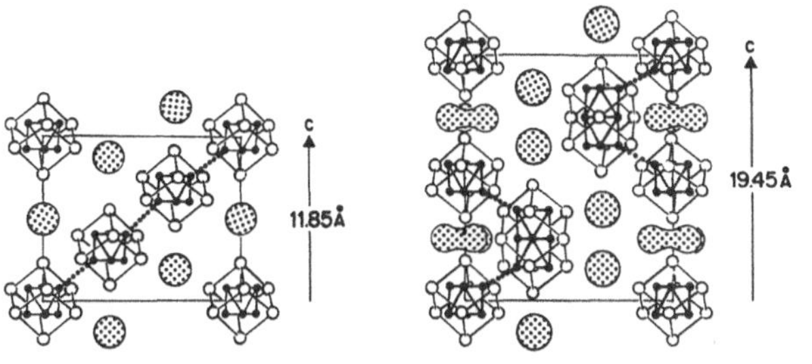

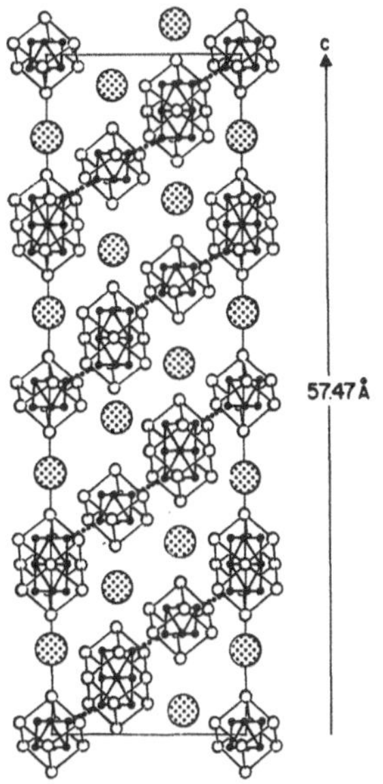

$InMo_6Se_8$ $\qquad$ $In_{\sim3}Mo_{15}Se_{19}$ $\qquad$ $In_2Mo_{15}Se_{19}$

Fig.3.5. The stacking of the [Mo_6Se_8] and [Mo_9Se_{11}] units in $InMo_6Se_8$, $In_{3.3}Mo_{15}Se_{19}$, and $In_2Mo_{15}Se_{19}$ [from 3.25]

The arrangement of the [$Mo_{3n}X_{3n+2}$] units in the crystal depends mainly on the nature and concentration of the ternary metal component, M (M = In, Tl, alkali metal, Ba). A list of structure types and corresponding crystal chemical formulas is given in Table 3.2. From the structural drawings shown in Fig.3.5 one can see that the arrangement of the [$Mo_{3n}X_{3n+2}$] building blocks in the crystal is similar to that of the [Mo_6X_8] units in MMo_6X_8. The general building principle of the different structural types can be described as follows. Each [$Mo_{3n}X_{3n+2}$] unit is connected to six other units by short Mo-X *intercluster* bonds which are formed between the Mo atoms of the terminal Mo_3X_3 subunit of one unit and the X atoms of the terminal Mo_3X_3 subunits of neighboring units. The atomic configurations and bond lengths are similar in all structures. Whereas the Mo and X atoms belonging to the terminal Mo_3X_3 subunits have, respectively, nine ligands (four Mo and four X atoms of the same, and one X atom of a different unit) and four ligands (three Mo atoms of the same, and one Mo atom of a different unit), those belonging to one of the median Mo_3X_3 subunits have, respectively, ten ligands (six Mo atoms and four X atoms) and four ligands (four Mo atoms) which all belong to the same unit. Similar to the Mo_6 clusters in MMo_6X_8, the Mo_6, Mo_9, and Mo_{12} units in the condensed cluster compounds interact only weakly via long metallic *intercluster* bonds (d^{inter}_{Mo-Mo} ~ 3.2 - 3.5 Å) which form a three-dimensional network of possible electronic conduction paths (dotted lines in Fig.3.5).

As a characteristic feature the concentration of Mo-Mo and Mo-X *intercluster* bonds in the crystals decreases as one goes from compounds containing small Mo_{3n}

Table 3.2. Mo-cluster chalcogenides: Structure, X/Mo ratio (X = S, Se, Te), number of Mo-Mo and Mo-X bonds per Mo atom, formal oxidation state (FOS) of Mo, and limiting ionic formula

Compound	Structure	Pearson symbol	X/Mo	Mo-Mo/Mo-X bonds	FOS (Mo)	Limiting ionic formula (□: electron hole)	References
$PbMo_6S_8$	$PbMo_6S_8$	hR15	1.33	2.0/5.0	+2.33	$Pb^{2+}[Mo_6S_8]^{4-}\square^{2+}$	3.24
$In_2Mo_{15}Se_{19}$	$In_2Mo_{15}Se_{19}$	hR216	1.27	2.2/4.8	+2.40	$In_2^+[Mo_6Se_8]^{4-}[Mo_9Se_{11}]^{4-}\square^{6+}$	3.25,26
$In_{\sim 3.3}Mo_{15}Se_{19}$	$In_{2.9}Mo_{15}Se_{19}$	hP74	1.27	2.2/4.8	+2.13	$In_{1.33}^{3+}In_2^+[Mo_6Se_8]^{4-}[Mo_9Se_{11}]^{4-}\square^{2+}$	3.27
$Tl_2Mo_9S_{11}$	$Tl_2Mo_9S_{11}$	hR132	1.22	2.3/4.1	+2.22	$Tl_4^+[Mo_6S_8]^{4-}[Mo_{12}S_{14}]^{4-}\square^{4+}$	3.28
$TlMo_3Se_3$	$TlFe_3Te_3$	hP14	1.00	3.0/4.0	+1.67	$In^+[Mo_3Se_3]^0$ 1e$^-$	3.29,30

clusters to those containing big Mo_{3n} clusters. In $InMo_3Se_3$, which has the $TlFe_3Te_3$ structural type [3.31] and is built up by a hexagonal array of quasiisolated $^1_\infty[Mo_3Se_3]$ fibers, such bonds no longer exist. As expected, this structural trend correlates with marked changes in physical properties. The electrical resistivity, for instance, increases; the critical magnetic field becomes more anisotropic; and the crystals can be more easily cleaved parallel to $\underline{c}$, as the concentration of the Mo-Mo and Mo-X *intercluster* bonds in the structure decreases. It is also apparent that compounds containing condensed $[Mo_{3n}X_{3n+2}]$ units are less favorable for superconductivity than those containing $[Mo_6X_8]$ units only.

3.3.2 Occurrence

Compared to MMo_6X_8 the condensed-cluster compounds have a much more restricted range of stability. All structures known so far form only in the presence of big metal cations of low valence, such as K^+, Rb^+, Cs^+, Ba^{2+}, Tl^+, and In^+. Although their homogeneity range is relatively small, it is not negligable, as can be seen from the significantly different lattice parameters which have been reported by different authors for compounds such as $In_2Mo_{15}Se_{19}$ [3.25,26] and $TlMo_3Se_3$ [3.29,30].

So far, only little is known about their structural properties at low temperature. In particular, no evidence for a dimerized "$Tl_2Mo_6Se_6$" compound [3.30] has been reported as yet, although such a modification is likely to exist (Chap.6).

3.3.3 Bonding

The bonding of the $[Mo_{3n}X_{3n+2}]$ building blocks has been discussed by HÖNLE et al. [3.29] and can be described as follows. Subtracting from the nine d^5sp^3 hybridized valence orbitals per T atom those necessary for the formation of bonds with the chalcogen ligands, there remain 15n-6 orbitals per $[Mo_{3n}X_{3n+2}]$ unit available for metal-metal bonds. Assuming that the overall distributions of the energy levels resemble those of the $[Mo_6^{2+}X_8^-]^{4+}$ halide or $[Mo_6^{2+}X_8^{2-}]^{4-}$ chalcogenide units, 6n orbitals should have bonding character. Thus, a structure of the $[Mo_{3n}X_{3n+2}]$ series is favored if there are approximately 12n valence electrons per Mo_{3n} unit available for metal-metal bonding, i.e., if the *cluster VEC* is close to four. This "12n rule" implies that the formal oxidation state of Mo in the *isolated* $[Mo_{3n}X_{3n+2}]$ building blocks (X = chalcogen) is +2, i.e., that these units all carry a fourfold negative formal charge, as indicated by the limiting ionic formula $[Mo_{3n}^{2+}X_{3n+2}^{2-}]^{4-}$.

Due to the metallic interactions between these units in the *crystal*, deviations from the above formal oxidation states have to be assumed. As can be seen from the different compounds and the corresponding limiting ionic formulas listed in Table 3.2, the negative formal charges on the isolated $[Mo_{3n}X_{3n+2}]$ units of finite length (n = 2,3,4) are only partially compensated in the crystal by the positive formal charges of the cations Tl^+, In^+, etc. Thus, the formal oxidation state of Mo in these compounds has probably increased above +2, and the conduction bands, which are presumably bonding, contain electron holes. On the other hand, the formal electric charge on the quasiinfinite isolated $^1_\infty[Mo_3X_3]$ units is zero. Thus, the valence electron donated by the M cation in the crystal decreases the formal oxidation state of Mo below +2 and populates a conduction band which is presumably antibonding [3.29].

It is interesting to estimate the electron requirement of the condensed Mo_{3n} clusters in the crystal from bond-order equation (3.1) and to compare the values obtained with those predicted from the "12n rule" for quasi-isolated $[Mo_{3n}X_{3n+2}]$ units. Using Pauling's single-bond distance for Mo and summing over all *intra-* and half the *intercluster* distances of the compounds listed in Table 3.3, the number of electrons on each Mo_{3n} cluster are 15-17 (Mo_6), 26 - 28 (Mo_9), ~ 40 (Mo_{12}), and ~ 12 ($^1_\infty Mo_3$). Clearly, the estimated electron concentrations on the Mo_6 octahedra are too low, as has already been noticed during the discussion of the $PbMo_6S_8$-type compounds. However, the values come closer to those predicted as the Mo_6 clusters condense into bigger Mo_{3n} units (n = 3,4,∞). In the $TlFe_3Te_3$-type compounds, for instance, the sum of bond orders per T atom are $\Sigma n \sim 3.9 - 4.2$ for the Mo-based representatives and $\Sigma n \sim 2.2$ for the Fe-based prototype [3.29]. These values correlate well with those calculated from the above MO description, which are $\Sigma n = (12-1)/3 \sim 4$ for the Mo based compounds and $\Sigma n = (12-7)/3 \sim 2$ for the Fe telluride (see Footnote 1).

On the other hand, if the modified single-bond distance of Mo used for the description of the $PbMo_6S_8$-type compounds ($d'(1) = 2.67$ Å) is taken as a reference, the electron counts for the small Mo_{3n} clusters (Mo_6: 20 - 23 $\bar{e}$, Mo_9: 35 - 38 $\bar{e}$), are close to those expected, whereas those of the big Mo_{3n} clusters (Mo_{12}: ~ 54 e^-, $^1_\infty Mo_3$: ~ 16 $\bar{e}$) are too high (see column $2\Sigma n'$ in Table 3.3). These calculations show that the Mo-Mo single-bond radii in this class of compounds not only differ from one structure type to another but also within one structure type from one cluster type to another. Clearly, the value of d(1) tends to decrease as the condensation of the $[Mo_{3n}X_{3n+2}]$ units increases. The most likely explanation for this behavior is the decrease in number of Mo-X and Mo-Mo *intercluster* bonds per Mo atom during cluster condensation. Both types of bonds tend to stretch the Mo-Mo intracluster bonds beyond their "normal" values, such as those found in the quasi-molecular $^1_\infty[Mo_3X_3]$ (X: chalcogen) or $[Mo_6X'_8]^{4+}$ (X': halogen) units in which Mo-X (X') and Mo-Mo *intercluster* bonds are absent.

As expected from the study of the $PbMo_6S_8$-type compounds, the electronic charge on the Mo_{3n} clusters also influences the dimensions of the lattice. The c parameter of the In-rich compound $In_{3.3}Mo_{15}Se_{19}$ (c = 19.40 Å), for instance, is shorter than that of the In-poor compound $In_{2.9}Mo_{15}Se_{19}$ (c = 19.49 Å) [3.27]. This shortening is partly due to the contraction of the Mo_6 and Mo_9 clusters along c resulting from the increase of their formal electronic charge. A similar trend pre-

Table 3.3. Mo-Mo bondlengths [Å], sum of bond orders, and average *cluster VEC* (electrons/Mo atom) in Mo-cluster chalcogenides

			d^{inter}_{Mo-Mo}		d^{inter}_{Mo-Mo}		$2\Sigma n$	$2\Sigma n'$	$<2\Sigma n>$	$<2\Sigma n'>$	average *cluster VEC*	References
$In_2Mo_{15}Se_{19}$	Mo_6	Δ	2.71	(6×)	3.49	(3×)	15.7	21.2	2.92	3.94	3.60	3.25
		Δ-Δ	2.70	(6×)								
	Mo_9	Δ	2.67	(6×)	3.49	(3×)	28.1	37.9				
		Δ^m	2.68	(3×)								
		$\Delta-\Delta^m$	2.67	(6×)								
		$\Delta-\Delta^m$	2.78	(6×)								
	Mo_6	Δ	2.689	(6×)	3.39	(3×)	14.5	19.6	2.71	3.66	3.60	3.26
		Δ-Δ	2.773	(6×)								
	Mo_9	Δ	2.673	(6×)	3.39	(3×)	26.2	35.3				
		Δ^m	2.681	(3×)								
		$\Delta-\Delta^m$	2.711	(6×)								
		$\Delta-\Delta^m$	2.808	(6×)								
$In_{2.9}Mo_{15}Se_{19}$	Mo_6	Δ	2.69	(6×)	3.50	(3×)	16.1	21.6	2.83	3.81	3.78	3.27
		Δ-Δ	2.71	(6×)								
	Mo_9	Δ	2.66	(6×)	3.50	(3×)	26.4	35.6				
		Δ^m	2.71	(3×)								
		$\Delta-\Delta^m$	2.73	(6×)								
		$\Delta-\Delta^m$	2.77	(6×)								
$In_{3.3}Mo_{15}Se_{19}$	Mo_6	Δ	2.69	(6×)	3.51	(3×)	17.0	22.9	2.84	3.83	3.87	3.27
		Δ-Δ	2.68	(6×)								
	Mo_9	Δ	2.66	(6×)	3.51	(3×)	25.6	34.5				
		Δ^m	2.77	(3×)								
		$\Delta-\Delta^m$	2.73	(6×)								
		$\Delta-\Delta^m$	2.77	(6×)								
$Tl_2Mo_9S_{11}$	Mo_6	Δ	2.693	(6×)	3.217	(3×)	14.5	19.6	3.02	4.08	3.78	3.28
		Δ-Δ	2.780	(6×)								
	Mo_{12}	Δ	2.658	(6×)	3.217	(3×)	39.9	53.8				
		Δ^m	2.688	(6×)								
		$\Delta-\Delta^m$	2.743	(6×)								
		$\Delta-\Delta^m$	2.771	(6×)								
		$\Delta^m-\Delta^m$	2.669	(6×)								
$InMo_3Se_3$	Mo_3	Δ	2.661	(3×)	-		11.9	16.1	3.97	5.37	4.33	3.29
		Δ-Δ	2.721	(6×)								

Δ indicate distances within, and $\Delta-\Delta(\Delta^m)$ distances between triangular Mo_3 groups. (Δ^m = median Mo_3 triangle). The sum of bond orders $2\Sigma n$ are calculated with d(1) = 2.592 Å and those of bond orders $2\Sigma n'$ with d(1) = 2.67 Å. The averages $<2\Sigma n>$ and $<2\Sigma n'>$ indicate average bond orders per Mo atom.

sumably also exists within the series of $In_2Mo_{15}Se_{19}$-type compounds. The $\underset{\sim}{c}$ parameter of the structure containing divalent Ba^{2+} (c = 57.60 Å) is anomalously short with respect to that containing, for instance, monovalent Tl^+ (c = 58.23 Å) [3.26]. Although atomic coordinates for the Ba-based compound are not yet available, it is likely that its Mo_6 and Mo_9 clusters are shorter along $\underline{c}$ than those in the Tl-based compound because of their higher electric charge.

Altogether, the above results further confirm the importance of metal-metal bonding for the stability of cluster chalcogenides. Although the number of such compounds known to date is still small, a few factors which could be favorable for T atom clustering already appear from their study. One is a low positive formal oxidation state of the T element. A lowering of the formal charge on T allows not only its valence electron shell to expand [3.10] but also its valence electron concentration to increase, and thus favors metal-metal bonding. Unfortunately, the correlation between the formal oxidation state (FOS) of T (or the average *cluster VEC*, Table 3.3) and the degree of cluster condensation is not always apparent (Table 3.2). In the selenides $M_2Mo_{15}Se_{19}$ (M = In, Tl, K), for instance, which contain Mo_6 octahedra and Mo_9 bioctahedra, the FOS of Mo is +2.4 (i.e., the average *cluster VEC* is 6.0-2.4 = 3.6). Yet, there exist many $PbMo_6S_8$-type selenides containing only Mo_6 octahedra in which the FOS of Mo is lower than +2.4 (i.e., the *cluster VEC* is higher than 3.6). The same situation also occurs with the sulfides $Mo_2Mo_9S_{11}$ (M = K, Tl). In these compounds, which are characterized by Mo_6 octahedra and Mo_{12} trioctahedra, the FOS of Mo is +2.22 (i.e., the average *cluster VEC* is 6.00-2.22 = 3.78), which is higher than that of most $PbMo_6S_8$-type sulfides.

This indicates that other factors such as the nonmetal-to-metal ratio also play a role for the stability of cluster chalcogenides. Clearly, low X/T ratios are favorable for T atom clustering, presumably because they imply an increase of the number of T-T bonds relative to that of the competing T-X bonds (Table 3.2). In the $PbMo_6S_8$-type compounds, for instance, there are two T-T and five T-X bonds per T atom, whereas in the $TlFe_3Te_3$-type compounds which have a lower X/T ratio, there are three T-T and four T-X bonds per T atom. The influence of the nonmetal-to-metal ratio on the cluster condensation has already been noticed in other classes of compounds. Examples are the reduced halides of the rare earths or lighter T elements containing condensed $[M_6X_8]$ or $[M_6X_{12}]$ units. As pointed out by SIMON et al. [3.32] the number of valence electrons per metal atom available for metal-metal bonding is the same (i.e., 1.5 electrons) for, say, M_2Cl_3 (M = Y, Gd, Tb) and Sc_7Cl_{12}). Yet, the former compounds (Cl/M = 1.5) contain chains of edge-sharing M_6 octahedra, whereas the latter compound (Cl/M = 1.71) contains discrete M_6 octahedra and isolated M atoms.

Finally, the formation of cluster chalcogenides which contain T elements such as Mo appears to be favored by the presence of an electropositive ternary metal component M. This is presumably due to the fact that such elements act as electron donors which fill T-T orbitals of mainly bonding character. On the other hand, they stabilize the structure from a geometrical point of view because their relatively large atomic size permits them to fill the large chalcogen holes between the condensed $[Mo_{3n}X_{3n+2}]$ units.

3.4 Ternary Borides MT_4B_4 (M: Metal, T: Transition Element)

Due to their unusual low-temperature properties, the MT_4B_4 borides (M: lanthanide, Y, Th, U; T = Co, Rh, Ir, Ru, Os) are currently subject of intensive research [Ref.3.33, Chap.4]. They resemble the above chalcogenides in the sense that they are cluster compounds in which the T element plays an important role for structural stability and electronic properties. Their crystal chemistry and bonding, however, are distinctly different from those of the chalcogenides. (For a discussion of the systematics of the occurrence of superconductivity in there materials, see [Ref.3.33, Chap.2]).

Depending on the nature of the metal constituents and the preparation conditions, several polytypes with closely related structures and similar properties occur.

3.4.1 $CeCo_4B_4$ Type (tP18, T_c^{max} = 12 K)[3]

Structure

The structure of $CeCo_4B_4$ has been analysed by KUZMA and BILONIZHKO [3.34] who derived approximate boron positions from geometrical considerations. Superconductivity has been found in the Rh-based analogues by MATTHIAS et al. [3.35], and metal atom parameters for YRh_4B_4 have been refined from X-ray powder diffraction analysis by VANDENBERG [3.36]. Significantly different atomic parameters, however, have been obtained for the same compound from a recent single crystal X-ray study [3.37] which was done to ascertain the B atom positions. The discrepancies affected mainly the Rh-Rh bond lengths, which differed by up to 0.1 Å with respect to those obtained from the powder study [3.36]. As will be shown later in the text the knowledge of accurate T atom parameters for these borides is necessary for a discussion of their bonding. It is also required for an analysis of their electronic properties, because theoretical calculations by FREEMAN and JARLBORG [Ref.3.33, Chap.6](based on the data of VANDENBERG [3.36]) have shown that the electronic energy bands at the Fermi level have rather fine structures which depend in a critical manner on the Rh atom sublattice.

---

3 Between the parentheses are listed the Pearson symbol [3.38], which consists of the crystal system, Bravais lattice, and number of atoms per unit cell, and the maximum critical temperature found for this structural type.

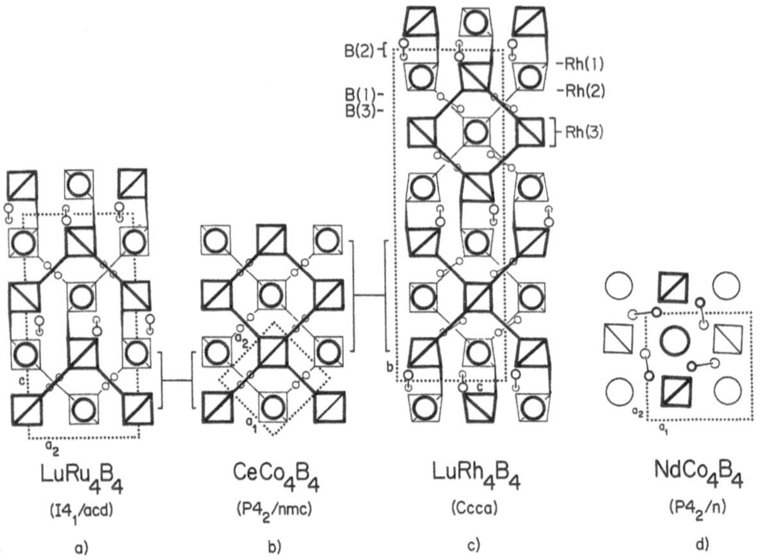

Fig.3.6a-d. The structures of the known MT_4B_4 polytypes. The T_4 tetrahedra and M atoms outlined by thick lines and thick circles and those outlined by thin lines and thin circles are centered on planes which are separated by half the lattice period perpendicular to the view direction. The boron atom pairs are shown by small circles. The $LuRu_4B_4$ (a) and $LuRh_4B_4$ (c) structure can be derived from $CeCo_4B_4$ (b) by a shift of structural slabs (containing, respectively, two and three sheets of T_4 tetrahedra) along $(a_1 + a_2 + c)/2$

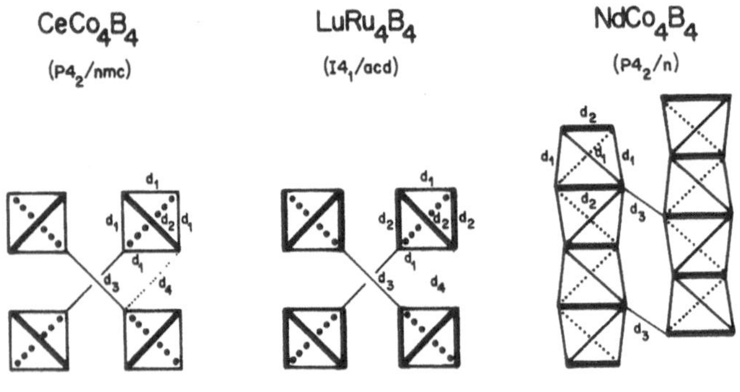

Fig.3.7. The T-T bond labels of the MT_4B_4 type borides (for the numerical values, see Table 3.4). Full (or dotted) lines of same thickness represent symmetry-equivalent bonds

The crystal structure can be described by a slightly distorted NaCl-type arrangement of tetrahedral T_4 clusters and M atoms (Fig.3.6). It has a tetragonal cell, the axial ratio of which is $c/a \sim \sqrt{2} = 1.41$ (Table 3.4). The T_4 tetrahedra are significantly elongated along $\underline{c}$ (see $\underline{d}_1$ and $\underline{d}_2$ in Fig.3.7 and Table 3.4) and are connected via relatively short T-T *intercluster* bonds ($\underline{d}_3$) to form quasi-two-dimensional infinite sheets ${}^2_\infty T_4$. The sheets are stacked along c and are connected to a three-dimensional network via relatively long, metallic T-T bonds ($\underline{d}_4$).

Table 3.4. Cell parameters and T-T distances [Å] in MT_4B_4 borides (for bond labels see Fig.3.7)

	a	c	c/a	V	d_1	d_2	d_3	d_4	Ref.
$CeCo_4B_4$ ($P4_2/nmc$)	5.059	7.063	1.396	180.8	2.59	2.50	2.56	3.01	3.34
YRh_4B_4	5.310	7.402	1.394	208.7	2.84	2.64	2.67	3.08	3.37
$SmRh_4B_4$	5.324	7.449	1.399	209.7	2.84	2.65	2.67	3.10	3.37
$ThRh_4B_4$	5.356	7.538	1.407	216.2	2.88	2.68	2.67	3.12	3.37
YRu_4B_4 ($I4_1/acd$)	7.443	14.990	2.014	830.3	2.79	2.71	2.75	2.98 (3.11)	3.37
$Y(Ru_{0.7}Rh_{0.3})_4B_4$	7.452	14.961	2.008	830.9	2.79	2.71	2.76	2.96 (3.12)	3.37
$Y(Ru_{0.5}Rh_{0.5})_4B_4$	7.459	14.941	2.003	831.3	2.80	2.72	2.74	2.96 (3.12)	3.37
$Y(Ru_{0.4}Rh_{0.6})_4B_4$	7.471	14.903	1.995	831.7	2.83	2.74	2.74	2.95 (3.11)	3.37
$Y(Ru_{0.15}Rh_{0.85})_4B_4$	7.478	14.887	1.991	832.4	2.84	2.76	2.71	2.94 (3.11)	3.37
$LaRu_4B_4$ ($P4_2/n$)	7.541	4.012	0.532	228.1	2.77	2.70	2.91	-	3.39
$LaIr_4B_4$	7.672	3.974	0.518	233.9	2.87	2.82	2.94	-	3.40
$NdOs_4B_4$	7.559	4.003	0.530	228.7	2.78	2.72	2.89	-	3.46

The B atoms form pairs (d[B-B] = 1.8 Å) which are aligned along the [110] direction and are bridging opposite faces of adjacent T_4 tetrahedra (Fig.3.6). They are binding different $^2_\infty T_4$ sheets together such that each B atom has three nearest T atom neighbors which belong to a T_4 tetrahedron of one $^2_\infty T_4$ sheet, and two which belong to two different T_4 tetrahedra of an adjacent $^2_\infty T_4$ sheet. The average T-B distance ($\bar{d}$[Rh-B] = 2.2 Å) is relatively short compared to the average M-B distance ($\bar{d}$[Y-B] = 3.1 Å), which suggests that boron interacts mainly with the transition element.

The coordination numbers of the metal atoms are high. The T atoms have 12 nearest neighbors (four T, three M, and five B atoms) and the M atoms have 24 (12 T and 12 B atoms). The T-M distances correspond approximately to the sum of metallic radii (r(Rh) + r(y) = 3.12 Å; $\bar{d}$[Rh-Y] = 3.2 Å), whereas many T-T distances ($\underline{d}_2$, $\underline{d}_3$) are shorter than those in the corresponding T metal (2r(Rh) = 2.68 Å). The crystal chemical formula of $CeCo_4B_4$ can be written as $Ce^{[12Co+12B]}\,{}^2_\infty[Co_4^{[6Co+3Ce+5B]}][B_2^{[1B+3Ce+5Co]}]_2$.

Occurrence

This structure has been found so far only with compounds containing T elements of the 9th group (Co, Rh, Ir) and trivalent metals of relatively small atomic size such as Y and the heavy lanthanides 3.34,36,41,42 . An apparent exception is

$ThRh_4B_4$, but its interatomic distances suggest that Th contributes only three valence electrons to the bonding. On the other hand, trivalent M metals of relatively large atomic size such as the light lanthanides (La, Pr, Nd, etc.) stabilize a different structure, typified by $NdCo_4B_4$ (see below). The switchover to this polytype occurs with different lanthanides in the Co(Sm)-, Rh(La)-, and Ir(Tb)-based alloys.

In general, the $CeCo_4B_4$-type compounds are metastable and form as high-temperature phases only. They transform into a low-temperature modification which is typified by a third polytype, $LuRh_4B_4$ (see below). Very recently a fourth MT_4B_4 polytype has been found among the Rh-based compounds containing heavy lanthanides. It is typified by $LuRh_4B_4$ (see below) and coexists with the $CeCo_4B_4$-type phases at high temperatures.

3.4.2 $LuRu_4B_4$ Type (tI72, T_c^{max} = 11 K)

Structure

This polytype was discovered by JOHNSTON [3.43] who derived atomic positions from X-ray powder data. The metal atom parameters were confirmed by single crystal analysis for isostructural YRu_4B_4 and a few pseudoternary $Y(Rh_{1-x}Rh_x)_4B_4$ compounds, and the exact B atom positions were determined [3.37]. As can be seen in Fig.3.6 the tetragonal structure is closely related to that of $CeCo_4B_4$. The T_4 tetrahedra and M atoms again form a slightly distorted NaCl-type structure having a cell parameter ratio $c/a \sim 2$, but the orientations of the T_4 tetrahedra are different from those in $CeCo_4B_4$. They are oriented in such a way that the structure of $LuRu_4B_4$ can be derived from $CeCo_4B_4$ by a shift of structural slabs along $(\underline{a}_1 + \underline{a}_2 + \underline{c})/2$ (Fig.3.6, caption). In this way the T_4 tetrahedra can no longer be connected to ${}^2_\infty T_4$ sheets but they form a three-dimensional ${}^3_\infty T_4$ network. The B atoms again form pairs (d[B-B] = 1.8 Å) which are bridging opposite faces of adjacent T_4 tetrahedra. They have practically the same atomic environment as those in $CeCo_4B_4$. The coordination numbers of the metal atoms are identical and the bond lengths very similar in both structure types.

Occurrence

This polytype occurs mainly with T elements of the 8th group (Ru, Os) [3.43,44] and as a low-temperature modification of the $CeCo_4B_4$-type phases with T elements of the 9th group (Rh, Ir).

Similar to the $CeCo_4B_4$ polytype, its stability range is rather limited. Isostructural compounds based on Ru, for instance, form with Y, U [3.43,44], and the lanthanides (except La) only. No compounds based on Fe and only one based on Os (UOs_4B_4 [3.44]) have been found so far. In many $LuRu_4B_4$-type compounds, Ru can be replaced by Rh without destabilizing the structure, for example, $Y(Ru_{1-x}Rh_x)_4B_4$ ($0 \le x \le 0.85$) [3.43].

3.4.3 $LuRh_4B_4$ Type (oC108, T_c^{max} = 6.3 K)

Structure

This orthorhombic polytype has been reported only recently [3.45]. As can be seen in Fig.3.6, its structure is again a shift variant of $CeCo_4B_4$. It contains a three-dimensional network of two different types of Rh_4 tetrahedra which are formed by three crystallographically independent Rh atoms. The Lu atoms and the Rh_4 tetrahedra are centered, respectively, on the cation and anion sites of a slightly distorted NaCl lattice with cell parameter relationships b/a $\sim$ 3 and c/a $\sim$ 1. The boron atoms form three types of pairs, of which one is unusual in that it contains two crystallographically independent B atoms. The average B atom separation of $\bar{d}$[B-B] = 1.4(1) Å is small compared to that of other compounds containing B atom pairs. As expected, the coordination numbers of the metal atoms are identical, and the bond lengths very similar to those of the two preceeding polytypes.

Occurrence

Similar to the $CeCo_4B_4$ structure type, it occurs for Rh-based compounds with lanthanides of small atomic size (Ho, Er, Tm, Lu). However, its structure is less favorable for superconductivity than that of $CeCo_4B_4$. The conditions of formation are not yet clarified, except that the structure could be stabilized by a deficiency of boron and coexist at high temperature with that of the $CeCo_4B_4$ polytype.

3.4.4 $NdCo_4B_4$ Type (tP18, T_c < 4 K)

Structure

Accurate structural parameters are only available for $LaRu_4B_4$ [3.39], $LaIr_4B_4$ [3.40] and $NdOs_4B_4$ [3.46]. As can be seen in Fig.3.6, the overall arrangement of the T_4 tetrahedra and M atoms differs from that of the preceeding polytypes. It corresponds no longer to a NaCl-type structure, but consists of a tetragonal array of quasi-one-dimensional infinite columns of edge-sharing tetrahedra ${}^1_\infty T_4$ running along $\underset{\sim}{c}$ which are separated by strings of M atoms (d[M-M] = c/2) and by pairs of B atoms (d[B-B] $\sim$ 1.7-1.8 Å). Formally the condensation of the T_4 atom tetrahedra into ${}^1_\infty T_4$ columns corresponds to that of the T_6 atom octahedra into ${}^1_\infty T_3$ columns in the $TlFe_3Te_3$-type chalcogenides. The situation in the borides, however, differs insofar as the ${}^1_\infty T_4$ columns are connected to each other by short T-T and T-B bonds. Similar to the $TlFe_3Te_3$ structure, this polytype is unfavorable for superconductivity, presumably because of its anisotropic character. Interestingly, the number of nearest neighbors for each atom in the structure is the same as that in the other polytypes, as can be seen from the crystal chemical formula

$$Nd^{[12Rh+8(+4)B]}{}^1_\infty[Co_4^{[5(+1)Co+3Nd+5B]}][B_2^{[1B+2(+1)Nd+5Ru]}]_2 .$$

Occurrence

The structure occurs for T elements of the 8th (Os) and 9th (Co, Rh, Ir) group and forms only in the presence of M metals of relatively large atomic size [3.39,40,47]. Within the lanthanide series of compounds, the switchover to this polytype occurs with different lanthanides for Co(Sm)-, Rh(La)-, Ir(Tb)-, and Os(Sm)-based borides. Thorium stabilizes the structure only in Os- and Ir-based borides [3.40]. So far, no Rh-based $NdCo_4B_4$-type borides has been reported.

3.4.5 Bonding of the MT_4B_4 Borides

Compared to the Mo cluster chalcogenides, the MT_4B_4 borides have considerably more metallic and less ionic bonding character. This can be seen from their structures which have features typical for alloys, such as high atomic coordination numbers and compact metal atom arrangements in which the nonmetal atoms occupy interstitial sites. On the other hand, the metal to nonmetal bonds in the borides are presumably less ionic than those in the chalcogenides because the electronegativity differences between boron and the metals are generally smaller than those between the chalcogens and the metals. In contrast to the chalcogenides in which the electronic state of the nonmetal component is known, that in the borides is unknown and cannot easily be derived from interatomic distances. This precludes a simple bonding description in terms of limiting ionic-covalent models such as those used for the chalcogenides.

The rather restricted range of existence of all the MT_4B_4 borides indicates that their stability depends in a critical manner on both geometrical and electronic factors. One which has been found to be less critical in the chalcogenides but obviously plays a major role in the borides is the size of the metal atoms. The $CeCo_4B_4$-, $LuRu_4B_4$-, and $LuRh_4B_4$-type compounds form predominantly with M metals of medium and relatively small atomic size ($1.75 < r_M < 1.8$ Å), whereas the $NdCo_4B_4$-type compounds form mainly with M metals of relatively large atomic size ($r_M > 1.8$ Å). Interestingly, the size constraint also applies to average atomic radii, as can be seen from the existence of pseudoternary $CeCo_4B_4$-type borides with mixed M atom occupancy, such as (Ca, Sc)Rh_4B_4 [3.41] and (Y, Lu)Ir_4B_4 [3.42]. The corresponding ternary borides based on one M component only do not exist, presumably because of the unfavorable atomic size of Sc(Lu), which is too small, and that of Ca (Y), which is too big.

The reason why the atomic size constraint is more effective in the borides than in the chalcogenides could be due to the different number and types of bonds involved in stabilizing their structures. Whereas the structures of the chalcogenides are stabilized by the interplay of three types of bonds, viz. one metal-metal (T-T) and two metal-nonmetal (T-X, M-X) bonds, the structures of the borides are stabilized by the interplay of five types of bonds, viz. two metal-metal

(T-T, M-T) and three metal-nonmetal (T-B, M-B, B-B) bonds. The balance of a large number of different bonds could be relatively sensitive to small changes of both geometrical origin (atomic size) and of electronic (valence electron concentration) origin. This can be seen, for instance, from a study of T-T bond lengths in $CeCo_4B_4$-type structures (Table 3.4). As the size of the M atoms (i.e., the cell volume) of the Rh-based compounds MRh_4B_4 (M = Y, Sm, Rh) is increased, the lengths of the T-T *intracluster* bonds ($\underline{d}_1$, $\underline{d}_2$, Fig.3.7) increases, whereas that of the T-T bonds between the T_4 clusters ($\underline{d}_3$) remains practically unchanged. Thus the volume expansion does not affect all metal-metal bonds equally. Yet the overall expansion of the structure is nearly isotropic, as can be seen from the almost constant c/a ratio. This is surprising because the bonding of this structure is relatively anisotropic. In particular there exist short T-T and B-B bonds only in directions perpendicular to the tetragonal axis but none parallel to it. The c/a ratio does also not change much if one substitutes T elements of different atomic size. Those of the Co (r = 1.25 Å) based compounds (1.38 < c/a < 1.40) are almost identical to those of the Rh (r = 1.34 Å) based compounds (1.39 < c/a < 1.40), whereas those of the Ir (r = 1.36 Å) based compounds (c/a ~ 1.35) are somewhat lower [3.42]. This suggests that the T-B and T-M bonds which both form a three-dimensional network of relatively isotropic character are more important for cohesion than the T-T and B-B bonds.

The stabilizing influence of the T-T bonds on the structure of the MT_4B_4 borides can be seen from the structural switchover to the $NdCo_4B_4$ polytype, which presumably occurs when these bonds are stretched by the M atoms beyond a certain critical value. The formation of condensed clusters of edge-sharing T_4 tetrahedra in this structural type apparently corresponds to an energetically more favorable T atom arrangement, presumably because it leads to an increase of the total number of T-T bonds in the structure ($NdCo_4B_4$: three short T-T bonds per T atom, other MT_4B_4 polytypes: two short T-T bonds per T atom). The fact that this switchover occurs in the Co-based borides with smaller M atoms than in the Rh-based borides supports this hypothesis. The behavior of the Ir-based compounds, however, is more complex, because they form this structure with smaller M atoms than both the Co- and Rh-based borides. The reason for this anomaly is unknown but could be related to the relatively low c/a ratio of the Ir-based $CeCo_4B_4$ representatives. Clearly, a single crystal X-ray study of such a compound would be of interest.

Another factor which plays an important role for the stability of the MT_4B_4 borides is the valency of the T element. This is not surprising from a valence bond theory point of view because the average number of valence electrons on the T_4 clusters (*cluster VEC*) determines the strength of the T-T bonds. The influence of the *cluster VEC* on the T-T bond lengths has been studied in the $Y(Ru_xRh_{1-x})_4B_4$ series of compounds [3.37]. As Ru (eight valence electrons) is replaced by Rh (nine valence electrons), which has practically the same atomic size, the $(Ru,Rh)_4$

tetrahedra expand, i.e., the T-T *intracluster* bonds become weaker, whereas the T-T bonds between these clusters contract, i.e., they become stronger (see $\underline{d}_1$, $\underline{d}_2$, and $\underline{d}_3$ in Table 3.4 and Fig.3.7). This behavior resembles that of the Mo_6 octahedra in the Mo cluster chalcogenides, except that the expansion of the T_4 clusters in the borides suggests that the electronic states at E_F are antibonding.

A similar trend also exists in the $NdCo_4B_4$-type compounds. The $^1_\infty T_4$ columns in the Ir-based (nine valence electrons) LaT_4B_4 boride show an expansion relative to those in the Ru-based (eight valence electrons) LaT_4B_4 boride (see $\underline{d}_1$ and $\underline{d}_2$ in Fig.3.7 and Table 3.4) which largely exceeds that expected from the atomic size difference between Ru and Ir.

Acknowledgement. The author thanks Professor H.G. von Schnering for many valuable discussions. He is also grateful to Miss. I. Kusel and Mrs. B. Künzler for their competent help in preparing the typescript and figures.

References

3.1 K. Yvon: "Structure and Superconductivity of Ternary Compounds", in *Proc. Int. Conf. Ternary Superconductors*, ed. by G. Shenoy (North Holland, Amsterdam 1981)
3.2 L. Pauling: *The Nature of the Chemical Bond*, 3rd ed. (Cornell University, Ithaca 1960)
3.3 Ø. Fischer: Appl. Phys. *16*, 1-28 (1978)
3.4 K. Yvon: "Bonding and Relationships between Structure and Physical Properties in Chevrel-Phase Compounds $M_xMo_6X_8$ (M = Metal, X = S, Se, Te)" in *Current Topics in Materials Science*, Vol.3, ed. by E. Kaldis (North Holland, Amsterdam 1979)
3.5 A. Grüttner, K. Yvon, R. Chevrel, M. Potel, M. Sergent, B. Seeber: Acta Crystallogr. B*35*, 285-292 (1979)
3.6 K. Yvon, R. Baillif, R. Flükiger: Acta Crystallogr. B*35*, 2859-2863 (1979)
3.7 K. Yvon, R. Chevrel, M. Sergent: Acta Crystallogr. B*36*, 685-687 (1980)
3.8 A. Lipka, K. Yvon: Acta Crystallogr. B*36*, 2123-2126 (1980)
3.9 O. Bars, J. Guillevic, D. Grandjean: J. Solid State Chem. *6*, 48-57 (1973)
3.10 F.A. Cotton, G. Wilkinson: *Advanced Inorganic Chemistry*, 3rd ed. (Interscience, New York 1972)
3.11 H. Schäfer, H.G. Schnering: Angew. Chem. *76*, 833- 868 (1964)
3.12 A. Simon: Z. Anorg. Allg. Chem. *355*, 311-322 (1967)
3.13 R. Flükiger, R. Baillif: Personal communication
3.14 E. Parthé: Acta Crystallogr. B*36*, 1-7 (1980)
3.15 A.M. Umarji, G.V. Subba Rao, M.P. Janawadkar, T.S. Radhakrishnan: J. Phys. Chem. Solids *41*, 421-429 (1980)
3.16 G.J. Dudley, K.Y. Cheung, B.C.H. Steele: J. Solid State Chem. *32*, 259-267 (1980)
3.17 R. Baillif, K. Yvon, R. Flükiger, J. Muller: J. Low Temp. Phys. *37*, 231-237 (1979)
3.18 A. Perrin, R. Chevrel, M. Sergent, Ø. Fischer: J. Solid State Chem. *33*, 43-47 (1980)
3.19 R. Flükiger, R. Baillif, J. Muller, K. Yvon: J. Less Common Met. *72*, 193-204 (1980)
3.20 A.M. Umarji, G.V. Subba Rao, V. Sankaranarayanan, G. Rangarajan, R. Srinivasan: Mater. Res. Bull. *15*, 1025-1031 (1980)

3.21 Ch. Perrin, R. Chevrel, M. Sergent, Ø. Fischer: Mater. Res. Bull. *14*, 1505-1515 (1979)
3.22 K. Yvon, A. Paoli: Solid State Commun. *24*, 41-45 (1977)
3.23 R. Chevrel, M. Sergent, J. Prigent: Mater. Res. Bull. *9*, 1487-1492 (1974)
3.24 M. Marezio, P.D. Dernier, J.P. Remeika, E. Corenzwit, B.T. Matthias: Mater. Res. Bull. *8*, 657-668 (1973)
3.25 A. Lipka, K. Yvon: Acta Crystallogr. B*36*, 2123- 2126 (1980)
3.26 R. Chevrel, M. Potel, M. Sergent, M. Decroux, Ø. Fischer: Mater. Res. Bull. *15*, 867-874 (1980)
3.27 A. Grüttner, K. Yvon, R. Chevrel, M. Potel, M. Sergent, B. Seeber: Acta Crystallogr. B*35*, 285-292 (1979)
3.28 M. Potel, R. Chevrel, M. Sergent: Acta Crystallogr. B*36*, 1319-1322 (1980)
3.29 W. Hönle, H.G. von Schnering, A. Lipka, K. Yvon: J. Less Common Met. *71*, 135-145 (1980)
3.30 M. Potel, R. Chevrel, M. Sergent: Acta Crystallogr. B*36*, 1545-1548 (1980)
3.31 K. Klepp, H. Boller: Monatsh. Chem. *110*, 677-684 (1979)
3.32 A. Simon, N. Holzer, H.-J. Mattausch: Z. Anorg. Allg. Chem. *456*, 207-216 (1979);
H.J. Mattausch, J.B. Hendricks, R. Eger, J.D. Corbett, A. Simon: Inorg. Chem. *19*, 2128-2132 (1980)
3.33 Ø. Fischer, M.B. Maple (eds.): *Superconductivity in Ternary Compounds, II*, Topics in Current Physics, Vol. 32 (Springer, Berlin, Heidelberg, New York 1982)
3.34 Yu.B. Kuzma, N.S. Biloniszko: Sov. Phys. Crystallogr. *16*, 897-898 (1972)
3.35 B.T. Matthias, E. Corenzwit, J.M. Vandenberg, H.E. Barz: Proc. Natl. Acad. Sci. USA *74*, 1334-1335 (1977)
3.36 J.M. Vandenberg, B.T. Matthias: Proc. Natl. Acad. Sci. USA *74*, 1336-1337 (1977)
3.37 K. Yvon, A. Grüttner: "The Influence of the Formal Electric Charge on the Size of the Transition Metal Atom Cluster in YRh_4B_4, YRu_4B_4 and $PbMo_6S_8$ Related Compounds", in *Superconductivity in d- and f- Band Metals*, ed. by H. Suhl, B. Maple (Academic, New York 1980)
3.38 W.B. Pearson: *A Handbook of Lattice Spacings and Structures of Metals and Alloys* (Pergamon, Oxford 1967)
3.39 A. Grüttner, K. Yvon: Acta Crystallogr. B*35*, 451-453 (1979)
3.40 P. Rogl: Monatsh. Chem. *110*, 235-243 (1979)
3.41 B.T. Matthias, C.K.N. Patel, H. Barz, E. Corenzwit, J.M. Vandenberg: Phys. Lett. *68*A, 119-121 (1978)
3.42 H.C. Ku, B.T. Matthias, H. Barz: Solid State Commun. *32*, 937-944 (1979)
3.43 D.C. Johnston: Solid State Commun. *24*, 699-702 (1977)
3.44 P. Rogl: Monatsh. Chem. *111*, 517-527 (1980)
3.45 K. Yvon, D.C. Johnston: Acta Crystallogr., B*38*, 246-250 (1982)
3.46 K. Hiebl, M.J. Sienko, P. Rogl: J. Less Common Met. *82*, 21-28 (1981)
3.47 Yu.B. Kuzma, N.S. Bilonizhko: Dopov. Akad. Nauk. Ukr. RSR Ser. A *3*, 275-277 (1978)

Additional References with Titles

D. Mamien, C.H. Novian, J. Gal: *Superconductivity in the Neptunium Chevrel Phase* $NpMo_6Se_8$, Solid State Commun. *38*, 437-440 (1981)

J.D. Corbett: *Chevrel Phases: An Analysis of Their Metal-Metal Bonding and Crystal Chemistry*, J. Solid State Chem. *39*, 56-74 (1981)

4. Metallurgy and Structural Transformations in Ternary Molybdenum Chalcogenides

R. Flükiger and R. Baillif

With 23 Figures

A brief review of phase diagram collections (e.g., [4.1-3]) reveals that metal sulfides and selenides have a general tendency to undergo structural transformations at temperatures around 300 K and below. This tendency is observed for superconducting as well as for nonsuperconducting compounds. This phenomenon is connected to the observation that in most cases, metal sulfides or selenides form structures which are far from being dense. This is particularly the case for ternary molybdenum chalcogenides of the formula $M_xMo_6X_8$ (M: metal, X: chalcogen, $0 \leq x \leq 4$) belonging to the space group R $\bar{3}$. These compounds exhibit remarkable physical properties, which arise from the unique configuration of Mo_6X_8 building blocks held together by the insertion of a weakly bounded third component, M. This particular arrangement can accomodate very different ions, which can be either magnetic or nonmagnetic. Depending upon the nature of M (transition or nontransition element, rare earth), the physical properties of the corresponding compound are very different.

In this paper, we will describe phenomenologically the structural transformations of ternary molybdenum chalcogenides in the temperature range between 60 and 500 K. We emphasize that the study of low-temperature relationships can only be carried out on well-defined samples. A review of the literature on ternary molybdenum chalcogenides shows, however, that very little attention has been given to the preparation of single-phase bulk samples. Different methods for synthesizing these compounds as well as for growing single crystals are thus reviewed. In addition, the high-temperature relationships for some selected systems are presented. The type of formation of the rhombohedral phase will be tentatively correlated to the nature of the M element.

4.1 Preparation Methods and Stoichiometry

Little is known about the metallurgy of ternary molybdenum chalcogenides. The high vapor pressure of the chalcogen X and of some of the M elements renders the preparation of homogeneous, single-phase samples very difficult.

4.1.1 Sintering

The most common method used to form the rhombohedral compounds $M_xMo_6X_8$ consists in reacting appropriate amounts of the three elements (or binary combinations of them) in quartz tubes at temperatures up to 1200°C. After this treatment, the samples, in general, are multiphased, the additional phases being usually Mo_2S_3, MoS_2, or free molybdenum. In rare earth (RE) compounds, the corresponding RE sulfide or selenide is also encountered. X-ray diffraction is the most convenient method for determining the amount of secondary phases in sintered samples. The absence of additional reflection lines in X-ray patterns is currently used to define the investigated sample as "single phases." However, microscopic observations on melted samples show that even in such cases, a few percent of additional phases may be present. The presence of free molybdenum is particularly hard to detect in sintered samples. The plastic deformation in the soft bcc molybdenum introduced during the powdering process prior to the X-ray analysis causes a broadening of the reflection lines: if the amount of Mo is small, the bcc lines are hard to distinguish from the background. They reappear after a "flash anneal" [4.4] of the powders. This consists in heating them for 1 or 2 min at 1000°C. This may be the principal reason for the discrepancies between different authors concerning the precise composition of ternary molybdenum chalcogenides. Additional phases in these systems can be eliminated by crushing the sintered material into a fine powder (40 μm size), compacting and reacting again. Sometimes the whole procedure has to be repeated several times.

4.1.2 RF Melting Under High-Pressure Argon

Dense samples allowing microscopic analysis can be obtained by hot pressing [4.5], CVD [4.6], melting in sealed Mo tubes [4.7], melting by the tube-in-tube method [4.8], or by melting under high-pressure argon [4.9,10]. We will focus our discussion on the melting in an RF furnace under a moderately high argon pressure (20-100 atm) which has been used in our laboratory. The higher pressure causes, in principle, a slight increase of the partial pressure of the volatile components of the compounds, but the drastic decrease of the diffusion speed of the evaporated atoms in the argon atmosphere at the vicinity of the surface largely counterbalances this effect, thus resulting in a lower evaporation rate. The RF melting device represented in Fig.4.1 is an improved version of that described in earlier reports [4.9,10]. The crucible material generally used was Al_2O_3. It was substituted by BeO for the higher melting compounds. Both ceramics showed no reaction with the Mo chalcogenides, except if the M components was a rare earth. In this case, boron nitride or thorium oxide had to be used, but a slight reaction with the crucible material could never be avoided. The melting temperatures of the intermediate phases Mo_2S_3 and MoS_2 were found to be substantially higher (> 1800°C) than those

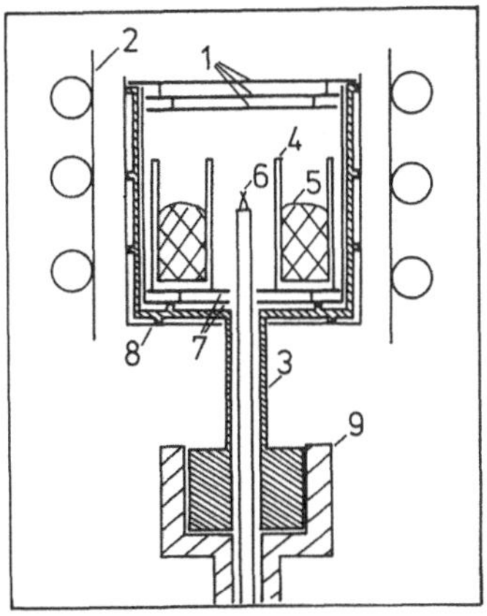

Fig.4.1. High pressure RF melting device: 1 upper radiation shield (molybdenum), 2 low-impedance RF coil, 3 tantalum support, 4 melting crucible, 5 sample, 6 thermocouple, 7 lower radiation shield (molybdenum), 8 susceptor, 9 molybdenum support [4.49]

of the ternary sulfides crystallizing in the rhombohedral phase, in agreement with MOH [4.8]. The melting temperature for ternary sulfides decreases with that of the M element, the lowest one being measured for M = Pb (around 1700°C). The melting temperatures of the selenides are markedly lower than those of the corresponding sulfides but show the same tendency if M varies. Typical microphotographs for several systems are reproduced in Figs.4.2-5.

From the study of grain growth in the analyzed systems, some conclusions about the formation of the rhombohedral phase can be drawn. In particular, the formation of large crystallites (of the order of several mm) is a strong indication for congruent formation. From our observations, we deduce that $Cu_xMo_6S_8$, $Ni_xMo_6S_8$, $Cu_xMo_6Se_8$, $PbMo_6Se_8$, $SnMo_6Se_8$, $AgMo_6Se_8$, and $ZnMo_6Se_8$ form congruently, while several sulfides, such as $PbMo_6S_8$, $SnMo_6S_8$, and $AgMo_6S_8$, form peritectically. The compounds containing rare earths with three valence electrons are thought to form peritectically.

There is some difference between the melting behavior of sulfides and selenides. The situation is best characterized by comparing the binary compounds Mo_2S_3(Mo_6S_8 is not stable) and Mo_6Se_8: the first compound shows practically no melting losses, while 1% Se or more is lost from the second. The same tendency of increased evaporation of Se is encountered in ternary compounds, thus suggesting a stronger bonding between Mo and S than between Mo and Se. In general, sulfides melt 50 to 100°C higher than the corresponding selenides. The melting losses in the sulfides is mostly due to the element M. In general, they reflect the elementary vapor pressure of the latter at the melting point of the rhombohedral phase. An interesting fact is that the crystals of the selenides are even more brittle than those of the sulfides: very small forces are sufficient to cleave the crystals.

The melting device described in Fig.4.1 can be used for growing single crystals (Bridgman-Stockbarger technique with fixed crucible and furnace). Important factors affecting directly the grain size were found to be the cooling speed (between 5 and 30°C/min) and the shape of the crucibles (a round or a conical bottom). A series

Fig.4.2

Fig.4.3

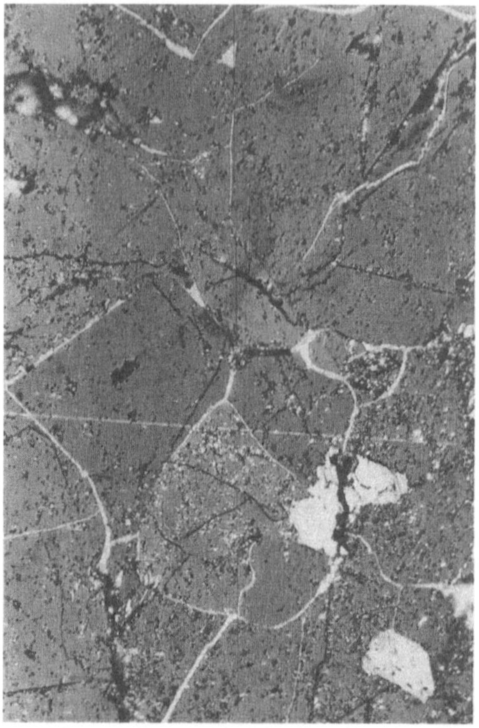

Fig.4.4

Fig.4.5

of single crystals [4.10] with sizes varying from 1 mm^3 ($LaMo_6S_8$, $GdMo_6S_8$, $GdMo_6Se_8$, $PbMo_6S_8$) to 5 mm^3 ($Cu_xMo_6S_8$, $Cu_xMo_6Se_8$, Mo_6Se_8, $PbMo_6Se_8$, $SnMo_6Se_8$, $AgMo_6Se_8$, $ZnMo_6Se_8$) were formed. A $Cu_{1.8}Mo_6S_8$ crystal greater than 3 cm^3 was obtained, which served for the measurement of the phonon spectrum [4.11]. Recently, ternary molybdenum sulfides were melted in an autoclave under argon pressures up to 3000 atm. Under these extreme conditions, a further decrease of the melting losses was observed [4.12]. Thermal analysis on $Cu_xMo_6S_8$ up to 1800°C has been performed in this device [4.13].

4.1.3 Stoichiometry

From considerations based on the crystal structure, it follows that the ideal formula describing ternary molybdenum chalcogenides would be $M_xMo_6S_8$. However, as a consequence of the metallurgical difficulties in preparing well-defined alloys, the precise composition for several systems is still controversial. A review of the literature data shows that the same compound has been characterized by different formulas. As examples, we just mention $Cu_xMo_5S_6$, $Cu_xMo_6S_7$, and $Cu_xMo_6S_8$ for the rhombohedral phase in the system Cu-Mo-S or $PbMo_6S_7$, $PbMo_{6.4}S_8$, and $Pb_xMo_6S_{8-y}$ for Pb-Mo-S. For the compounds containing rare earths, the formula $(RE)_{1.2}Mo_6X_8$ alternates with $(RE)_1Mo_6X_8$.

The problem of choosing between very similar formulas for the characterization of systems with more than two components is a general one and is also encountered in other systems with different crystal structures. The problem of determining the precise composition of a given phase in ternary of multinary systems can only be solved by a combined *metallurgical and crystallographic* analysis. This may be illustrated by the system Pb-Mo-S. MAREZIO et al. [4.14] proposed for this system the formula $Pb_{0.92}Mo_6S_{7.5}$, while CHEVREL [4.6] used the expression $PbMo_{6.4}S_8$. Both formulas denote the same overall composition, since the former one can be rewritten as $Pb_{0.98}Mo_{6.4}S_8$. However, the initial formulas [4.6,14] lead to fundamentally different interpretations when the crystallographic structure is taken into account. $PbMo_{6.4}S_8$ suggests the presence of excess Mo, while in the other case, each sixteenth sulfur site should be vacant. As will be seen later, the answer can be given on the basis of density measurements on single crystals. In all cases where single

Fig.4.2. Optical micrograph of the single-phase alloy $Cu_2Mo_6Se_8$. The parallel lines are cleavage planes. 40X

Fig.4.3. Optical micrograph of the alloy $AgMo_6S_8$, showing two phases, the rhombohedral phase (dark) and free Mo (white). 40X

Fig.4.4. Optical micrograph of the single-phase alloy $SnMo_6Se_8$. The cleavage planes are very marked, as for all selenide compound. 40X

Fig.4.5. Optical micrograph of the alloy with the nominal composition $PbMo_{6.8}S_8$. The rhombohedral phase (dark) is in equilibrium with the ($Mo + Mo_2S_3$) eutectic. 160X crystals were available, the stoichiometric ratio 6:8 between Mo and the chalcogen has been confirmed, i.e., in the systems $Cu_xMo_6S_8$ [4.10,13,15], $Cu_xMo_6Se_8$ [4.10,13], Mo_6Se_8 [4.10,13], and $PbMo_6S_8$ [4.10,16]. From X-ray refinements on single crystals, stoichiometry was also found to occur in rare earth molybdenum sulfides with M = La and Gd [4.16].

Although S or Se defects have been discussed in the literature [4.17], no experimental proof has so far revealed their presence in ternary molybdenum chalcogenides. The estimated amount of S or Se vacancies on the analyzed single crystals is, if any, less than 1% [4.10]. This suggests that the only variable in the composition of ternary molybdenum chalcogenides is the parameter x in the formula $M_xMo_6X_8$.

4.2 High-Temperature Phase Fields in Some Selected Systems

4.2.1 The Binary System Mo-S

The phase diagram of the binary system Mo-S is represented in Fig.4.6 and was established on the basis of the high-temperature data of MOH [4.8] and our own results [4.12]. All solidus temperatures are approximate. The main features of this diagram are a) no solubility of S in Mo, b) two eutectics between Mo_2S_3 and Mo, c) the rhombohedral phase "Mo_6S_8" is not stable in the binary Mo-S system at equilibrium, d) the Mo_2S_3 phase (not superconducting above 1 K) shows a low-temperature phase transformation at 195 K, e) the phases Mo_2S_3 and MoS_2 exhibit very high melting points, 1950 and 2375°C, respectively. In ternary systems, the Mo_2S_3 phase is exceedingly stable and competes with the rhombohedral phase. The problem of preparing single-phase rhombohedral samples is almost identical to the problem of avoiding the Mo_2S_3 phase.

There have been several attempts to form the metastable binary compound Mo_6S_8. Only one was successful, the "leaching technique," which was first used by CHEVREL et al. [4.18]. It consists in preparing $Ni_2Mo_6S_8$ or $Cu_2Mo_6S_8$ which are subsequently etched in HCl acid. The resulting compound, "Mo_6S_8," is superconducting at 1.8 K, regardless whether the starting material was Cu or Ni based [4.18]. Since chemical analysis and neutron activation experiments revealed less than 0.1 wt% Cu, Ni, or Cl, a stabilization of "Mo_6S_8" by these impurities is practically excluded [4.13].

It was recently shown by FLÜKIGER et al. [4.13] that the density of "Mo_6S_8" as reported by CHEVREL et al. [4.18] is approximately 2% smaller than expected from a linear extrapolation of the density values in $Cu_xMo_6S_8$ (Fig.4.7). From a comparison with the corresponding selenides showing a linear variation of density, it is thus thought that the bonding in "Mo_6S_8" and Mo_6Se_8 may be different, leading to the instability of the former. It is interesting that the partial substitution of S by Br, I, or O stabilizes the rhombohedral phase: the compounds $Mo_6S_6Br_2$, $Mo_6S_6I_2$

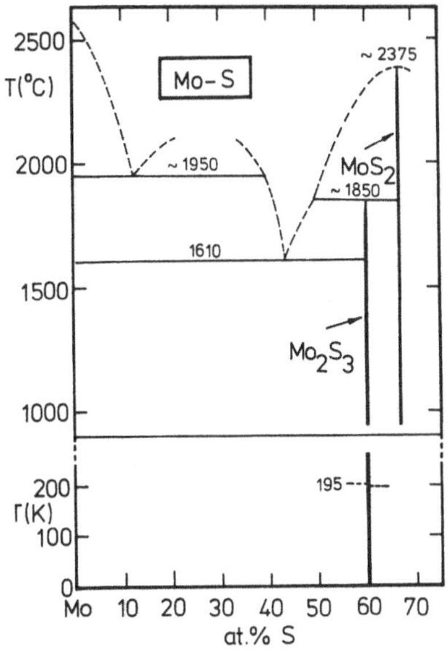

Fig.4.6

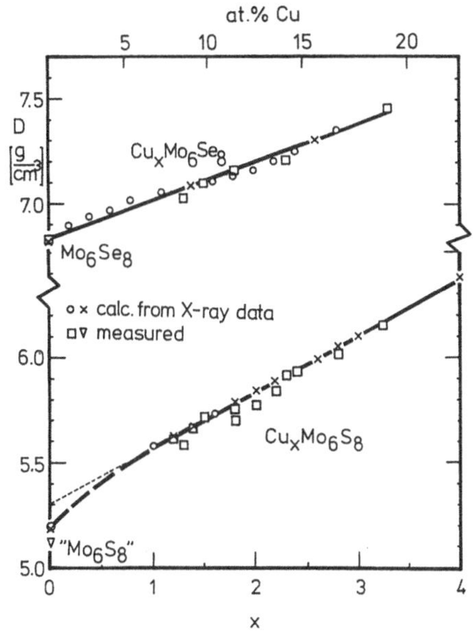

Fig.4.7

Fig.4.6. The Mo-S phase diagram after MOH 4.8 , modified by FLÜKIGER and BAILLIF [4.12]

Fig.4.7. Densities for the systems $Cu_xMo_6S_8$ and $Cu_xMo_6Se_8$, calculated from X-ray data (o,x) and measured (□,∇) [4.13]. The variation of the Cu content in the equilibrium phase fields is essentially linear [4.13]

[4.19], and $Mo_6S_6O_2$ [4.20] have been synthesized. The question of stability in Mo_6S_8 is of particular interest, since in other structures it has been observed that metastable compounds can exhibit very high T_c values, i.e., Nb_3Ge [4.21].

These systems have been produced by nonequilibrium methods, as coevaporation, sputtering, CVD, or ultrafast quenching. A particularity of rhombohedral compounds with respect to A15-type compounds resides in the fact that none of these techniques was successful in stabilizing new metastable compounds, or even to extend the homogeneity range of the rhombohedral phase. The only exception is $Cu_{1.2}Mo_6S_8$, which will be discussed below.

4.2.2 The System $Cu_xMo_6S_8$

The only ternary molybdenum chalcogenide for which the phase diagram has been extensively studied so far is the system $Cu_xMo_6S_8$ [4.10,13,15]. The complete phase diagram from 11 to 2000 K [4.13] is represented in Fig.4.8. We first discuss the region above 800°C. The rhombohedral phase in the system Cu-Mo-S forms congruently at 1750 ± 30°C and at a composition close to x = 2 (12 at % Cu). The phase limits are strongly temperature dependent: at 1500°C the phase is stable within the range $1.2 \le x \le 3$ ($7.9 \le x \le 17$ at % Cu), whereas at 850°C the limits are shifted to

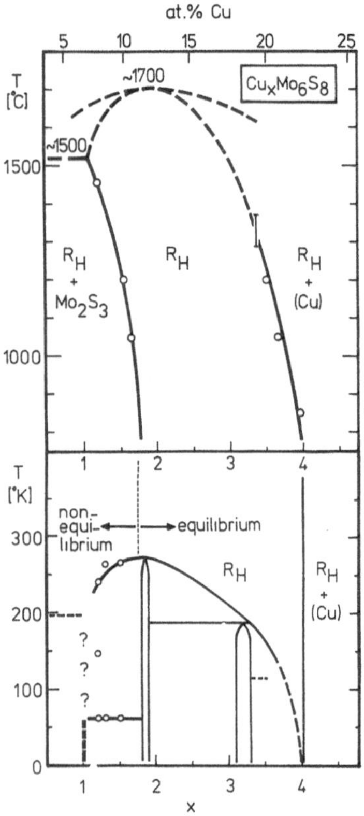

Fig.4.8

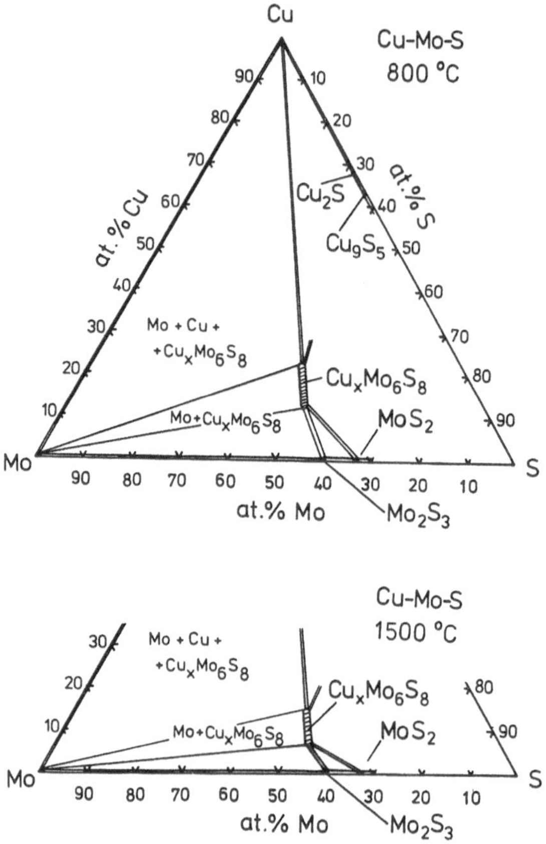

Fig.4.9

Fig.4.8. The $Cu_xMo_6S_8$ phase diagram in the range $10 < T < 2000$ K. The limits of the rhombohedral phase are strongly temperature dependent. The compositions $1.2 < x < 1.8$ are metastable and can only be retained by quenching. At 60 K, these metastable compositions undergo a phase transformation into a new phase [4.13]

Fig.4.9. The ternary phase diagram Cu-Mo-S at 800 and 1500°C, showing the shift of the $Cu_xMo_6S_8$ phase limits with temperature. For better clarity, the two-phase regions have been drawn wider than reality [4.49]

$1.8 \leq x \leq 4$ ($11 \leq x \leq 22$ at% Cu) [4.13]. On the low Cu side there is a three-phase region $R_H + Mo_2S_3 + Mo$, whereas on the Cu rich side the rhombohedral phase is in equilibrium with Cu. These phase limits were found by analyzing the variation of the lattice parameters after argon jet quenching from different temperatures [4.13]. The ternary Cu-Mo-S phase diagram at 1500° and 700°C resulting from our data [4.13,15] is represented in Fig.4.9. For better clarity, the width of the two-phase regions Mo + Mo_2S_3, Mo_2S_3 + MoS_2, Cu + Mo, ... has been drawn larger than reality. The homogeneity range of the $Cu_xMo_6S_8$ phase extends from x = 1.2 to x = 4, corresponding to a width of 14 at % Cu. An investigation on a series of single crystals has shown that the ratio Mo:S is invariantly 6:8 over the whole homogeneity range [4.13].

4.2.3 The System $Cu_xMo_6Se_8$

The binary rhombohedral phase Mo_6Se_8 is stable at equilibrium, in contrast to the corresponding sulfide. The system $Cu_xMo_6Se_8$ appears to be a solid solution of Mo_6Se_8, the solubility limit of Cu being x = 2.8 [4.13,22]. Over the entire homogeneity range, the width of the rhombohedral phase is very narrow, i.e., the ratio Mo:Se is invariantly 6:8. Copper molybdenum sulfides and selenides have many things in common. Both form by a congruent reaction and exhibit several modifications at low temperature. The variation of the lattice parameters a and c as a function of the Cu content is also very similar for both cases: as follows from Fig.4.10, the parameter c shows its strongest variation for x → 0 and exhibits a minimum.

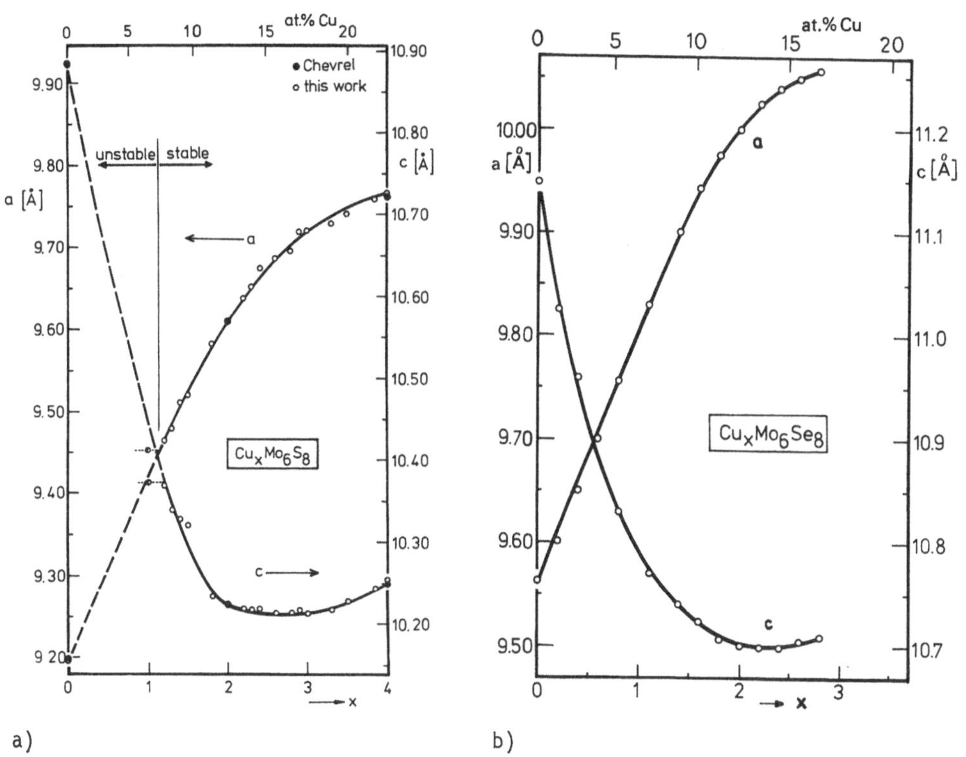

Fig.4.10a,b. Variation of the lattice parameters c and a of the rhombohedral phases $Cu_xMo_6S_8$ and $Cu_xMo_6Se_8$ with the Cu content, x. Circle-data from [4.13]. Dots-data from [4.22]

4.2.4 The System Pb-Mo-S: $Pb_xMo_6S_{8-y}$ or $Pb_xMo_6S_8$?

The detailed high-temperature phase diagram of the system Pb-Mo-S is not known. However, metallographical analysis and crystal growing experiments strongly support a peritectic formation of the rhombohedral phase in this system [4.10]. The peritectic temperature is situated between 1500° and 1700°C [4.7,10]. Another indication for the peritectic formation of this phase is furnished by the crystal growing process. Indeed, appreciably large Pb-Mo-S single crystals (up to 1 mm^3 size) could only be obtained if the initial Mo content exceeded the stoichiometric ratio 6:8. For Pb, the largest crystals were obtained for nominal compositions ranging between $Pb_{1.2}Mo_{6.9}S_8$ and $Pb_{1.2}Mo_{7.2}S_8$ [4.10]. Melted alloys of these compositions show two distinctly different zones, one containing the eutectic Mo_2S_3 + Mo, the other being $PbMo_6S_8$ (Fig.4.5). This crystal growing process, the "traveling solvent zone process," is typical for the peritectic phase formation. During the slow cooling after melting, the interface (rhombohedral + liquid/(rhombohedral) is moved from the warmer to the cooler region of the crucible, crystallization occurring at the cooler side.

The average density of eight small single-crystal pieces originating from the same crystal was determined to be 6.38 ± 0.08 g/cm^3 (Table 4.1). In [4.10] we concluded that a slight Mo excess was present, which was also confirmed by the chemical analysis yielding the composition $Pb_{1.02}Mo_{6.5}S_8$. However, some traces of undissolved Mo (2 to 4 vol.%) were always present in the analyzed crystals, so that the composition of the rhombohedral phase was corrected to $PbMo_{6.2}S_8$ [4.10]. In light of later microscopic observation on the same samples [4.12], the correct formula appears to be even closer to the stoichiometric composition $PbMo_6S_8$.

Table 4.1. Lattice constants, calculated and measured densities and T_c for several Chevrel compounds. The data of HAUCK [4.7] have been added for comparison

Compound	a_{hex} [Å]	c_{hex} [Å]	D_{calc} [g/cm^3]	D_{meas} [g/cm^3]	T_c [K]
$Cu_{1.3}Mo_6S_8$	9.479	10.340	5.67	5.64	10.3/5.6
$Cu_2Mo_6S_8$	9.605	10.225	5.86	5.82	10.8
$Cu_3Mo_6S_8$	9.720	10.213	6.08	6.05	5.5
$AgMo_6S_8$	9.307	10.832			7.5
$PbMo_{6.2}S_8$	9.193	11.473	6.30	6.38 ± 0.08	12.0
$PbMo_6S_7$ [4.7]	9.174	11.431	5.99		12.0
Mo_6Se_8	9.567	11.176	6.80	6.81	6.3
$Cu_{1.4}Mo_6Se_8$	9.865	10.800	7.04	7.04	5.7
$Cu_{2.6}Mo_6Se_8$	10.001	10.746			<1.2
$Pb_{1.2}Mo_6Se_8$	9.533	11.936			6.75

On the basis of a study on sintered samples, HAUCK [4.7] established the rhombohedral phase field of the system Pb-Mo-S at 1100°C. He found a variation in both the Pb and the S content and proposed the formula $Pb_xMo_6S_{8-y}$, with $0.85 \leq x \leq 1.05$ and $0.6 \leq y \leq 1.2$. This range does not include the stoichiometric composition and suggests the presence of a large amount of S defects. However, this diagram is in contradiction to the above results, which may be due to the above mentioned difficulties in analyzing the correct amount of additional phases in sintered samples. At least, the phase field must be extended up to the stoichiometric composition, characterized by y = 0 [4.10,16]. Unfortunately, single-phase samples with other compositions could not be prepared, so that the question of whether or not vacancies occur cannot be answered with certainty. HAUCK [4.7] and SERGENT et al. [4.23] have reported a variation of the rhombohedral angle from $\alpha = 89.2^\circ$ to 89.55° within the homogeneity range of lead molybdenum sulfide. The correlation between α and T_c reported by both authors [4.7,23] has been plotted in Fig.4.11. Taking into account shielding effects due to inhomogeneities, particularly in the low T_c region, both sets of data agree.

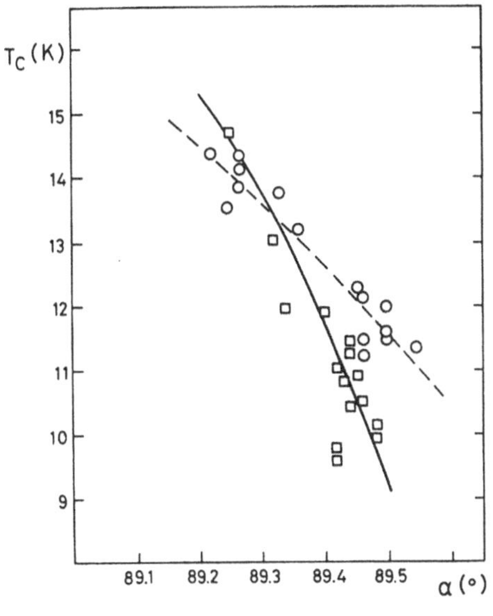

Fig.4.11. Correlation between T_c and the rhombohedral angle for different Pb-Mo-S samples, originating from HAUCK [4.4] (circles) and SERGENT et al. [4.23] (squares). Both sets of data are contained in a relatively narrow area. From [4.36]

From a series of experiments with the system Sn-Mo-S [4.12] it was found that this system has metallurgical properties very similar to Pb-Mo-S. In particular, it is also formed by a peritectic reaction. Recently, the homogeneity range of the rhombohedral phase in Sn-Mo-S was established by WAGNER et al. [4.24], who proposed the formula $Sn_xMo_6S_{8-y}$. In their diagram, the stoichiometric composition $SnMo_6S_8$ is not included. This is in contrast to the result of single-crystal X-ray refinements [4.25]. It is interesting that exactly the same problem is encountered for both systems, Pb-Mo-S and Sn-Mo-S. The reason is probably the same and resides in an incomplete analysis of all components after the sintering process, in particular the sample portions adhering at the surface of the quartz tube.

4.2.5 The Systems Ag-Mo-S, Ag-Mo-Se, Pb-Mo-Se, Sn-Mo-Se, ...

The rhombohedral phase $Ag_xMo_6S_8$ has a very narrow phase field, essentially centered on x = 1 [4.12], and is fully described by the stoichiometric formula $AgMo_6S_8$. In spite of long heat treatments at 1200°C, it has not been possible to produce microscopically single-phase alloys of this phase, which is thought to form peritectically (Fig.4.3). This behavior is in contrast to that encountered in $Cu_xMo_6S_8$. The question arises if this difference is due to a possible change of the valency of Ag from +1 to +2 in the environment of the rhombohedral phase. There is only a limited solubility of Ag in the system $(Cu,Ag)Mo_6S_8$. The formation of the rhombohedral phase in Mo selenides is less complex than for the analogous sulfides: as well in Pb-Mo-Se as in Sn-Mo-Se (Fig.4.4) the rhombohedral phase is found to form congruently [4.10]. The formation temperatures have so far not been determined with precision but are situated between 1550 and 1650°C. In all cases, single crystals can be easily prepared by the Bridgman technique [4.10].

4.3 Low-Temperature Relationships in $Cu_xMo_6S_8$ and $Cu_xMo_6Se_8$

The available information about a low-temperature phase transformation is usually restricted to the transformation temperature, T_M. This can be understood for two reasons a) it requires much more work to determine the formation conditions and the homogeneity range of the analyzed low-temperature phase, b) the knowledge of these properties has for a long time been considered to be of little help for the understanding of problems in solid-state physics. The only ternary molybdenum chalcogenide system for which the low-temperature phase relationships have been reported so far, is $Cu_xMo_6S_8$ [4.13,15,26].

In this section both the systems $Cu_xMo_6S_8$ and $Cu_xMo_6Se_8$ [4.12] will be reviewed. The experimental methods leading to these phase diagrams will briefly be discussed.

4.3.1 Experimental Methods

The small energies involved in phase transformations of intermetallic compounds at temperatures below 300 K do not favor important rearrangements of atoms by site exchange or diffusion. In the great majority of cases, these transitions are diffusionless, the major change being a change in symmetry. The term "martensitic transformation" for low-temperature phase transformations is current, but should not be used before the transformation mechanism for each individual case is known. The techniques of establishing high-temperature and low-temperature phase diagrams are somewhat different. The impossibility of performing observations by optical microscopy below 300 K is a major obstacle to working in this temperature range. The first step in detecting structural transformations usually consists in measuring the electrical resistivity as a function of temperature $\rho(T)$; a change in slope of $\rho(T)$ is usually an indication of a phase transformation. In such anomalously brittle phases, however, this change can also be caused by the sudden appearance of internal cracks during the heating or the cooling runs. The confirmation of a phase transition by additional X-ray measurements is thus necessary. The presence of microcracks can be detected by measuring $\rho(T)$ after different temperature cycles between 4.2 and 300 K: in contrast to other brittle phases, i.e., A15 compounds, the $\rho(T)$ values of rhombohedral phases sometimes show an increase after repeated cycles [4.12].

As shown earlier [4.15], specific heat measurements are very sensitive to transformations at temperatures below 300 K: they indicate the transformation temperature and the heat of transformation and give qualitative information about whether the transition corresponds to a horizontal or to an ablique line in the equilibrium phase diagram. The combination of electrical resistivity, specific heat and X-ray diffraction measurements for determining phase diagrams below 300 K will be discussed for the system $Cu_xMo_6S_8$, where it was first applied.

4.3.2 The System $Cu_xMo_6S_8$

The study of the phase relationships of $Cu_xMo_6S_8$ below 300 K reveals the presence of four different low-temperature modifications of the rhombohedral phase [4.13]. Since each modification has different superconducting properties, it is of particular importance to establish the correct low-temperature phase relationships. The system $Cu_xMo_6S_8$ can be considered as a pilot system for establishing phase diagrams below 300 K, based on the commonly used phase rules. In Figs.4.12 and 4.13, specific heat curves for compounds with different Cu contents are shown. The structural transformations show up as large peaks, the top of which coincides with the discontinuity in the slope of the electrical resistivity (Fig.4.12). The detailed low-temperature phase diagram shown in Fig.4.14 for the composition range $1.5 \leq x \leq 3.5$ has been established on the basis of these measurements [4.15], in combination with additional resistivity and X-ray diffraction data. At $x = 1.8$ in Fig.4.12 there is a single peak, corresponding to the congruent formation of the triclinic $Cu_{1.8}Mo_6S_8$ from the high-temperature rhombohedral phase [4.26,27]. A small bump in the specific heat of the sample with $x = 2$ at 187 K indicates that below this temperature, the alloy sits in a two-phase field, $Cu_{1.8}Mo_6S_8 + Cu_{3.2}Mo_6S_8$. When heating the alloy with $x = 2.7$ from the lowest temperatures, where the phase ratio according to Fig.4.14 is 1/3 $(Cu_{1.8}Mo_6S_8)$ + 2/3$(Cu_{3.2}Mo_6S_8)$, the first peak occurs at the peritectoid, 187 K. At this point, the phase with $x = 3.2$ transforms

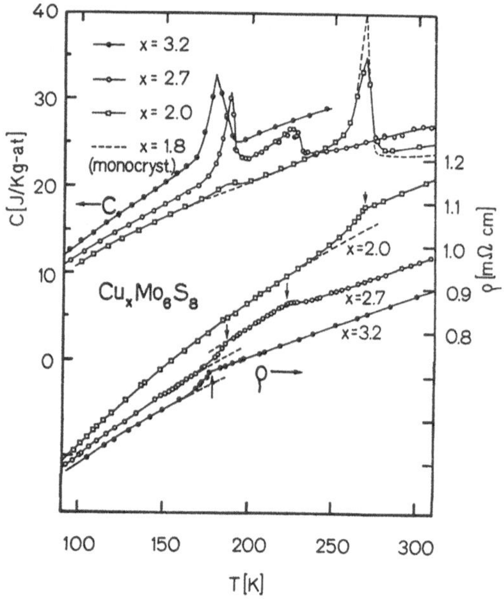

Fig.4.12. Low-temperature specific heat and electrical resistivity curves for compounds $Cu_xMo_6S_8$ with x=1.8, 2, 2.7, and 3.2. The lower transition temperature at 187 K for the composition x=2 can only be detected by specific heat measurements. From [4.15]

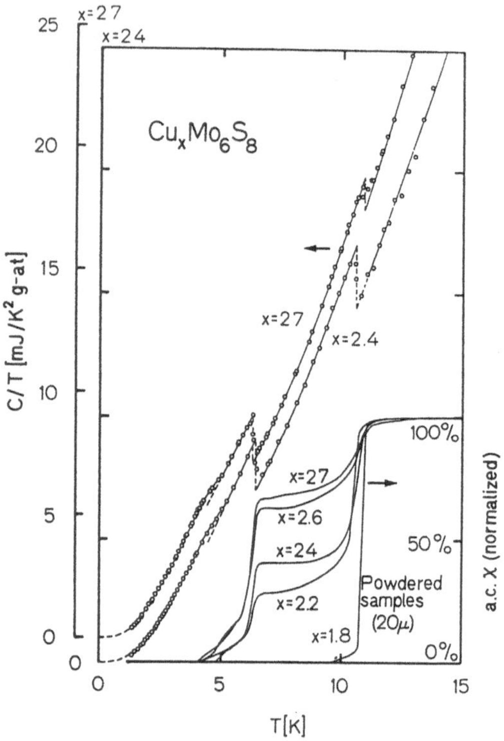

Fig.4.13. Specific heat curves for the compounds $Cu_{2.4}Mo_6S_8$ and $Cu_{2.7}Mo_6S_8$, showing two superconducting transitions, confirming inductive measurements of T_c on powdered samples. From [4.15]

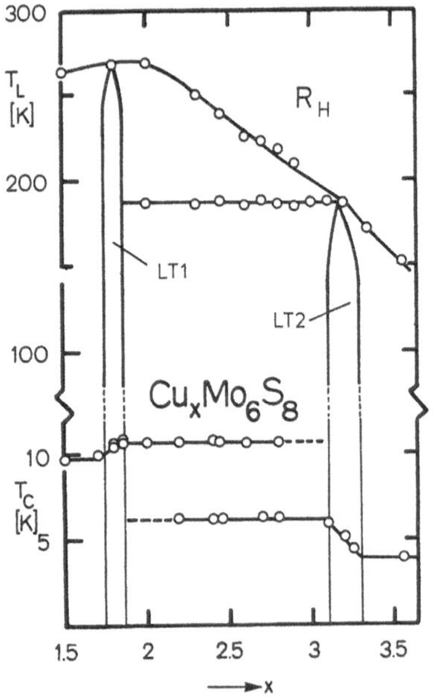

Fig.4.14. Portion of the low-temperature phase diagram of the system $Cu_xMo_6S_8$. This diagram serves as a basis for the understanding of the specific heat data in Figs.4.12 and 4.13. After [4.15]

into the high-temperature rhombohedral phase, R_H, and the heat of transformation effectively amounts to two-thirds of that observed in $Cu_{3.2}Mo_6S_8$.

Between 187 and 230 K, the phase with x = 1.8 gradually transforms into the rhombohedral phase, following the $(Cu_{1.8}Mo_6S_8 + R_H)/R_H$ phase boundary. The associated heat of transformation drops to zero when this process is over, i.e., near 230 K. The very broad peak in specific heat at 230 K thus signifies that an oblique line in the equilibrium phase diagram is crossed. Since both of the phases $Cu_{1.8}Mo_6S_8$ and $Cu_{3.2}Mo_6S_8$ are superconducting, the jump of the specific heat at T_c in Fig.4.13 allows the limits of these phases to be determined by applying the "lever law."

Inductive measurements of T_c on powdered samples confirmed the phase limits obtained by specific heat measurements. This allowed the low-temperature phase field of the phase with x = 1.8 to be determined within a very narrow region, $1.75 < x < 1.85$ [4.15]. The corresponding X-ray patterns at 293 and 5 K are shown in Fig.4.15. The determination of T_M by resistivity measurements is shown in Fig. 4.16: from this figure, it can be seen that the rhombohedral $Cu_xMo_6S_8$ phase is really stable up to 1300 K [4.13]. Differential thermal analysis up to the melting point confirms the stability of the rhombohedral phase in this temperature range [4.28]. The composition with the smallest Cu content, $Cu_{1.2}Mo_6S_8$, is metastable at room temperature but can be obtained by argon jet quenching. It is remarkable that at 60 K this metastable composition undergoes a transformation into a new low-temperature phase (Fig.4.14). Below 300 K, the rhombohedral $Cu_xMo_6S_8$ phase thus decomposes into four different modifications (Table 4.2). The low-temperature phase relationships for the equilibrium part of the diagram ($x \geq 1.8$ in Fig.4.8) can be drawn following the phase rules under the assumption of first-order phase transformations [4.13]. The region $x < 1.8$ is metastable (nonequilibrium region in Fig.4.8).

The superconducting transition temperatures of the four modifications are T_c = 5.6 K for x = 1.2, T_c = 11 K for x = 1.8, T_c = 6.4 K for x = 3.2, while the phase with x = 4, the untransformed rhombohedral phase, was found to be normal down to 0.8 K. The phase with x = 1.8 is triclinic [4.26,27]: the value T_c = 11 K is the highest value found so far for a triclinic phase.

4.3.3 $Cu_xMo_6Se_8$

The same principles as for the sulfide were applied for determining the low temperature phase relationships in $Cu_xMo_6Se_8$, represented in Fig.4.17 [4.12]. The homogeneity range of the rhombohedral phase in this system is confined within the limits $0 \leq x \leq 2.8$ [4.22]. At T_M = 167 K, there is a congruent transformation from the rhombohedral phase into a low temperature modification, $Cu_{1.7}Mo_6Se_8$ (the crystal structure of this phase has not been determined). Due to the toxicity of the evaporated Se, the investigation had to be restricted to a small number of samples. The picture in Fig.4.17 is thus incomplete, in particular at high Cu contents,

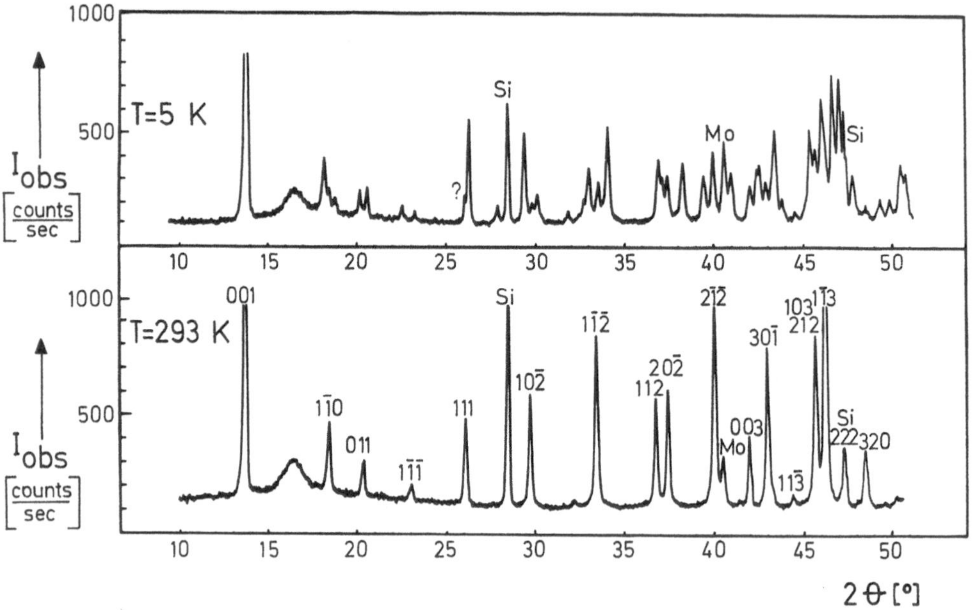

Fig.4.15. Powder X-ray diffraction data for $Cu_{1.8}Mo_6S_8$ at 5K (top) and 293 K (bottom). From [4.26]

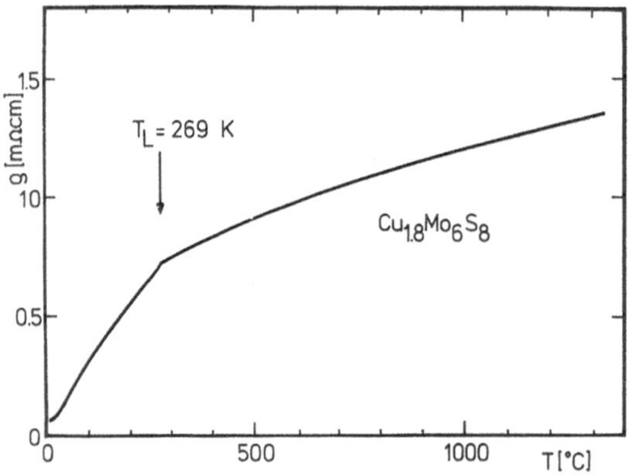

Fig.4.16. The electrical resistivity of $Cu_{1.8}Mo_6S_8$ in the temperature range $10 < T < 1350$ K. From [4.13]

where an additional low-temperature modification at $x \simeq 2.5$ is expected, based on the horizontal line at 118 K. The value T_c = 2 K [4.9] observed in this region has to be attributed to the phase with $x \simeq 2.5$. The usually reported value of T_c = 6 K corresponds to the $Cu_{1.7}Mo_6Se_8$ phase.

The untransformed rhombohedral phase with x = 2.8 is normal down to 1 K. The copper molybdenum selenide phase corresponding to the metastable $Cu_{1.2}Mo_6S_8$ (Fig. 4.8) is missing, due to the stability of the rhombohedral phase in this composition range. A comparison between the systems $Cu_xMo_6S_8$ and $Cu_xMo_6Se_8$ shows that in both

Table 4.2. Low-temperature modifications of the rhombohedral phase in the systems $Cu_xMo_6S_8$ ($1.2 \leq x \leq 4$) and $Cu_xMo_6Se_8$ ($0 \leq x \leq 2.8$) and their superconducting temperature T_c

$Cu_xMo_6S_8$	T_c[K]	Low-temperature modification
Phase x = 1.2	5.6 1.5	Unknown
Phase x = 1.8		Triclinic
Low copper limit x = 1.75	9.9	
High copper limit x = 1.85	11.0	
Phase x = 3.2		Unknown
Low copper limit x = 3.1	6.4	
High copper limit x = 3.3	4.0	
Phase x = 4		Rhombohedral
$Cu_xMo_6Se_8$	T_c[K]	Low-temperature modification
Phase $0 \leq x \lesssim 1.0$	6.3	Rhombohedral
Phase $x \simeq 1.7$	5.7	Unknown
Phase $x \simeq 2.5$	2.0	Unknown
Phase x = 2.8		Rhombohedral

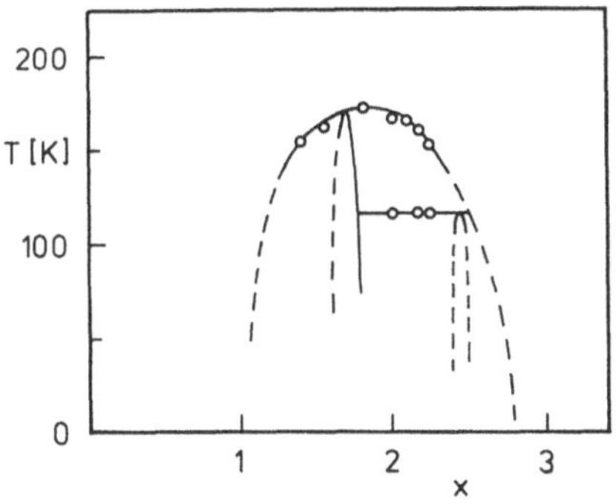

Fig.4.17. Portion of the low-temperature phase diagram of the system $Cu_xMo_6Se_8$. The three low-temperature modifications are situated at $x \simeq 1.7$, $x \simeq 2.5$, and $x \simeq 2.8$. From [4.12]

cases, the low-temperature modifications occur within a composition range limited by a miscibility gap of the rhombohedral phase.

The transformation temperatures for the Cu selenides are lower than for the sulfides, T_M = 167 K and 118 K, compared with T_M = 269 and 187 K, respectively. The resistive transformation at T_M = 167 K for the alloy $Cu_{1.7}Mo_6Se_8$ is reproduced in Fig.4.18. A comparison with Fig.4.12 shows that the change of $\rho(T)$ for $Cu_xMo_6Se_8$ at the transition is very different from the change in the corresponding sulfide. These transformations are all of the first order: for the selenide at x = 1.7, a hysteresis of 3 K is observed [4.18]; the corresponding hysteresis for the sulfide

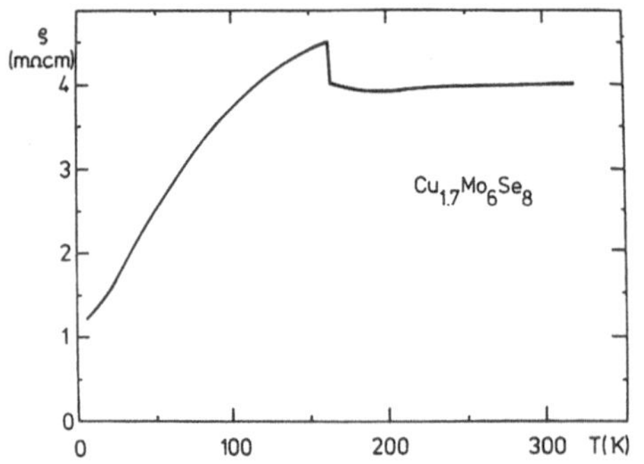

Fig.4.18. Low-temperature electrical resistivity of the compound $Cu_{1.7}Mo_6Se_8$, showing a lattice transformation at 167 K. From [4.9]

with x = 1.8 is less than 1 K [4.12]. For the latter, the mean atomic volume at T_M changes by 1.5% [4.26]. Like in the system $Cu_xMo_6S_8$, the untransformed rhombohedral phase in melted samples of $Cu_xMo_6Se_8$ for $x \approx 2.8$, the Cu-rich limit, can only be obtained after heat treatments around 800°C [4.12]. For comparison, the low-temperature relationships in $Cu_xMo_6S_8$ and $Cu_xMo_6Se_8$ have been summarized in Table 4.2.

4.4 The Type of Formation of the Triclinic Phase in Ternary Molybdenum Chalcogenides

4.4.1 Existence of a Miscibility Gap in the Rhombohedral Phase

From the phase diagrams represented in Figs.4.8,17 it follows that the low-temperature modifications occur within a miscibility gap of the rhombohedral phase. This is particularly evident for the system $Cu_xMo_6Se_8$ but is also visible for $Cu_xMo_6S_8$, where the metastable region $x < 1.8$ has to be retained by quenching (nonequilibrium region in Fig.4.8). The observation of the miscibility gap in these two systems is greatly facilitated by the excessively large width of their rhombohedral phase fields, i.e., $1.2 \leq x \leq 4$ for $Cu_xMo_6S_8$ and $0 \leq x \leq 2.8$ for $Cu_xMo_6Se_8$, i.e., ~ 14 at % Cu.

4.4.2 "Small" M Atoms: M = Ni, Zn, Co, Cr, Mn, Ti, V, Fe

The existence of a miscibility gap is quite common in intermetallic systems, a well-known example being Cu-Au. However, the unique feature of ternary molybdenum chalcogenides is the occurrence of a miscibility gap at temperatures close or even below room temperature. The metallurgical characterization of the phase transformations in this class of compounds, shown in Fig.4.19, can be summarized as follows: a) if first-order transformations are assumed, the presence of a new modification at low temperatures is always connected with the simultaneous presence of the untransformed rhombohedral phase at a different composition, and b) for a system in thermal equilibrium, there are compositions at both sides of the miscibility gap

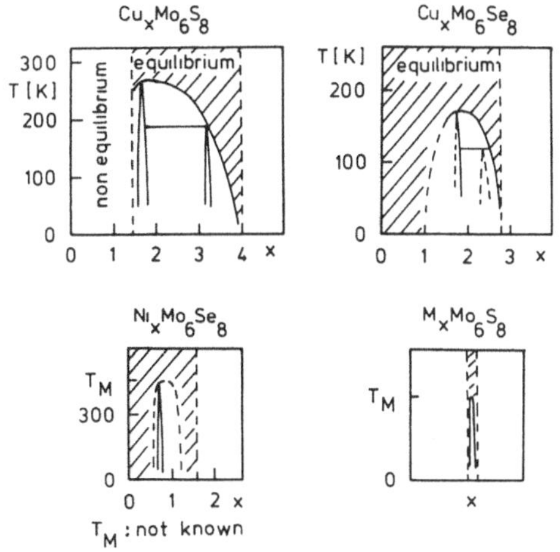

Fig.4.19. The miscibility gap in the systems $Cu_xMo_6S_8$, $Cu_xMo_6Se_8$, $Ni_xMo_6Se_8$ (tentative phase diagram), and for a general compound $M_xMo_6S_8$, where M is a "small" atom and X = S, Se. It is seen how the miscibility gap contains at least two phases for the Cu-based systems which have very large phase fields. Other systems, as $Ni_xMo_6Se_8$ with a narrow phase field have only one low-temperature modification at $T = T_M$

which remain untransformed. These two statements can be verified very easily by studying the literature. Indeed, in the first papers dealing with structural transformation in rhombohedral phases, the presence of both the triclinic and the untransformed rhombohedral phase was reported [4.22,29]. However, the metallurgical aspect of the structural transformation was not recognized at this time. We will now critically review all known data on structural transformations of $M_xMo_6X_8$ (x = Se, S) compounds with "small" M atoms.

a) *The Systems* $Ni_xMo_6Se_8$ *and* $Ni_xMo_6S_8$

The system $Ni_xMo_6Se_8$ has been studied by SERGENT and CHEVREL [4.22], who found a phase field confined within the limits $0 \leq x \leq 1.6$. The same authors reported a triclinic phase x = 0.6 and x = 1.2 [4.22] at 300 K. The refinement of a single crystal with the composition $Ni_{0.66}Mo_6Se_8$ yielded a triclinic structure [4.30] with the lattice parameters indicated in Table 4.3. By analogy with $Cu_xMo_6Se_8$ (Fig. 4.17) and $Cu_xMo_6S_8$ (Figs.4.8 and 4.14) the low temperature portion of the $Ni_xMo_6Se_8$ phase diagram can be provisionally drawn as shown in Fig.4.19.

The corresponding sulfide $Ni_xMo_6S_8$ has been analyzed at 300 K only [4.29]. At this temperature, it was found to be rhombohedral over the whole stability range [4.29]. No X-ray measurements were performed to our knowledge at T < 300 K, where a phase transformation is expected, in analogy to M = Co and Cu. At x = 2, LAWSON et al. [4.31] did not find any resistive indication for a phase transformation.

In order to verify the thermodynamical model presented above [4.32], we have recently investigated the system $Ni_xMo_6S_8$ in the region $1 \leq x \leq 2$ and found the expected new phase transition at $x \approx 1.4$ (the precise composition has not yet been determined) [4.12].

Table 4.3. Lattice parameters of some Chevrel phase compounds at different temperatures T. The rhombohedral compounds transform into a triclinic structure at the temperature T_M

Compound	T[K]	T_M[K]	Structure	Lattice parameters [Å], [°]	Reference
$Cu_{1.8}Mo_6S_8$	293	269	rhombohedral	a= 6.503, α=94.93	4.26
	5		triclinic	a= 6.450, b= 6.533, c= 6.521 α=96.81 , β=93.21 , γ=95.74	4.26
$Fe_2Mo_6S_8$	500	~ 400	rhombohedral	a= 6.50 , α=94.9	4.37
	300		triclinic	a= 6.502, b= 6.466, c= 6.481 α=95.94 , β=97.37 , γ=91.33	4.25
$Ni_{0.66}Mo_6Se_8$	300		triclinic	a= 6.727, b= 6.582, c= 6.751 α=90.61 , β=92.17 , γ=90.98	4.30
$Co_2Mo_6S_8$	300	220	rhombohedral	a= 6.48 , α=95.2	4.33
	220		triclinic		4.25
$ZnMo_5S_6$	320	320	rhombohedral	a= 6.50 , α=94.6	4.31
	300		triclinic		4.25
$In_{1}Mo_6S_8$	300	?	rhombohedral	a= 6.52 , α=93.2	4.25
	10		triclinic		4.25
$Cu_{1.4}Mo_6Se_8$	300	165	rhombohedral	a= 6.76 , α=94.2	4.31
	165		triclinic		4.25
$ZnMo_5Se_6$	300	130	rhombohedral	a= 6.70 , α=94.6	4.31
	130		triclinic		4.25
$Eu_{1.1}Mo_6S_8$	293	109	rhombohedral	a= 6.552, α=88.91	4.34
	10		triclinic	a= 6.472, b= 6.616, c= 6.573 α=89.26 , β=88.10 , γ=89.25	4.34
$BaMo_6S_8$	293	~ 175	rhombohedral	a= 6.640, α=88.77	4.35
	10		triclinic	a= 6.572, b= 6.691, c= 6.651 α=89.05 , β=88.10 , γ=88.90	4.35

b) *The Systems* $Zn_xMo_6Se_8$ *and* $Zn_xMo_6S_8$

The system $Zn_xMo_6Se_8$ was first studied by SERGENT and CHEVREL [4.22], who found a very narrow phase field centered at x = 1.2. At 300 K the compound $Zn_{1.2}Mo_6Se_8$ is rhombohedral [4.22]. LAWSON and SHELTON [4.31] observed by means of electrical resistivity measurements a lattice transformation at T_m = 130 K for a Zn-Mo-Se alloy of nominal composition $ZnMo_5Se_6$. In Fig.4.20, we have reproduced our own electrical resistivity measurements on a melted sample with the nominal composition $Zn_{1.2}Mo_6Se_8$ [4.12]. Hysteresis of ~1 K at the transition is observed, which indicates that the transition is of the first order, in analogy to the chalcogenides with M = Cu. The corresponding sulfide $Zn_xMo_6S_8$ was studied by CHEVREL et al. [4.29], who observed both the triclinic and the rhombohedral modification at 300 K. LAWSON and SHELTON [4.31] found for the nominal composition $ZnMo_5S_6$ a lattice transformation at 320 K

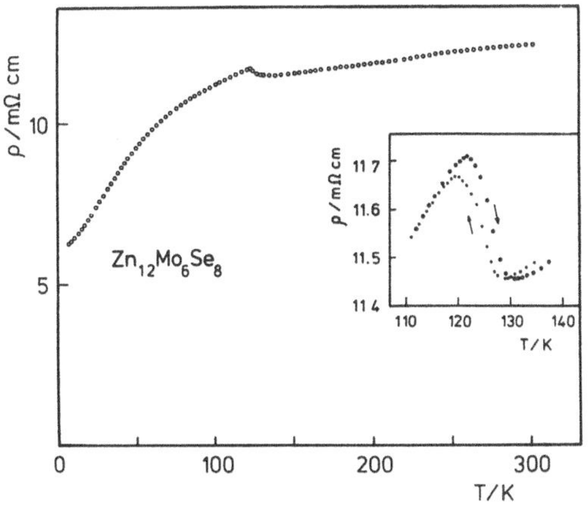

Fig.4.20. Electrical resistivity of $Zn_{1.2}Mo_6Se_8$ as a function of temperature. The insert shows a hysteresis of ~1 K at the transition

(Table 4.3). From the available data, it can be concluded that the low-temperature phase diagram of the systems $Zn_xMo_6S_8$ and $Zn_xMo_6Se_8$ is of the same type as that of $Ni_xMo_6Se_8$, represented in Fig.4.19.

c) *Ternary Molybdenum Chalcogenides with* M = Co, Cr, Mn, Ti, V *and* Fe

The selenides with M = Co, Cr, Mn, Ti, V, and Fe were first studied by SERGENT and CHEVREL [4.22]. The system $Co_xMo_6Se_8$ with $0 \le x \le 1.4$ was found to be rhombohedral in the whole composition range at 300 K. For the compound of the nominal composition $CoMo_3S_4$, LAWSON [4.33] reported a phase transition at 220 K. This transition is also of the type rhombohedral → triclinic [4.25]. The phase field in the selenides with M = Cr, Mn, Ti, V, and Fe was determined by SERGENT and CHEVREL [4.22] by means of density measurements. For all these systems, the homogeneity range is included within the limits $1 < x < 1.4$. The crystal structure at 300 K for the selenides with M = Fe, Cr, and V as determined on X-ray refinement of single crystals with x = 1.2 were found to be triclinic [4.22]. The same result was obtained on powders of the selenides with M = Mn and Ti, which are isotypical [4.22]. Unfortunately, the original paper does not clearly indicate if, and at what compositions, the rhombohedral phase was observed at 300 K. The corresponding sulfides with M = Co, Fe, Mn, and Cr were analyzed by CHEVREL et al. [4.29] (the rhombohedral phase in the sulfides with M = V and Ti could not be found). All of these systems were found to exhibit an extended rhombohedral phase field, but its limits were not determined with precision [4.29]. Nevertheless, for M = Fe and Mn, both the triclinic and the rhombohedral modifications were found at 300 K [4.29], the latter for the composition x = 2. For the sulfide with M = Mn, LAWSON [4.33] reported a phase transformation at ~190 K. The analyzed compound $Cr_2Mo_6S_8$ was found to be triclinic at 300 K [4.29].

4.4.3 "Large" M Atoms

For "large" M atoms, the homogeneity range of the rhombohedral phase $M_xMo_6X_8$ is usually very narrow and is restricted to a region close to x = 1. This is also the case for the compounds with M = Eu and Ba, where a structural transformation has been recently found by BAILLIF et al. [4.34,35] (Sect.4.4.3). The difficult preparation of these compounds together with the narrow rhombohedral phase field are actually the main obstacles in deciding whether a miscibility gap exists.

In the case of $Eu_{\sim 1.1}Mo_6S_8$, BAILLIF et al. [4.35] found a hysteresis of 0.4 K in the electrical resistivity curve at T_M, which suggests a first-order transition. Nevertheless, the type of formation of triclinic phases with "large" M atoms (involving no mobility of the M atom) must be considered as unknown.

4.5 Mechanism of Structural Transformations in Ternary Molybdenum Chalcogenides

4.5.1 The Delocalization of the M Atom

The search for factors leading to low-temperature structural transformations is of general interest and is not only restricted to ternary molybdenum chalcogenides. Due to the lack of general criteria, the usual way to proceed is to establish empirical correlations. It has been repeatedly proposed [4.6,25,29,36] that the instability of the rhombohedral phase may be correlated with a geometrical factor, i.e., the size of the M cation. Indeed, rhombohedral compounds with smaller M atoms (M = Cu, Ni, Co, Zn, ...) seem to have a pronounced tendency to undergo structural transformations [4.15,22,29]. Based on single-crystal X-ray refinements on a large number of rhombohedral compounds, YVON [4.16,18] has recently introduced a new quantity for classifying this class of materials, the "delocalization" of the M atom. This quantity was derived from the fact that the atoms do not occupy the exact center of the chalcogen hole at the origin but are found at a certain distance from it. Significant changes of the thermal motion in the plane perpendicular to the ternary axis are thought to be correlated with the observed delocalization of the M atom. As shown in Fig.4.21, the value of the delocalization is largest for the M atoms with the smallest size. Since the presence of small M atoms seems to favor structural transformations, Fig.4.21 could suggest a correlation between the degree of delocalization and phase stability. However, it is easy to show that delocalization alone cannot account for the observed phase transformations. Indeed, it can be seen from Fig.4.21 that the degree of delocalization for the sulfides with M = Cu and x = 1.8, 2.76, 2.94, and 3.66 is essentially the same. It is unlikely that dramatic changes occur up to x = 4, where the rhombohedral phase is stable down to the lowest temperatures. Therefore, a transformed and a nontransformed phase are observed for almost the same degree of delocalization. This clearly indicates that this quantity is not alone responsible for the ability of a rhombohedral compound

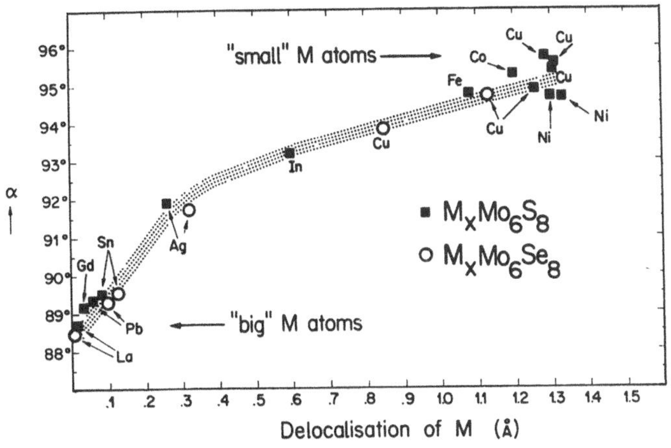

Fig.4.21. The delocalization of the atoms M from the inversion center as a function of the rhombohedral angle α in $M_xMo_6X_8$ (X=S,Se). From [4.25]

to undergo low-temperature lattice transformations. It is interesting that in all cases known so far (Table 4.3), the low-temperature modification is triclinic (in three compounds, $K_2Mo_6S_8$, $Rb_2Mo_6S_8$, and $Cs_2Mo_6S_8$ [4.29], a tetragonal phase was reported at 300 K, but it is not known if the rhombohedral phase in these compounds forms at all.)

In Table 4.3, we have summarized the corresponding lattice parameters of the triclinic phase, showing that the deformation relative to the high temperature rhombohedral lattice is quite small. From the known cases of $Fe_2Mo_6S_8$ [4.37] and $Cu_{1.8}Mo_6S_8$ [4.26,27], this rhombohedral → triclinic transformation may appear as an order-disorder transition of the Fe or Cu atoms.

4.5.2 The Mechanism of Structural Transformations in $M_xMo_6S_8$ Compounds (M = Cu, Fe, ...)

From the analysis of the metallurgical and crystallographical changes at $T = T_M$, the formation temperature of the triclinic phase, it follows that the mobility of the M atom, even at low temperature, must be extraordinarily high [4.30]. This follows immediately from Fig.4.22, which represents the Cu positions of both modifications as reported by YVON et al. [4.27] for $Cu_xMo_6S_8$. This system, together with $Fe_xMo_6S_8$, can be considered as a model for other ones with "small" M atoms.

a) *Intrasite Mobility*

At 300 K, there are 12 possible Cu positions, 6 equivalent inner positions (site 1) and 6 equivalent outer positions (site 2), indicated as Cu(1) and Cu(2) in Fig. 4.22. The closest distance between the sites Cu(1) and Cu(2) varies between 1.12 and 1.42 Å [4.27], depending on the Cu content, x, which is much less than 1.92 Å, corresponding to the sum of two neighboring Cu^+ radii. There is thus an overlap of the two sixfold crystallographic sites, excluding the simultaneous oc-

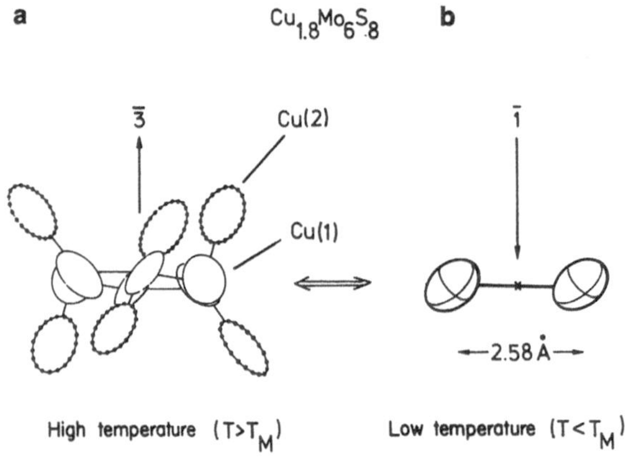

Fig.4.22a,b. The rhombohedral high-temperature and triclinic low-temperature structures of $Cu_{1.8}Mo_6S_8$, a) Cu(1) and Cu(2) sites for the rhombohedral phase, b) change of Cu(1) positions at the structural transformation temperature T_M = 269 K. From [4.25]

cupation of two closest positions. Different degrees of probability for the occupancy on both sites have been found. YVON et al. [4.38] called this effect "dynamic" disorder, in contrast to the usual "static" disorder.

Below T_M, there are only two single Cu sites, as shown in Fig.4.22. The transition mechanism from the rhombohedral to the triclinic phase involves translations between Cu(1) sites. The smallest translational distances for the composition x = 1.8 are observed for Cu atoms arising from the Cu(1) sites and are 1.26 Å [4.27]. The distance between two Cu(2) sites is somewhat larger, 1.70 Å. However, it is more probable that the diffusion path between two Cu(2) sites includes the intermediate Cu(1) sites.

It is interesting to compare the above described transformation with classical order-disorder transformations observed, for example, in Cu-Au. In the Cu-Au system the transition implies site exchanges of the Cu and Au atoms through interdiffusion at $T \gtrsim 1/2\ T_F$, where T_F is the formation temperature of the high-temperature phase. The fundamental lattice retains its symmetry. In the case of $Cu_{1.8}Mo_6S_8$ (as a representative of other rhombohedral phases) the transition occurs at $T \gtrsim 1/12\ T_F$ (Fig.4.17) and does not imply interdiffusion. At all temperatures, the M atoms remain in the "channels" separating the Mo_6S_8 building blocks. Furthermore, the symmetry of the fundamental lattice is altered by a distortion at the transformation. The term "generalized" order-disorder transformation would therefore be more appropriate to describe this particular kind of lattice instability.

b) *Metallurgical Arguments: Intersite Mobility*

The mobility of Cu ions between different $Cu_xMo_6S_8$ cells can be directly concluded from the equilibrium low-temperature phase diagram of this system, represented in Figs.4.8,14. As an example, we will follow the alloy with x = 2.5 during the cooling process from 300 K to liquid He temperature. At T ~ 230 K, the oblique LT1 + R_H/R_H phase limit is crossed. At T = 200 K, the sample contains the two phases,

LT1, with x = 1.85 (57%), and R_H, with x ≃ 2.9 (43%). At T = 100 K, equal parts of the sample crystallize in the LT1 phase (with x = 1.85) and in the LT2 phase (with x ~ 3.1). There is thus a shift of the chemical composition of the observed phases. However, at these low temperatures, the motion of grain boundaries as a consequence of nucleation or interdiffusion is not expected to take place; consequently, other mechanisms must determine the observed shift of chemical composition as well as the structural transformation. A mechanism accounting for both effects must thus be based on the mobility of the Cu ions between two or more $Cu_xMo_6S_8$ clusters.

From [4.27] it follows that the minimum distances between Cu(2) sites of two neighboring clusters vary between 2.07 and 2.44 Å, depending on the Cu content. This distance is larger than the minimum intrasite distance. On the other hand, there are less Cu ions changing their site. In our example, the transition from x = 2.5 (at 300 K) to x_1 = 1.85 and x_2 = 3.2 at T < 180 K involves an average motion of 0.3 Cu ions per cluster. The distribution of the two phases LT 1 and LT 2 in the crystal, i.e., the "grain sizes" at low temperatures, cannot be obtained directly. The narrow X-ray diffraction lines of the LT 1 and LT 2 phases, however, suggest that grain sizes are larger than 100 or even 1000 Å. As a consequence, the total migration distance for the Cu ions would be of the order of several hundreds of unit-cell lengths.

There are other effects indicating the extraordinary mobility of the Cu (or other "small" cations), such as the ionic conductivity described by DUDLEY et al. [4.39] or the topotactic redox reaction reported by SCHÖLLHORN et al. [4.40]. However, the most striking effect remains the possibility of obtaining the metastable "Mo_6S_8" by the "leaching" technique introduced by CHEVREL [4.18]. From the size of the crystallites, ~0.1 mm, a total migration distance of the same order of magnitude can be deduced.

4.5.3 The Mechanism of Structural Transformations in Compounds with "Large" M Atoms

Mössbauer effect measurements of KIMBALL et al. [4.41] on $SnMo_6S_8$ have shown that the vibrational behavior of the Sn atoms is anisotropic, anharmonic, and has a typical soft mode behavior. Similar measurements on $Sn_xEu_{1-x}Mo_6S_8$ compounds [4.42] confirmed the existence of an anharmonic potential around the Sn atoms and also showed that irregularities of the Debye-Waller factor occur in the temperature region between 50 K and 110 K. These anomalies were attributed to a structural phase transformation involving vibrational modes of the Mo_6S_8 building blocks. However, no crystallographical evidence for the occurrence of a structural transformation have been obtained. The first crystallographical proof for a structural transformation of a rhombohedral compound with a "large" M atom was recently found by BAILLIF et al. [4.34,35] for M = Eu and Ba. Both compounds were found to transform into a triclinic distorted low-temperature structure at T = 110 and 175 K, respectively. In both cases, the distortion does not involve the movement of the M atoms between two clusters. In contrast to the cases with M = Cu, Ni, ..., the transformation is thus not correlated to a particular mobility of Eu or Ba: the electronic instability of the Mo_6X_8 clusters seems to be rather responsible for this transformation.

After the detection of a structural transformation in compounds with M = Eu and Ba, it is probable that other compounds with "large" M atoms also transform. The question arises if these transformations even occur in all Chevrel phases. But these transformations have not been observed so far, the triclinic deformation being too small to be detected with the resolution of conventional diffraction methods.

4.6 Superconductivity and Structural Transformations in Ternary Rhombohedral Compounds

According to FISCHER [4.36] and to band-structure calculations [4.43-46], superconductivity in ternary molybdenum chalcogenides is due to the Mo 4d electrons which occupy narrow bands very close to the Fermi level. However, all available band-structure calculations have been performed for the untransformed rhombohedral structure. The corresponding band structure for the triclinic phase is unknown. The lower symmetry of the triclinic cell is thought to affect the position of the narrow d band with respect to E_F, thus leading to a decreased density of states. Unfortunately, there is no system where the transformed and the untransformed rhombohedral phase can be observed at the same composition. Quenching experiments undertaken to prevent the low-temperature modification have so far failed [4.12].

A possible way of affecting the low-temperature phase equilibrium is to apply a high external pressure. Pressure effects on T_c have been reported for $Cu_xMo_6S_8$ [4.47], $PbMo_6S_8$ [4.47], and $EuMo_6S_8$ [4.48]. In the first two cases, an increase in T_c was observed, while a dramatic change in T_c has been reported for the latter, which is usually normal but exhibits T_c values up to 11 K under external pressure. In the case of $PbMo_6S_8$, it is not excluded that a possible distortion is hidden by insufficient instrumental resolution, T_c being more sensitive to a phase change. This would also be supported by the fact that additions of rare earths to ternary $Pb_xMo_6S_8$ increase T_c [4.23]. The volume change correlated to the increase of T_c [4.23] could in turn influence the conditions of a possible phase transformation. However, this remains a speculation as no crystallographic data of higher resolution are available. The suppression of the low-temperature phase transformation by pressure or substitution still remains to be proved. Due to the experimental uncertainty in detecting structural transformations in cases where the triclinic deformation is very small, it could be that most compounds reported as "rhombohedral" at low temperature are in reality triclinic. There are not enough data in order to decide if the degree of triclinic deformation affects T_c. At this point, it is tempting to compare the rhombohedral compounds with Nb_3Sn, the latter crystallizing in the

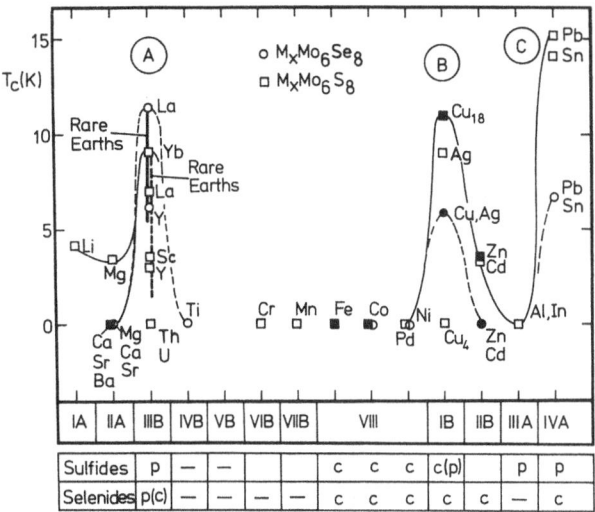

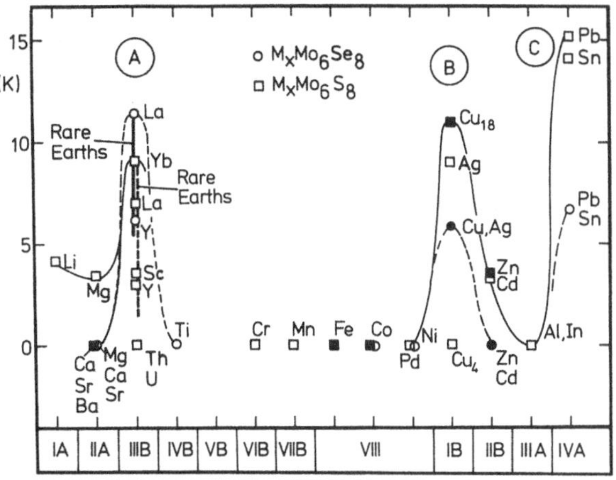

	IIIB	IVB	VB	VIB	VIIB	VIII			IB	IIB	IIIA	IVA
Sulfides	p	—	—			c	c	c	c(p)		p	p
Selenides	p(c)	—	—	—	—	c	c	c	c	c	—	c

Fig.4.23. Superconducting transition temperature T_c for rhombohedral sulfides and selenides $M_xMo_6X_8$ (X=S,Se) as a function of the atomic number of the M atom. Superconductivity occures in three regions: A (earth alkaline, rare earth), B (Cu, Ag), and C (Sn, Pb). The type of formation of the rhombohedral phase is indicated below (P = peritectic, c = congruent). Blackened symbols denote compounds with known lattice transformation

A15-type structure. For this compound T_c is maximum for the cubic case and is gradually depressed with increasing value of tetragonality (1-c/a), as shown by FLÜKIGER et al. [4.49].

Figure 4.23 shows the variation of T_c as a function of the atomic number of the M element in the formula $M_xMo_6X_8$, where X = S and Se. There are essentially three stability regions for the rhombohedral phase for both S- and Se-based compounds. The first region (A) includes the alkalies, alkaline earths, rare earths, and actinides. The second region (B) contains transition elements with nearly filled electron shells (noble metals are excluded) and elements of the IB and IIB group of the periodic system. The third region (C) is essentially limited to M = Pb, Sn, Al, and In. The highest transition temperatures in region A are reached by the RE Mo_6X_8 compounds, the values of selenides being systematically higher than those of sulfides [4.36]. The situation is reversed in both regions B and C where sulfides exhibit higher values of T_c than selenides. $PbMo_6S_8$ and $SnMo_6S_8$ show the highest T_c values, close to 15 K. The type of formation of the rhombohedral phase is tentatively added in Fig.4.23 (c stands for congruent and p for peritectic). It is seen that in region A, the peritectic formation is favored for both, sulfides and selenides. In region B, the congruent formation is observed again in both cases sulfides and selenides. The main difference is observed in region C, where the sulfides form peritectically, in contrast to the selenides, which form congruently. It follows from Fig.4.23 that the type of formation (congruent or peritectic) is determined by the M element. This is analogous to the case of A15-type compounds A_3X, where the minority X atom has a dominant effect on the phase formation [4.49]. For both phase types, regularities in the type of phase formation are encountered when increasing the group number of the M or the X element, respectively.

The width of the rhombohedral phase is different for the regions A, B, and C. In region A (RE's, actinides), the rhombohedral phase field must be very narrow, since no variation of T_c as a function of composition has been reported so far. The widest homogeneity ranges occur in region B, where they reach up to 14 at % in the case of $Cu_xMo_6S_8$. As shown in Fig.4.11 the homogeneity range of $Pb_xMo_6S_8$ is narrow but has an appreciable width, with important variations of T_c and the lattice parameters as a function of composition. The different regions A, B, C in Fig.4.23 correspond to a different behavior of the electronic properties, as shown by a comparison of the compounds $La_{1.2}Mo_6S_8$(A), $Cu_{1.8}Mo_6S_8$(B), and $PbMo_6S_8$(C). The T_c values of these three compounds are similar, T_c = 12, 11, and 15 K, respectively, the electronic specific heat coefficients γ 79, 54, and 105 mJ K^{-2} $mole^{-1}$, respectively [4.36].

References

4.1 M. Hansen, K. Anderko (eds.): *Constitution of Binary Alloys* (McGraw-Hill, New York 1958)
4.2 R.P. Elliott (ed.): *Constitution of Binary Alloys* (McGraw-Hill, New York 1965)
4.3 F.A. Shunk (ed.): *Constitution of Binary Alloys*, 2nd Suppl. (McGraw-Hill, New York 1969)
4.4 A. Taylor, N.J. Doyle, B.J. Kagle: J. Less Common Metals *4*, 436 (1962)
4.5 D.C. Johnston, R.N. Shelton, J.J. Bugaj: Solid State Comm. *21*, 949 (1977)
4.6 R. Chevrel: Thesis No. *B112*, Université de Rennes, France (1974)
4.7 J. Hauck: Mater. Res. Bull. *12*, 1015 (1977)
4.8 G.H. Moh: In *Aspects of Molybdenum and Related Chemistry*, Topics in Current Chemistry, Vol.76 (Springer, Berlin, Heidelberg, New York 1978) p.107
4.9 R. Flükiger, H. Devantay, J.L. Jorda, J. Muller: IEEE Trans. MAG *13*, 818 (1978)
4.10 R. Flükiger, R. Baillif, E. Walker: Mater. Res. Bull. *13*, 743 (1978)
4.11 P. Schweiss, B. Renker, R. Flükiger: In *Ternary Superconductors*, ed. by G.K. Shenoy, B.D. Dunlap, F.Y. Fradin (North Holland, Amsterdam 1981) p. 29
4.12 R. Flükiger, R. Baillif: Unpublished results
4.13 R. Flükiger, R. Baillif, J. Muller, K. Yvon: J. Less Common Metals *72*, 193 (1980)
4.14 M. Marezio, P.D. Dernier, J.P. Remeika, E. Corenzwit, B.T. Matthias: Mater. Res. Bull. *8*, 657 (1973)
4.15 R. Flükiger, A. Junod, R. Baillif, P. Sptizli, A. Treyvaud, A. Paoli, H. Devantay, J. Muller: Solid State Commun. *23*, 699 (1977)
4.16 K. Yvon: Solid State Comm. *25*, 327 (1978)
4.17 K. Yvon, A. Paoli: Solid State Comm. *24*, 41 (1977)
4.18 R. Chevrel, M. Sergent, J. Prigent: Mater. Res. Bull. *9*, 1487 (1974)
4.19 M. Sergent, Ø. Fischer, M. Decroux, C. Perrin, R. Chevrel: J. Solid State Chem. *22*, 87 (1977)
4.20 A.M. Umarji, G.V. Subba Rao, V. Sankaranarayanan, G. Ragarajan, R. Srinivasan: Mater. Res. Bull. *15*, 1025 (1980)
4.21 J.R. Gavaler: Appl. Phys. Lett. *23*, 480 (1973)
4.22 M. Sergent, R. Chevrel: J. Solid State Chem. *6*, 433 (1973)
4.23 M. Sergent, R. Chevrel, C. Rossel, Ø. Fischer: J. Less Common. Metals *58*, 179 (1978)
4.24 H.A. Wagner, H.C. Freyhardt: To be published
4.25 K. Yvon: In *Current Topics in Materials Science*, Vol.3, ed. by E. Kaldis (North-Holland, Amsterdam 1979) p.55
4.26 R. Baillif, K. Yvon, R. Flükiger, J. Muller: J. Low Temp. Phys. *37*, 231 (1979)

4.27 K. Yvon, R. Baillif, R. Flükiger: Acta Crystallogr. B*33*, 3066 (1977); B*35*, 2859 (1979)
4.28 J.L. Jorda: Unpublished results
4.29 R. Chevrel, M. Sergent, J. Prigent: J. Solid State Chem. *3*, 515 (1971)
4.30 O. Bars, J. Guillevic, D. Grandjean: J. Solid State Chem. *6*, 3 (1973)
4.31 A.C. Lawson, R.N. Shelton: Mater. Res. Bull. *12*, 375 (1977)
4.32 R. Flükiger: To be published
4.33 A.C. Lawson: Mater. Res. Bull. *7*, 773 (1972)
4.34 R. Baillif, A. Dunand, J. Muller, K. Yvon: To be published
4.35 R. Baillif, A. Junod, B. Lachal, J. Muller, K. Yvon: To be published
4.36 Ø. Fischer: Appl. Phys. *16*, 1 (1978)
4.37 K. Yvon, R. Chevrel, M. Sergent: Acta Crystallogr. B*36*, 685 (1980)
4.38 K. Yvon, A. Paoli, R. Flükiger, R. Chevrel: Acta Crystallogr. B*33*, 3066 (1977)
4.39 G.J. Dudley, K.Y. Cheung, B.C.H. Steele: J. Solid State Chem. *32*, 259 (1980)
4.40 R. Schöllhorn, M. Kümpers, J.O. Besenhard: Mater. Res. Bull. *12*, 781 (1977)
4.41 C.W. Kimball, L. Weber, G. van Landmyt, F.Y. Fradin, B.D. Dunlap, G.K. Shenoy: Phys. Rev. Lett. *36*, 412 (1976)
4.42 J. Bolz, J. Hauck, F. Pobell: Z. Phys. B*25*, 351 (1976)
4.43 T. Jarlborg, A.J. Freeman: Phys. Rev. Lett. *44*, 178 (1980)
4.44 O.K. Andersen, W. Klose, H. Nohl: Phys. Rev. B*17*, 1209 (1978)
4.45 L.F. Mattheiss, C.Y. Fong: Phys. Rev. B*15*, 1760 (1977)
4.46 D.W. Bullet: Phys. Rev. Lett. *39*, 664 (1977)
4.47 R.N. Shelton, A.C. Lawson, D.C. Johnston: Mater. Res. Bull. *10*, 297 (1975)
4.48 C.W. Chu, S.Z. Huang, C.H. Lin, R.L. Meng, K. Wu, P.H. Schmidt: Phys. Rev. Lett. *46*, 276 (1981);
D.W. Harrison, K.C. Lim, J.D. Thompson, C.Y. Huang, P.D. Hambourger, H.L. Luo: Phys. Rev. Lett. *46*, 280 (1981)
4.49 R. Flükiger: In *Superconductor Materials Science: Metallurgy, Fabrications and Applications*, ed. by S. Foner, B.B. Schwartz (Plenum, New York 1981) p.511

5. Thin-Film Ternary Superconductors

J. A. Woollam, S. A. Alterovitz, and H.-L. Luo

With 12 Figures

Physical properties and preparation methods of thin film ternary superconductors, (mainly molybdenum chalcogenides) are reviewed. Properties discussed include the superconducting critical fields and critical currents, resistivity and the Hall effect. Experimental results at low temperatures, together with electron microscopy data are used to determine magnetic flux pinning mechanisms in films. Flux pinning results, together with an empirical model for pinning, are used to get estimates for possible applications of thin film ternary superconductors where high current densities are needed in the presence of high magnetic fields. The normal state experimental data is used to derive several Fermi surface parameters, e.g. the Fermi velocity and the effective Fermi surface area.

5.1 Preliminary Comments

There are only a few systems of ternary compound superconductors, as opposed to the large number of three (or more) element pseudobinary superconductors [5.1]. The first high-temperature ternary superconductors to be found were the molybdenum chalcogenides [5.2], followed soon after by the discovery of superconductivity in rare-earth-metal rhodium borides [5.3], in the perovskite- and spinel-structure oxides [5.4,5], and most recently in germanides and silicides [5.6,58].

Most of the ternary superconductors have been prepared in bulk form, for example by sintering. The first successful preparation of any ternary superconductor in thin-film form was for the molybdenum sulfides, by BANKS et al. [5.7]. Subsequently, other successful attempts were reported for both sputtered and evaporated molybdenum sulfide compounds [5.8-14], as well as for perovskite oxides [5.15] and for rhodium borides [5.16,17].

Several of the ternary superconductors have transition temperatures T_c above 8 to 10 K. For example, $PbMo_6S_8$ has $T_c \lesssim 14$ K, and an upper critical field B_{c2} of $\lesssim 60$ Tesla at 0 K. This upper critical field compares favorably with that of the A15-structure superconductors such as Nb_3Ge, where T_c is approximately 23 K and

B_{c2} is $\lesssim$ 35 Tesla [5.18]. With high B_{c2} and high T_c it is reasonable to expect high critical currents J_c at high fields. Thus the ternary superconductors are of both scientific and practical interest [5.19]. The study and potential use of ternary compounds in bulk form can be difficult. (Sintered molybdenum chalcogenides are highly permeable, for example). Thin films offer a convenient geometry for the study of transport properties and especially for evaluation of the critical currents found in these materials. Thin films also offer a useful geometry for potential applications.

In this chapter we shall describe the preparation and properties of thinfilm ternary superconductors. Preparation methods include both sputtering and evaporation. Properties discussed include:

1) superconducting critical fields B_{c2}, B_{c1}, B_c, critical temperatures T_c, and critical currents J_c;
2) phenomenological parameters such as coherence length ξ_0, penetration depth λ, and Ginzburg-Landau constant κ;
3) normal state properties like magnetoresistance $\Delta\rho/\rho$, Hall effect R_H, heat capacity γ, electrical resistivity ρ, and derived values for: mean free path ℓ, number of carriers n, Fermi surface area S, effective masses m^*, Fermi velocity v_F, scattering times τ, and density of states $N(E)$.

Emphasis will be on thin-film molybdenum chalcogenides, where the most extensive studies have been made. For a discussion of bulk samples see also [Ref.5.20, Chap.3].

5.2 Preparation and Structure

Even though ternary superconductors have quite different constituents, compositions, and crystal structures, there are some features common to all ternary-compound film preparation. Firstly, stoichiometries are complex, and preparation by sputtering and evaporation ("physical" methods) are favored over methods such as chemical vapor deposition. This is not surprising since physical methods often provide good control and versatility. Secondly, substrate conditions always play an important role in thin-film deposition, and ternary superconductors are no exception. In general, the most critical factors are substrate material and temperature during deposition. Ternary films prepared on substrates held at room temperature are usually highly disordered or even amorphous.

To promote crystallization and improve crystalline order, substrates can be held at fixed elevated temperatures related to the solid-state reaction temperatures for bulk-form synthesis. To do this, substrates must be able to withstand elevated temperatures and simultaneously remain chemically inert. The relative thermal expansion coefficient is also a consideration.

Also common to most thin-film ternary-compound superconductors is that few surface effects have been reported. Films of thickness $\lesssim$ 200 Å exhibit properties strongly influenced by the nature and condition of the surface. Ternary superconductor films prepared to date have all been much thicker than this (~ 0.4 to 17 μm) and have been prepared on substrates at elevated temperatures. Thus, film properties approach those of bulk samples.

As an example of the above general considerations, we will now describe the sputter deposition of the ternary molybdenum chalcogenides $Cu_xMo_6S_8$ and $PbMo_6S_8$, as described by KAMMERDINER and LUO [5.21]. Samples were RF sputtered using a base target of MoS_2 which was overlayed with whatever additional constituents were desired:copper for $Cu_xMo_6S_8$, and PbS or Pb for $PbMo_6S_8$. To change the Mo-to-S ratio, additional Mo was overlayed. The ternary phase was identified (using X-ray diffraction) for $Cu_xMo_6S_8$ only when the substrates (sapphire or molybdenum) were in the range 750° to 900°C. For other temperatures only other phases were found, including MoS_2, free Mo, etc. For $PbMo_6S_8$ the ternary was formed only after additional annealing in Pb vapor at temperatures between 750° and 1100°C. At the higher Pb-vapor annealing temperatures, the superconducting transition temperature was sharper, but film adherence to sapphire became a problem and films became powdery. A compromise was found for an annealing temperature of 850° to 950°C.

Figure 5.1 shows T_c as a function of Mo-to-S target ratio for $PbMo_6S_8$. About a 20% variation in T_c is found for a factor of greater than two change in the Mo-to-S ratio. Each data point represents the average onset T_c for a large number of samples, however. The highest T_c was found for Mo to S in the desired 6 to 8 ratio for the ternary compound.

Composition is a very important parameter influencing T_c for the ternaries. For this reason KAMMERDINER and LUO [5.21] investigated T_c versus rhombohedral angle α of the Chevrel ternary structure, using X-ray determined lattice constants. This is dramatically shown in Fig.5.2, where T_c is found to go from above 14 K to near zero with an α change of less than 0.5°. Data from bulk samples are also included [5.21-24]. Critical current is related to composition also, as will be discussed below. Thus, composition and structure can be controlled by preparation conditions, and both influence the resultant superconducting properties.

Other thin-film molybdenum chalcogenide ternary superconductors have been sputtered, including $Pb(Gd)Mo_6S_8$, $SnMo_6S_8$, $Cu_xMo_6S_8$, and $AgMo_6S_8$ (these are nominal compositions only). Table 5.1 gives a general comparison of some of the properties of these films (as well as two non-Chevrel phase films), including a comparison with bulk sample data. Table 5.2 is a condensed summary of data on twelve representative sputtered $Cu_xMo_6S_8$ specimens [5.25]. The formation of $Cu_xMo_6S_8$ is an important example since it can be prepared with a wide range of x [5.26]. The phase diagram for the $Cu_xMo_6S_8$ system is shown in Fig.5.3 [5.27]. Note the presence of two phases LT1 and LT2. By sputtering, values of x from 1.5 to 2.0 were formed [5.7,21]. A wider range of x (1.8 to above 3.1) was obtained for $Cu_xMo_6S_8$ prepared by evaporation.

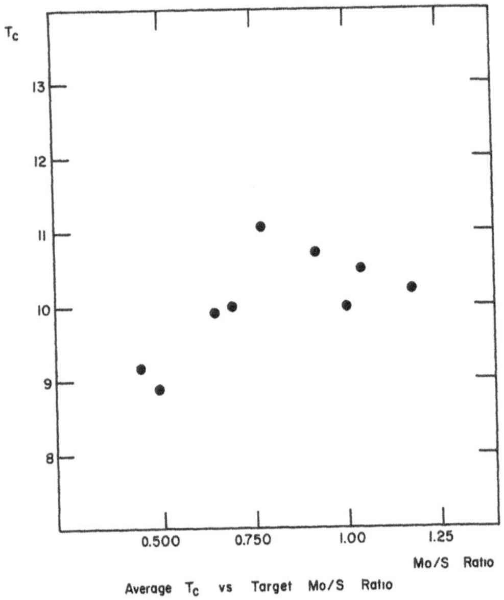

Fig.5.1. Superconducting transition temperature T_c vs molybdenum-to-sulfur volume ratio of sputtering target [5.21]

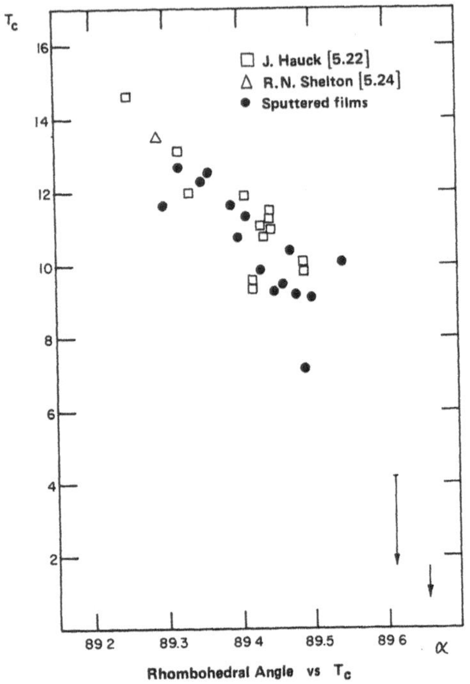

Fig.5.2. Superconducting transition temperature T_c vs rhombohedral angle for sputtered $PbMo_6S_8$ and several sintered samples [5.21,22]

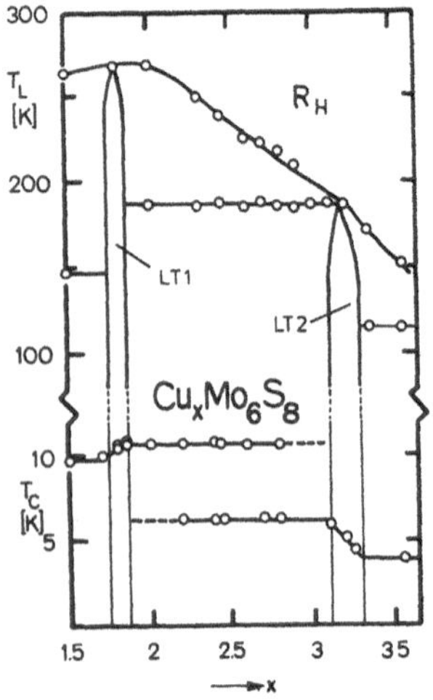

Fig.5.3.
Phase diagram of the $Cu_xMo_6S_8$ system [5.27]

Table 5.1. Summary of thin-film ternary superconductor results

Ternary compound	Film configuration		Substrate			Bulk form		Reference
[a]	Preparation method	Thickness [μm]	Material	Temperature [°C]	T_c [K]	Solid state reaction temperature [°C]	T_c [K]	
$PbMo_6S_8$	sputtering	0.4∼4	Mo or	b	13.3	900°-1250°	12.5-13.2[f]	This work[g]
$SnMo_6S_8$			sapphire		11.0		10.9-11.3	and [5.7]
$CuMo_8S_8$	sputtering	0.4∼4	Mo or	750°- 900°	10.7	900°-1250°	10.8-10.9	This work[g]
$AgMo_6S_8$			sapphire	850°-1050°[c]	8.8		8.4- 8.9	and [5.7]
$CuMo_6S_8$	evaporation	5-7	sapphire	~815	10.1-10.8	-	-	[5.11,12,33]
$ErRh_4B_4$	evaporation	0.4	sapphire	980°-1130°	8.3	arc melting	8.9	[5.16,17]
$BaPb_{0.7}Bi_{0.3}O_3$[d]	evaporation	-	glass	e	13.3	900°-1000°	13	[5.15]

[a]Nominal composition only. [b]Deposited at ambient temperature; then annealed in closed space at 850°-1050°C in the presence of Pb or Sn vapor as the case may be. [c]Deposited at ambient temperature; then annealed in closed space at 850°-1050°C. [d]Recently doubt has been cast upon the true nature of superconductivity of this pseudo ternary compound because no specific-heat anomaly was detected at T_c (C.E. Methfessel, et al., Proc. Nat. Acad. Sci., in press). [e]Deposited at ambient temperature, then annealed at 350°-550°C. The upper temperature was limited by the substrate material. [f]This represents the range of values reported by numerous authors. Each transition is normally sharper than the "range" given here. [g]Some of the numbers in these rows of the table have not been previously published.

Table 5.2. A condensed summary of data on sputtered $Cu_xMo_6S_8$ specimens [5.25]

Sample no. series A	T_{sub}[°C]	T_c[K]	$J_c \times 10^4$ [a] [A/m^2]	a[Å]	c[Å]	$R_{RT}/R_{T>T_c}$	Relative amounts of precipitated Mo[b]
177	810°	7.6	0.20	9.47	10.30	1.82	vvw
179	820°	8.0	0.42	9.44	10.30	2.01	vw
180	840°	9.0	0.72	9.49	10.29	2.28	w
188	840°	8.5	1.50	9.54	10.28	-	w
191	850°	9.5	2.40	9.57	10.24	-	m
202	860°	9.4	2.70	9.55	10.30	2.1	m
196	865°	9.9	3.5	9.52	10.24	2.7	m
194	870°	9.3	4.2	9.67	10.22	2.6	s
199	875°	8.5	4.2	9.53	10.27	2.4	s
201	870°	9.6	4.7	9.55	10.22	2.8	m
197	870°	9.4	5.7	9.61	10.23	2.6	s
198	865°	9.3	7.8	9.56	10.25	2.2	s

[a] In zero external field at 4.2 K. [b] vvw, very very weak; vw, very weak; w, weak; m, medium; s, strong.

Films of $Cu_xMo_6S_8$ were prepared by evaporation by CHI et al. [5.11,12] in a standard high-vacuum evaporator, and these procedures will now be described. Copper and molybdenum were evaporated simultaneously using resistively heated and electron-beam heated sources, respectively, and a base pressure of 10^{-5} torr. Sulfur was provided by bleeding high-purity H_2S gas (Table 5.3a) or hot sulfur gas (Table 5.3b) into the system to a pressure of $\sim 10^{-3}$ torr. Sapphire substrates were used and were mounted 15 cm above the centerline between the molybdenum and copper sources. Substrates were heated from the back side to 760° to 850°C by radiation from a tantalum filament. Two techniques, reactive evaporation and activated reactive evaporation (ARE), were used. In the ARE, a blue plasma was generated by putting a probe between substrates and the molybdenum billet. The probe was biased at about 70 V and drew a current of 1.5 A. Tables 5.3a,b give preparation conditions and some pertinent results for coevaporated $Cu_xMo_6S_8$. Note that x ranges from 1.8 to above 3.1. This is a wider span of the phase diagram than was found for sputtered samples. As a result samples with either LT1 phase or LT2 phase or both were found. Both the ternary and pure molybdenum phases were found. Values of x for the ternary were determined by X-ray lattice parameters and by microprobe analysis. The presence of more than just the pure ternary phase was found to influence values of critical current J_c.

Since the most extensively prepared and evaluated thin-film superconductors have been $PbMo_6S_8$ and $Cu_xMo_6S_8$, detailed results on these systems will now be presented. Further description of film properties will be made in the context of particular experimental results.

Table 5.3a. Evaporation conditions and characterization of copper molybdenum films[a]

Run	Substrate temperature [°C]	Working pressure [10^{-2} Pa]	Process[b]	Phases[c]	Lattice parameter a[Å]	α[deg]	x	$T_c \pm \Delta T_c$
60	760°	10	ARE	T,M	6.55	95.6	$\geq$ 3.1	5.1 ± 0.65
61	760°	10	ARE	T,M	6.48	94.8	1.6	9.9 ± 0.40
62	760°	12	ARE	T,M	6.45	94.4	1.2	8.9 ± 1.0
55	830°	2.2	CE	T,M	-	-	2.7	8.1 ± 0.4

[a] H_2S gas was used as sulfur source for samples listed in this table.

[b] CE: coevaporation, ARE: activated reactive evaporation.

[c] T: $Cu_xMo_6S_8$; M: molybdenum.

Table 5.3b. Evaporation conditions and properties of copper molybdenum sulfide films[a]

Run	ARE Process	Substrate Temperature [°C]	Approx. Thickness [μm]	Phases[b]	Ternary lattice parameters a[Å]	α	Upper limit of x	T_c [c]	ΔT_c [c]	R_{300}/R_{11K} [d]
15	NO	715°	12	M,T	-	-	-	6.2	0.7	~ 2.5
18	NO	720°	17	M,T,B,C	6.61	95.6°	> 4	N	-	1.72
23	NO	715°	4	M,T,A	6.61	95.6°	> 4	N	-	2.08
32	YES	810°	7	M,T	6.52	95.3°	2.2 ± 0.3	10.1	0.50	2.50
33	YES	820°	7	M,T,C	6.58	95.6°	3.4 ± 0.4	N	-	2.33
34	YES	815°	4	M,T,C	6.57	95.8°	3.4 ± 0.4	N	-	2.30
36	YES	815°	5	M,T,C	6.58	95.5°	3.2 ± 0.7	N	-	2.36
39A	NO	815°	5	M,T	6.54	95.4°	2.5 ± 0.3	10.8	0.40	2.86
39B	NO	815°	5	-	-	-	-	10.8	0.50	2.60

[a] Hot S_2 gas was used as the sulfur source for samples in this table.
[b] Designations are $A \equiv MoS_2$, $B \equiv Cu_{1.96}S$, $C \equiv Cu$, $M \equiv Mo$, $T \equiv Cu_xMo_6S_8$.
[c] N denotes lack of a superconducting transition T_c above 4.2 K.
[d] Ratios of resistance at temperatures indicated [K].

5.3 Critical Fields

The upper critical field B_{c2} and its temperature dependence for the ternary molybdenum chalcogenides (TMC) is of great interest. In $PbMo_6S_8$, for example, the critical field is exceedingly high ($\gtrsim$ 60 Tesla): higher at 0 K than the much higher T_c binary compound Nb_3Ge (38 Tesla) or pseudo binary $Nb_3Al_xGe_{1-x}$ (45 Tesla) [5.18]. With such high B_{c2} and yet lower T_c, the slope $(dB_{c2}/dT)_{T_c}$ is also exceptionally high. Among all superconducting materials, only the lower dimensional materials (such as intercalated MoS_2 [5.28,29]) have such high values for dB_{c2}/dT, and even in the intercalated MoS_2 case, dB_{c2}/dT is high only for certain magnetic field orientations. Thus the high critical fields achievable in TMC are both of technological and scientific interest.

Critical fields have been studied in both sintered (bulk) [5.30,31] and thin-film samples of the TMC [5.32-34]. We will discuss only results in sputtered and evaporated thin films in this review, while the critical fields of bulk samples will be reviewed in Ref.5.20, Chap.4 . Nearly all studies of B_{c2} in thin-film TMC were done on $Cu_xMo_6S_8$, and only a few attempts have been made to form thin-film TMC with magnetic ions. Thus a summary for this system will be made.

Copper molybdenum sulfide has a relatively low $B_{c2}(0)$ so most of the $B_{c2}(T)$ curve could be observed using steady-state magnetic fields, rather than more electrically "noisy" pulsed fields. Critical fields can be measured in several ways. Commonly, either magnetic inductance changes or four probe resistance transitions are measured. To get $B_{c2}(T)$, either the field is held constant and temperature changed, or vice versa. At high fields superconducting transitions broaden. Thus the best choice of method takes the sample through the superconducting transition in a path normal to the $B_{c2}(T)$ curve, resulting in the sharpest transitions. Also, to define the $B_{c2}(T)$ curve, a choice of standards needs to be made: the onset, the 50%, the 90%, etc. point in the transition. For comparison with theory, use of several definitions, and more than one type (conductive or resistive) of observation is wise. For most of the thin-film work reported below, the midpoint of the resistive transition was most often chosen. [A final word of caution is that for comparison of critical current (J_c) data with theory, another criterion for determining B_{c2} had to be invoked, and this will be described in Sect.5.4 on J_c].

For samples containing *no* magnetic impurities, $B_{c2}(T)$ closely follows (with a small deviation, discussed below) the theoretical predictions of WERTHAMER, HELFAND, and HOHENBERG (WHH) for an isotropic, dirty, weak-coupled type II superconductor [5.35]. In the WHH theory the paramagnetic limiting parameter α and the spin-orbit parameter λ_{so} are variables to be determined by experiment. The value of α is determined by the slope, dB_{c2}/dT near T_c, and λ_{so} is then varied to fit the high-field, lower-temperature data. Figure 5.4 shows examples of fits of data on $Cu_xMo_6S_8$ ($1.5 \leq x \leq 2.0$) to the WHH theory. Data from two sputtered and one sintered sample

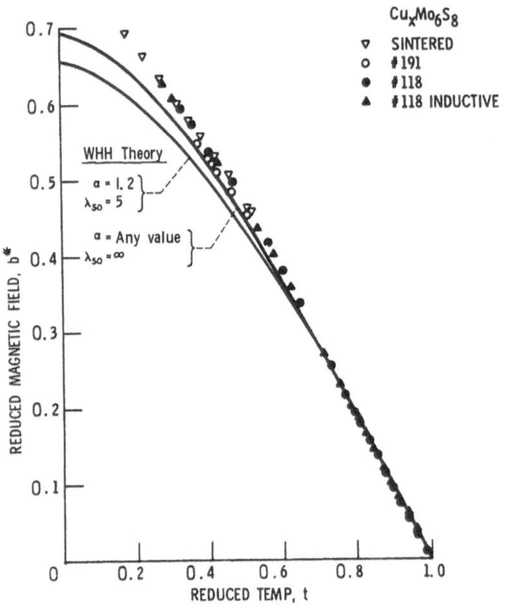

Fig.5.4. Reduced magnetic field b^* vs reduced temperature t for several $Cu_xMo_6S_8$ samples. Solid lines are from WHH theory [5.35]

are shown. Both inductive and resistively measured transitions are shown for sputtered sample number 118. Solid lines show examples of WHH theory for two sets of α and λ_{SO} values. A "reduced field" parameter b^*, given by

$$b^*(T) \equiv [B_{c2}(T)/T_c(-dB_{c2}/dT)_{T_c}] \quad , \tag{5.1}$$

is plotted as a function of "reduced temperature" $t \equiv T/T_c$. Experimental data were plotted by assuming a "WHH fit" in the range $0.7 \leq t \leq 1.0$. Figure 5.4 shows that for any value of t, the experimental data are above the theoretical curve for $\lambda_{SO} = \infty$ and α being any value. These parameters give the highest $b^*(T)$ within the limits of the theory, so an explanation is needed.

There are three possible mechanisms for b^*(low T) enhancement: 1) strong coupling effects [5.36], 2) purity effects [5.37], and 3) anisotropy of the electron-phonon interaction [5.38]. The first possibility can be ruled out by considering the known phonon spectra responsible for superconductivity in these compounds. According to BADER et al. [5.39], these phonons are concentrated at $\gtrsim 5\, k_B T_c$. According to theory for strong coupling [5.36], this range of phonon energies will probably reduce $b^*(0)$ slightly, rather than increase it. The purity dependence of b^*(low T) assuming isotropy in the electron-phonon coupling is small but not negligible [5.37]. Using experimental data and a few resonable assumptions, it was determined that $\xi_0 \sim \ell$, where ξ is the BCS coherence length and ℓ is the low-temperature mean free path. The ratio ξ_0/ℓ can be defined as a "purity parameter". Using this parameter, the theoretical value of $b^*(0)$ can be increased enough to explain the discrepancy between WHH theory and experiment. However, the anomalously high $b^*(0)$ could also be due to anisotropy of the electron-phonon coupling, even in the limit of short mean free path ℓ. Moreover, the cleaner the material (longer ℓ and lower ξ_0/ℓ), the greater is the anisotropy enhancement of b^*. Thus both purity and anisotropy can effect b^* enhancement, and it is difficult at this point to distinguish between the two effects. Indeed, anisotropy effects have been detected in some Chevrel materials ($PbMo_6S_8$ and $SnMo_6S_8$) [5.31].

The above discussion has been only for single-phase ($1.5 \lesssim x \lesssim 2.0$) $Cu_xMo_6S_8$. A fascinating phenomenon occurs for samples ($2.0 \lesssim x \lesssim 3.0$) where two superconducting phases coexist (Fig.5.3). The atomic structure of these two phases is not known. However, since they exist at low temperature and for other reasons, we believe that these phases are of a weakly distorted Chevrel structure [5.40]. In any case, the two low-temperature phases have $T_c \simeq 11$ K and 6 K. Samples exhibiting either superconducting transition were prepared in thin-film form by coevaporation [5.11,12, 32,34]. Samples with $x \geq 3.1$ had T_c's near 5 K. With $x < 3$ both superconducting phases coexisted but the lower T_c transitions were not observed due to shielding and shorting by the higher T_c phase. Both $B_{c2}(T)$ in high and low T_c phases again resulted in $b^*(0)$ above the WHH limits. Comparison with HELFAND and WERTHAMER [5.37] theory for "clean-limit," isotropic, nonparamagnetically limited material gave a similar result: experimental $b^*(0)$ is higher than theory, but the fit is better than with WHH theory. It appears that the enhanced $b^*(0)$ is due to both purity and anisotropy effects.

5.4 Critical Currents

As discussed in Sect.5.3, the ternary molybdenum chalcogenides have the highest upper critical fields known. This suggests that they may be useful as conductor material for high-field superconducting magnets. The critical parameters for practical uses are critical current density J_c as a function of external field B and temperature T, as well as the mechanical properties. Because of its high T_c and B_{c2} the compound $PbMo_6S_8$ and closely related pseudoternary compounds are likely candidates for applications. In this section we will describe J_c measurements on thin films of $PbMo_6S_8$, $Pb(Gd)Mo_6S_8$, $M_xPb_{1-x}Mo_{6+y}S_8$, and $Cu_xMo_6S_8$ [5.9,19,24,33,41-43]. Thin films are generally nonporous and have reliable geometries for critical current I_c to critical current density J_c conversions. Films are also easy to examine by scanning electron microscope and electron microprobe.

Central to the study of mechanisms for critical currents is fluxoid pinning on defects, precipitates, boundaries, etc. Flux pinning force densities are vectorial and are given by

$$\underline{P} = \underline{J}_c \times \underline{B} \quad , \tag{5.2}$$

where J_c is the critical current measured in a field B. Pinning force densities "scale" if they can be described by [5.44]

$$|P| = g(T)f(b) \quad , \tag{5.3}$$

where T is temperature, b is the reduced field = B/B_{c2}, and g and f are single variable functions (of T and b, respectively). Lack of scaling can occur when there is paramagnetic limiting, matching of a flux line lattice with an array of pinning centers, size effects, multiple species of pinning centers, or inhomogeneous material. Thus the observation of pinning force density scaling suggests that a material is homogeneous with only one effective pinning center. The general applicability of scaling laws is discussed in [5.44].

The most widely used theoretical prediction for the forms of g(T) and f(b) is due to KRAMER [5.45,46]. In general f(b) = 0 at b = 0 and b = 1, therefore a peak in f(b) occurs at some value of $f(b_{max})$. For $b > b_{max}$ Kramer's model predicts

$$P = 2\, C_s\, B_c^2(T)\, B_{c2}^{\frac{1}{2}}(T) b^{\frac{1}{2}}\, (1 - b)^2 \quad , \tag{5.4}$$

where $B_c(T)$ is the thermodynamic critical field and C_s is a constant having values in the range $0.14 \leq C_s \leq 0.56$ depending on the pin density. The general form of (5.4) has been found to be obeyed for numerous cases [5.46]. However, it should be pointed out that (5.4) may be followed in cases not really justified by the assumptions of the original theory. Examples of when the form of (5.4) is followed are for normal-state σ phase precipitates in Nb_3Ge and grain boundaries in Nb_3Sn. Below, we will show that in most cases pinning in the TMC follow (5.4). Thus (5.4) can be used to give an empirical upper estimate to P and therefore to J_c in the ternary molybdenum chalcogenide compounds.

For $Cu_xMo_6S_8$ the upper critical field B_{c2} appropriate for use in (5.4) was found using a fit to f(b) of the form

$$f(b) \propto b^{\frac{1}{2}}(1 - b)^2 \quad . \tag{5.5}$$

Experimental plots of total pinning force F versus f(b) were made, with B_{c2}^* used as a parameter to result in straight-line plots. An example is shown in Fig.5.5 for a series of $Cu_xMo_6S_8$ sputtered and evaporated films with x ranging from 1.85 to 2.5 [5.43]. These data confirm the applicability of the field dependent form (5.5) in Kramer's equation (5.4), for the copper molybdenum sulfide system.

In Figs.5.6-8 the separable field and temperature dependencies (5.3) are demonstrated to be of the form of (5.4) for a sputtered film [5.25]. In Fig.5.6, P is plotted versus b for a series of measured temperatures from 1.56 to 4.19 K. In Fig.5.7 the data of Fig.5.6 have been normalized by dividing by the values of P_{max} for each temperature. In Fig.5.8 the temperature dependence given by (5.4) is demonstrated to be followed. Equation (5.4) predicts a slope of 2.5 on the log P_{max} versus $B_{c2}(T)$ curve, within the limits of the observed 2.35 ± 0.2.

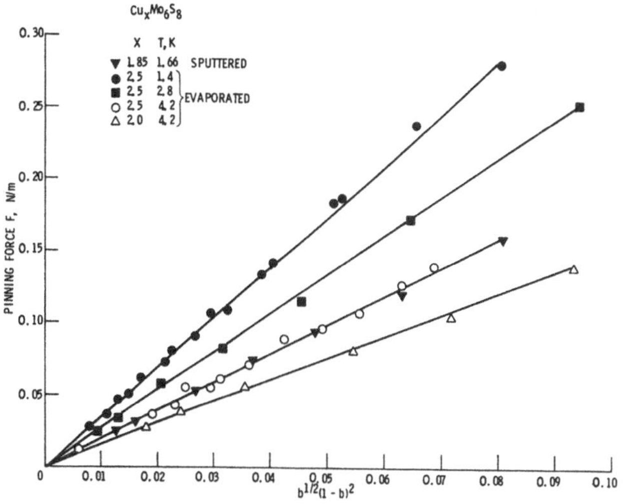

Fig.5.5. Pinning force F vs "Kramer parameter" $b^{\frac{1}{2}}(1-b)^2$, where b is the reduced field B/B_{c2}, for several $Cu_xMo_6S_8$ samples and temperatures [5.43]

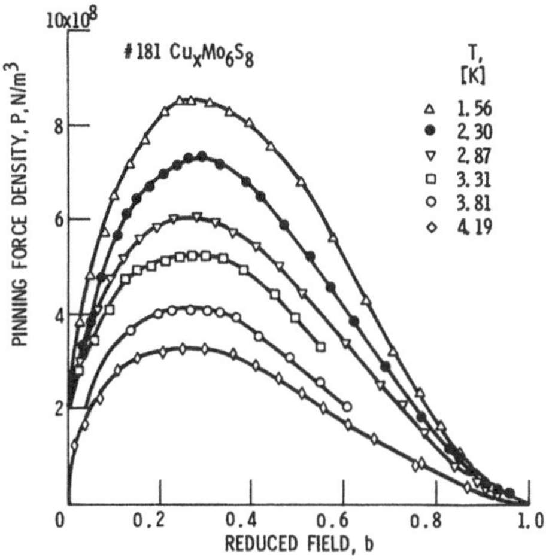

Fig.5.6. Pinning force density P versus reduced field b for a $Cu_xMo_6S_8$ film at six temperatures

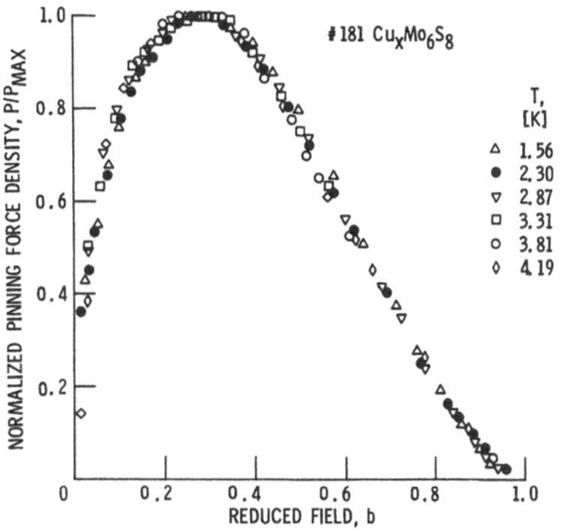

Fig.5.7. Normalized pinning force density P/P_{max} versus reduced field b for the same sample as in Fig.5.6

The form given by (5.4) for f(b) was found for all $Cu_xMo_6S_8$ samples except three inhomogeneous evaporated films. Thus, with only three exceptions, data on both sputtered and evaporated $Cu_xMo_6S_8$ demonstrated that there was only one pinning mechanism and that the pinning source was homogeneously distributed throughout the film. In the sputtered films a simple proportionality was found between P_{max} and excess Mo metal precipitate. The higher amounts of Mo were found in samples with higher substrate temperatures during deposition, whereas grain size of the ternary was found to be nearly independent of substrate temperature. It is concluded, then, that the single pinning mechanism in sputtered $Cu_xMo_6S_8$ is small particles of Mo precipitates. This is in agreement with the observation that P

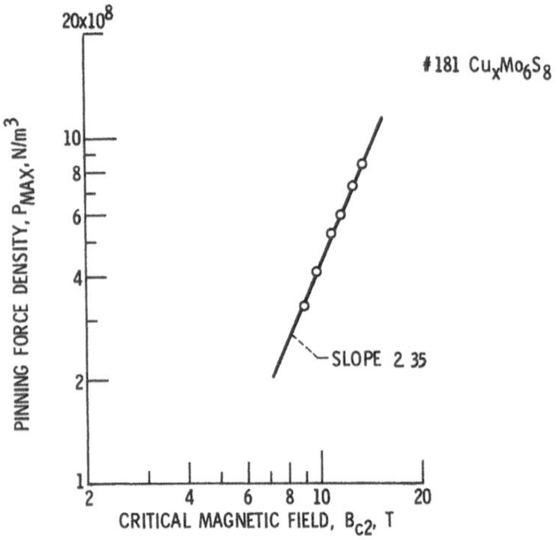

Fig.5.8. Log pinning force density P_{max} versus log critical field B_{c2} for the same sample as in Figs.5.6,7

was independent of whether the field was perpendicular or parallel to the film plane (in both cases $\underline{J}$ and $\underline{B}$ were perpendicular to each other).

In evaporated films a direct correlation of P_{max} with Mo was not attempted, but other evidence [5.47] supports the conclusion that Mo was the pinning source for both evaporated and sputtered films of $Cu_xMo_6S_8$.

As shown in Fig.5.3, the phase diagram predicts two superconducting phases in $Cu_xMo_6S_8$ for $2.0 \lesssim x \lesssim 3.0$. An evaporated sample with $x = 2.5$ was extensively studied [5.43], and the presence of two phases caused a lack of scaling of pinning forces. By careful analysis, it was possible to separate the total pinning force into two parts, each having a separate f(b) of the shape shown in Figs.5.6,7 with two different B_{c2}'s. Thus a separation of the form

$$F(B,T) = F_1(B,T) + F_2(B,T) \tag{5.6}$$

was found, where F_1 and F_2 are the pinning forces for the two phases. The form of pinning in (5.4,5) were found for each phase.

For $Cu_xMo_6S_8$ with $x \geq 3.1$ there is only a single phase present, that having the lower T_c and lower B_{c2}. Such a sample was prepared by evaporation (but not by sputtering). The data on B_{c2} are presented in Sect.5.3. Critical current measurements were performed on this single phase, low T_c material. The results show scaling, and are consistent with values of B_{c2} for the low T_c phase.

Critical current data versus applied field B are shown in Fig.5.9 for three sputtered films of $PbMo_6S_8$ [5.42]. The effective upper critical field for these measurements is $B^*_{c2} \sim 25$ Tesla, low compared with inductively determined B_{c2} in sintered material. Midpoint transition temperatures for these films were $9\ K \leq T_c \leq 13\ K$, which are again low compared with the maximum field for sintered material. When analyzed for scaling, less than half of the sputtered films were found to obey pinning force scaling laws, and those were found to have no free

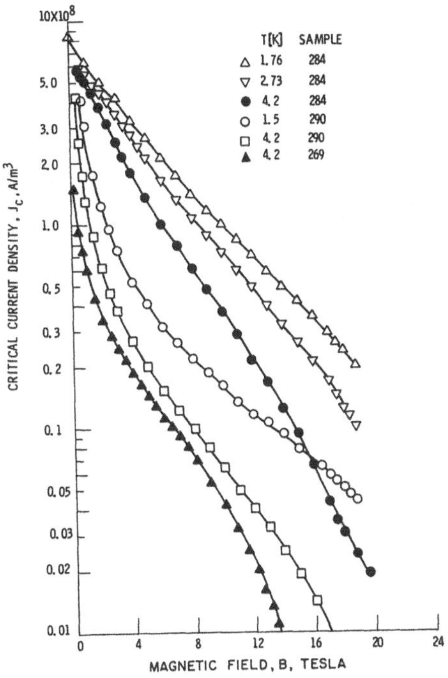

Fig.5.9. Critical current J_c versus magnetic field B for three sputtered $PbMo_6S_8$ samples at several temperatures [5.42].

lead, yet had free molybdenum. Evidence was presented [5.42,47] which suggested that free molybdenum was the source of flux pinning.

Films of $PbMo_6S_8$ having high J_c were found to not obey scaling laws and were inhomogeneous, with two pinning mechanisms. Scaling was found at high fields in these films, and evidence suggested that superconducting (granular) free lead shares some of the current at low fields and acts as a pinning center when in the normal state at high fields. Films of nominally $Pb(Gd)Mo_6S_8$ were also sputtered, but J_c was generally lower than for $PbMo_6S_8$ and no scaling of pinning was found. Since the system was complicated, no further work was done.

A recent paper on $J_c(B)$ for $M_xPb_{1-x}Mo_{6+y}S_8$ films was published [5.9]; where M = La, Nd, Pr, Gd, Y or U; x = 0.1 and 0.2; and y = 0 and 0.35. Films were produced by sputtering on a cold substrate, followed by annealing to as high as 1100°C in an argon atmosphere. Exact composition of the films is unknown, and the composition quoted is that of the target, likely to be different than that of the films. X-ray data on these pseudoternary compounds were generally characteristic of the Chevrel phase, but peaks were broad compared with the simple ternaries. Sharp drops in $J_c(B)$ in fields of about 2 to 3 Tesla suggest that these films may contain free lead also. Values of J_c for the pseudoternary were about an order of magnitude lower than those found for $PbMo_6S_8$ (Fig.5.9). Thus controlled deposition of pseudoternaries by sputtering appears to be difficult at this time.

We would now like to make an estimate of the highest $J_c(B)$ possible at 4.2 K for $PbMo_6S_8$. To do this we will use Kramer's formulas (5.4,5) in an empirical fit,

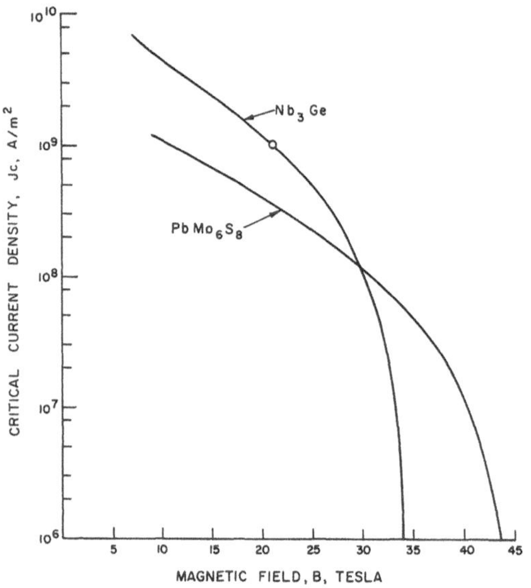

Fig.5.10. Log of projected critical current versus magnetic field B, comparing Nb_3Ge to $PbMo_6S_8$ projections [5.19]. Explanation of projections is given in text

using optimistic estimates of parameters [5.19,47]. The basis for this is the fact that all $PbMo_6S_8$ films obey a scaling law with the form of (5.5) for $b \geq 0.5$. According to theory and experiment, $0.14 \leq C_s \leq 0.56$ is expected [5.45]. In a few cases, values of C_s higher by a factor of two have been found. To project upper limits to J_c in $PbMo_6S_8$, Kramer's equations are used in an empirical numerical fit only. That is, no dependence on first- principles theory is assumed. Taking $C_s = 0.56$, $\gamma = 6400$ erg cm^{-3} K^{-1} [5.48], $B^*_{c2} = 45$ Tesla, $T = 4.2$ K, and $T_c = 13$ K, values of $J_c(B)$ were calculated and are shown in Fig.5.10. A similar calculation was made for the A15 structure Nb_3Ge with $B_{c2} = 35$ Tesla and $T_c = 21$ K, using an experimentally observed value of 10^9 A/m^2 at 20 T, rather than assuming a value for C_s. Plots of both predictions are shown in Fig.5.10. Note that a crossover is found at $B \simeq 30$ Tesla and $J_c \sim 1.2 \times 10^8$ A/m^2. If $C_s = 1.12$ is used, this crossover moves to 24 Tesla and $J_c \simeq 5 \times 10^8$ A/m^2. An increase in T_c to 14 K will move the crossover to B = 23 Tesla and 8×10^8 A/m^2 [5.59].

Since excess molybdenum and lead can be produced by sputtering, along with the ternary, good pinning centers are easily formed (a result found by at least two groups). Perhaps the pseudoternaries can increase B_{c2} significantly while still maintaining good flux pinning centers. If so, even higher $J_c(B)$ values might be expected. One of the greatest problems for achieving useful high $J_c(B)$ values in the lead-based ternary molybdenum chalcogenide films is the sometimes poor adhesion to substrates. To our knowledge, no one has yet attacked this problem. Compared with the stage of development of A15-structure conductors, the development of TMC is in its infancy. There are, however, a number of areas worthy of further research in an attempt to find a practical high-field conductor.

5.5 Electronic Properties

Above T_c the TMC exhibit a temperature-dependent electrical resistivity, a Hall effect, and a magnetoresistance [5.49-53].

In bulk samples the heat capacity γ can be measured, whereas in films the substrate mass is too large as compared with the film mass. From these γ measurements, coupled with measured superconducting properties, many electronic properties of the TMC can be deduced [5.49,52]. In this section experimental results are discussed which lead to evaluation of the number of carriers, the coherence length ξ_0, the actual to free electron Fermi surface area ratio S/S_f, the effective mass m^*/m_0, the Fermi velocity v_F, the transport relaxation time τ_{tr}, the London penetration depth λ_L, the Ginzburg-Landau constant $K_1(0)$, the thermodynamic critical field $H_c(0)$, and the density of states at the Fermi surface per spin direction N(E).

The electrical resistance of both bulk and film TMC samples exhibit an unusual temperature dependence. In most films there is a T^2 dependence at low temperature, and a slow tendency towards saturation to a constant value at high temperature. The low-temperature T^2 dependence is shown in Fig.5.11 for two sputtered TMC films, one evaporated film, and one sintered sample [5.49,53]. The T^2 dependence from T_c to above 35 K is evident. In Fig.5.12 the magnetoresistance of sintered $Cu_{1.6}Mo_6S_8$ to 14 Tesla is shown [5.51]. From the fractional exchange in resistivity with field B,

$$\Delta\rho/\rho = \mu^2B^2 \qquad (5.7)$$

one obtains the average carrier mobility μ. And from the Hall effect, the carrier density n can be estimated assuming a simple band model [5.49,51,52].

If the saturation of resitivity at high T is assumed to be due to the mean free path becoming comparable to lattice spacing, then a low-temperature mean free path ℓ can be obtained from the ratio of ρ at high temperature to its value near T_c from

$$\ell = (\rho_{max}/\rho_0)\ell_{min} \qquad (5.8)$$

From HAKE [5.54], WIESMANN et al. [5.55], and GHOSH et al. [5.56] we obtain the following set of simultaneous equations:

$$\ell = 1.27 \times 10^4[\rho n^{2/3}(S/S_F)]^{-1} \quad , \qquad (5.9)$$

$$\xi_0 = 7.93 \times 10^{-17}\, n^{2/3}(S/S_F)(\gamma T_c)^{-1} \quad , \qquad (5.10)$$

$$\rho = 2.16 \times 10^{-5}(dH_{c2}/dT)_{T_c}[\eta\gamma(1 + 1.3\ell/\xi_0)]^{-1} \quad , \qquad (5.11)$$

where ρ is in ohm cm, n is in cm^{-3}, the heat capacity coefficient γ in erg cm^{-3} K^{-2}, the coherence length ξ_0 in cm, and dH_{c2}/dT in G/K. S/S_F is the ratio of the actual to the free electron Fermi surface areas, and η is a correction factor for strong coupling effects. Since ℓ, n, γ, $(dH_{c2}/dT)_{T_c}$, and T_c were obtainable from experiment,

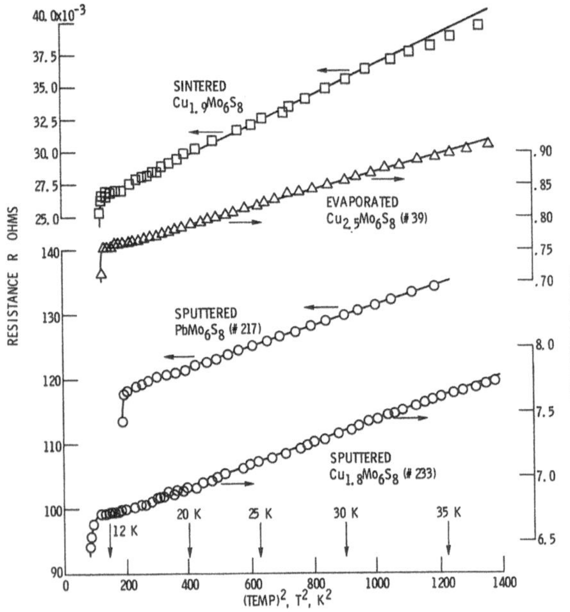

Fig.5.11. Resistance versus temperature squared for several films of $Cu_xMo_6S_8$ and $PbMo_6S_8$ [5.49,53]

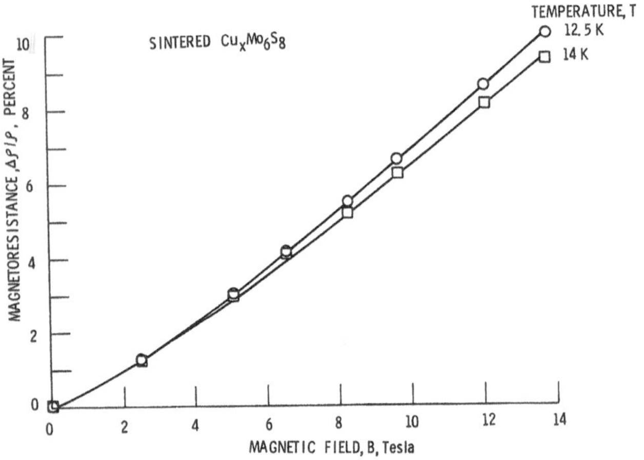

Fig.5.12. Magnetoresistance $\Delta\rho/\rho$ (in %) versus magnetic field B for sintered $Cu_xMo_6S_8$ for two temperatures just above T_c. Similar, but lower magnitude, changes are found in films [5.50]

(5.9-11) could be solved for ρ, ξ_0, and S/S_F. (Due to possible errors in thickness measurement, it was found best to find ρ from equations. The numbers are consistent with those measured directly). In addition, HAKE [5.54] gave

$$v_F = 5.76 \times 10^{-5}\, n^{2/3}(S/S_F)\gamma^{-1} \quad , \tag{5.12}$$

$$m^*/m_0 = 6.21 \times 10^4 \gamma n^{-1/3} \quad , \tag{5.13}$$

$$\tau_{tr} = 2.21 \times 10^8 \gamma[\rho n^{4/3}(S/S_F)^2]^{-1} \quad , \tag{5.14}$$

and

$$N(E) = 8 \times 10^{30}\gamma \tag{5.15}$$

from which v_F, m^*/m_0, $N(E)$, and τ_{tr} can be obtained.

Table 5.4 gives values of measured quantities for two films, as well as values for derived quantities for these films. Note that the low-temperature mean free path is comparable to or only slightly smaller than the coherence length. This reinforces the conclusion reached in Sect.5.3 (where critical field versus T data were analyzed) that sputtered and evaporated films may be neither completely "dirty" nor "clean" limit superconductors. Sintered samples are thus even more in the "clean" limit. This is an important fact since a dirty limit is often assumed for analysis of superconductors. The low mobility, high resistivity, high effective masses, and low Fermi velocities are consistent with $PbMo_6S_8$ and $Cu_xMo_6S_8$ being d-band metals. All of the above analysis is highly model dependent, but does give reasonable and consistent numbers. Band calculations [5.57] and other evidence [5.31] suggest that the Fermi level should lie in a d band originating from molybdenum d bands. Thus our results support this interpretation.

Table 5.4. Measured quantities for two films

Measured quantities	Sputtered $Cu_{2.1}Mo_6S_8$ (No. 233)	Sputtered $PbMo_6S_8$ (No.217)
ρ_{max}/ρ	7	3.6
Hall coefficient R_H	2.3×10^{-24} esu	0.9×10^{-24} esu
Transition temperature T_c	9.8 K	11.5 K
Critical field slope	20 kG/K	60 kG/K
Heat capacity coefficient γ, [5.48]	4000 erg cm^{-3} K^{-2}	6400 erg cm^{-3} K^{-2}
Derived quantities		
Low temperature mean free path, ℓ	46 Å	23 Å
Carrier mobility, $\bar{\mu}$	80 cm^2/Vs	42 cm^2/Vs
Resistivity, ρ	65 µΩ cm	125 µΩ cm
Number of carriers, n	3×10^{22} cm^{-3}	7×10^{22} cm^{-3}
Coherence length, ξ_0	86 Å	48 Å
Fermi surface ratio, S/S_F	0.44	0.26
Effective mass, m^*/m_0	8	9.6
Fermi velocity, v_F	6×10^6 cm/s	4 $10^6\times$cm/s
Transport relaxation time, τ_{tr}	8×10^{-14} s	5.8×10^{-14}
Density of states, N(E)	3.2×10^{34} states erg^{-1} cm^{-3}	5.1×10^{34} states erg^{-1} cm^{-3}

5.6 Final Statements and Conclusions

Although a significant amount of work has been done on ternary molybdenum chalcogenide films, most of this has been on $PbMo_6S_8$ and $Cu_xMo_6S_8$. Thus there is room for considerably more work in preparing and evaluating new films of not only the ternary molybdenum chalcogenides, but other ternaries as well. Even within the TMC "family" of materials there is need for clarification of the state of "purity" (that is, values of l compared with coherence length) and how this affects certain properties. For practical reasons it is worth investigating critical currents in pseudoternaries based on $PbMo_6S_8$, in search for a high-field, high-current-density superconductor. For the same reason the mechanical properties should be studied, and if useful $J_c(B)$ values are found, then improvement of mechanical properties should be sought.

There are certain properties of materials which can not easily be found using thin films. Examples are the heat capacity and anisotropy effects. Thus close comparison between bulk (even single crystal) and film properties should be made. As mentioned in Sect.5.2 on preparation, films much thicker than a few hundred Angstroms will exhibit few "surface" effects. Thus most of the intrinsic properties (such as effective mass) will be the same for both films and bulk samples. One of the main advantages of thin-film ternaries is then the control of geometry, morphology, and microstructure. These have an especially important influence on superconducting critical currents.

References

5.1 B.W. Roberts: J. Phys. Chem. Ref. Data *5*, 581 (1976)
5.2 B.T. Matthias, M. Marezio, E. Corenzwit, A.S. Cooper, H.E. Barz: Science *175*, 1465 (1972)
5.3 B.T. Matthias, E. Corenzwit, J.M. Vanderberg, H.E. Barz: Proc. Nat. Acad. Sci. (USA) *74*, 1334 (1977)
5.4 A.W. Sleight, J.L. Gillson, P.E. Bierstedt: Solid State Commun. *17*, 27 (1975)
5.5 D.C. Johnston, H. Prokash, W.H. Zachariasen, R. Viswanathan: Mater. Res. Bull. *8*, 777 (1973)
5.6 H.F. Braun, C.V. Segre: Solid State Comm. *35*, 735 (1980)
5.7 C.K. Banks, L. Kammerdiner, H.-L. Luo: J. Solid State Chem. *15*, 271 (1975)
5.8 P. Przyslupski, R. Horyn, J. Szwasyek: Solid State Commun. *28*, 869 (1978)
5.9 P. Przyslupski, R. Horyn, B. Gren: J. Low Temp. Phys. *38*, 93 (1980)
5.10 P. Przyslupski, B. Gren: *Ternary Superconductors*, ed. by G.K. Shenoy, B.D. Dunlap, F.Y. Fradin (North Holland, New York 1981) p.125
5.11 K.C. Chi, R.O. Dillon, R.F. Bunshah, Samuel Altervitz, John A. Woollam: Thin Solid Films *54*, 259 (1978)
5.12 K.C. Chi, R.O. Dillon, R.F. Bunshah, S. Alterovitz, D.C. Martin, J.A. Woollam: Thin Solid Films *47*, L9 (1977)
5.13 M. Decroux, Ø. Fischer, R. Chevrel: Cryogenics *17*, 291 (1977)
5.14 N.E. Alexseevski, M. Glinski, N.M. Dobrovolskii, V.I. Tsebro: JETP Lett. *23*, 412 (1976)
5.15 L.R. Gilbert, R. Messier, R. Roy: Thin Solid Films *54*, 129 (1978)
5.16 J.M. Rowell, R.C. Dynes, P.H. Schmidt: Solid State Comm. *30*, 191 (1979)

5.17 G.L. Christner, B. Bradford, L.E. Toth, R. Cantor, E.D. Dahlberg, A.M. Goldman, C.Y. Huang: J. Appl. Phys. *50*, 5820 (1979)
5.18 S. Foner: In *Superconductivity in d- and f-Band Metals*, ed. by D.H. Douglass (Plenum, New York 1976) p.161
5.19 S.A. Alterovitz, J.A. Woollam: Cryogenics *19*, 167 (1979)
5.20 Ø. Fischer, M.B. Maple (eds.): *Superconductivity in Ternary Compounds II*, Topics in Current Physics, Vol.34 (Springer, Berlin, Heidelberg, New York, to be published)
5.21 L. Kammerdiner, H.-L. Luo: Bull. Amer. Phys. Soc. *23*, 383 (1978)
5.22 J. Hauck: Mater. Res. Bull. *12*, 1015 (1977)
5.23 M. Sergent, R. Chevrel, C. Rossel, Ø. Fischer: J. Less Common Met. *58*, 179 (1978)
5.24 R.N. Shelton: Private communication to H.-L. Luo
5.25 S.A. Alterovitz, J.A. Woollam, L. Kammerdiner, H.-L. Luo: J. Low Temp. Phys. *30*, 797 (1978)
5.26 R. Flükiger, H. Devantay, J.L. Jorda, J. Muller: IEEE Trans. MAG-13, 818 (1977); D.C. Johnston, R.N. Shelton, J.J. Bugaj: Solid State Commun. *21*, 949 (1977); M. Tovar, L.E. DeLong, D.C. Johnston, M.B. Maple: Solid State Commun. *30*, 551 (1979); R. Flükiger, R. Baillif, J. Muller, K. Yvon: J. Less Common Met. *72*, 193 (1980)
5.27 R. Flükiger, A. Junod, R. Baillif, P. Spitzli, A. Treyvaud, A. Paoli, H. Devantay, J. Muller: Solid State Commun. *23*, 699 (1977)
5.28 J.A. Woollam, R.B. Somoano: Mater. Sci. Eng. *31*, 289 (1977)
5.29 R.B. Somoano, J.A. Woollam: In *Intercalated Layered Materials*, ed. by F. Levy (D. Reidel, Dordrecht 1980) p.307
5.30 Ø. Fischer, M. Decroux, R. Chevrel, M. Sergent: In *Superconductivity in d- and f-Band Metals*, ed. by D.H. Douglass (Plenum, New York 1976) p.175
5.31 Ø. Fischer: Appl. Phys. *16*, 1 (1978)
5.32 S.A. Alterovitz, J.A. Woollam: Solid State Commun. *25*, 141 (1978)
5.33 J.A. Woollam, S.A. Alterovitz, K.C. Chi, R.O. Dillon, R.F. Bunshah: J. Appl. Phys. *49*, 6027 (1978)
5.34 S.A. Alterovitz, J.A. Woollam: Philos. Mag. B*40*, 497 (1979)
5.35 R.N. Werthamer, E. Helfand, P.C. Hohenberg: Phys. Rev. *147*, 295 (1966)
5.36 D. Rainer, G. Bergman: J. Low Temp. Phys. *14*, 501 (1974)
5.37 E. Helfand, R.N. Werthamer: Phys. Rev. *147*, 288 (1966)
5.38 P. Entel, M. Peter: J. Low Temp. Phys. *22*, 613 (1976)
5.39 S.D. Bader, S.K. Sinha, R.H. Shelton: In *Superconductivity in d- and f-Band Metals*, ed. by D.H. Douglass (Plenum, New York 1977) p.209
5.40 R. Flükiger, K. Yvon: Private communication
5.41 S.A. Alterovitz, J.A. Woollam, L. Kammerdiner, H.-L. Luo: Appl. Phys. Letts. *31*, 233 (1977)
5.42 S.A. Alterovitz, J.A. Woollam, L. Kammerdiner, H.-L. Luo: Appl. Phys. Letts. *33*, 264 (1978)
5.43 S.A. Alterovitz, J.A. Woollam: J. Low Temp. Phys. *32*, 839 (1978)
5.44 S.A. Alterovitz, J.A. Woollam: Philos. Mag. B*38*, 619 (1978)
5.45 E.J. Kramer: J. Appl. Phys. *44*, 1360 (1973)
5.46 E.J. Kramer: J. Electron. Mater. *4*, 839 (1975)
5.47 S.A. Alterovitz, J.A. Woollam: In *Ternary Superconductors*, ed. by G.K. Shenoy, B.D. Dunlap, F.Y. Fradin (North Holland, New York 1981) p.113
5.48 F.Y. Fradin, G.S. Knapp, S.D. Bader, G. Cinader, C.W. Kimball: In *Superconductivity in d- and f-Band Metals*, ed. by D.H. Douglass (Plenum, New York 1976) p.297
5.49 J.A. Woollam, S.A. Alterovitz: Phys. Rev. B*19*, 749 (1978)
5.50 J.A. Woollam, S.A. Alterovitz, E.J. Haugland: Phys. Lett. *68*A, 122 (1978)
5.51 J.A. Woollam, S.A. Alterovitz: Solid State Commun. *27*, 669 (1978)
5.52 J.A. Woollam, S.A. Alterovitz: J. Magn. Magn. Met. *11*, 177 (1979)
5.53 J.A. Woollam, S.A. Alterovitz: Solid State Commun. *27*, 571 (1978)
5.54 R.R. Hake: Phys. Rev. *158*, 356 (1967)
5.55 H. Wiesmann, M. Gurvitz, A.K. Ghosh, H. Lutz, O.F. Kammerer, M. Strongin: Phys. Rev. B*17*, 122 (1978)
5.56 A.K. Ghosh, M. Gurvitch, H. Wiesmann, M. Strongin: Phys. Rev. B*18*, 6116 (1978)
5.57 O.K. Anderson, W. Klose, H. Nohl: Phys. Rev. B*17*, 1209 (1978)

5.58 H.F. Braun: "Superconducting Ternary Silicides and Germanides", in *Ternary Superconductors*, ed. by G.K. Shenoy, B.D. Dunlap, and F.Y. Fradin (North Holland, New York 1981) p. 225
5.59 E.H. Brandt: "The Maximum Loss-Free Current of Type II Superconductors with Flux Pinning", Phys. Letts. *77A*, 484 (1980). The estimate for maximum J_c is a factor of about 30 higher than implied using Kramer's model and $C_s = 0.56$. If Brandt's results could be achieved it would raise J_c to about 2×10^{10} A/m^2. The crossover for J_c for Nb_3Ge and $PbMo_6S_8$ (see discussion p. 158) would remain at about 23 T. *See also*: B. Seeber, K. Rossel, and Ø. Fischer: "$PbMo_6S_8$: A New Generation of Superconducting Wires?", in *Ternary Superconductors*, ed. by G.K. Shenoy, B.D. Dunlap, and F.Y. Fradin (North Holland, New York 1981) p. 119

6. Band Structures of $M_xMo_6X_8$- and $M_2Mo_6X_6$-Cluster Compounds

H. Nohl, W. Klose, and O. K. Andersen

With 23 Figures

New detailed calculations of the electronic structure of the Chevrel phases Mo_6S_8, Mo_6Se_8, Mo_6Te_8, $PbMo_6S_8$, $PbMo_6Se_8$, $BaMo_6S_8$, $EuMo_6S_8$ and $GdMo_6S_8$ and of the pseudo-one-dimensional compounds of type $TlMo_3Se_3$ are presented. The results are discussed in terms of simple models and are related to observed structural and physical properties.

6.1 General Considerations

This chapter deals with the electronic structure, as obtained from one-electron calculations, of the Chevrel phases and of the chain compounds described in Chaps. 2 and 3.

The input to such calculations is the stoichiometry and the crystal structure, and the output is the one-electron energy levels and wave functions which, in principle, determine the total energy. By investigating how the calculated electronic structure depends on the input, an understanding of the *chemical* properties, the relation between stoichiometry and structure, may therefore be obtained. It turns out that the role of the third component, M, is essentially to donate electrons to the valence band composed of the Mo 4d molecular orbitals of the Mo_6 octahedron. We shall try to answer questions concerning the stability of the Mo_6X_8 cluster. Why do compounds containing separated Mo_6X_8 units usually form when from $\sim$ 20 to 24 electrons are available for the Mo_6 d orbitals? Why are the octahedra elongated when less than 24 electrons are available, and why do tellurium (X = Te) compounds with 24 electrons not exist? What is the cause of the seemingly strange mutual arrangement of the Mo_6X_8 clusters in the Chevrel phases? When more than 24 electrons are available, why do the octahedra condense into face-sharing bioctahedra, trioctahedra, or even infinite chains? The answer to these questions is connected with the existence of a gap in the one-electron spectrum for the Mo_6X_8 cluster at 24 electrons, and it follows from a description of the valence band in terms of bond orbitals.

The *physical* properties, in particular the occurrence of superconductivity, depend on the density and character of the one-electron states at the Fermi level.

We shall see that these, in turn, depend sensitively on the number of electrons available, the nature of the X atoms, and the positions of the atoms, first of all on the mutual arrangement of the octahedra and of the distortion of the octahedron. In the Chevrel phases the conduction band, being the topmost band of the valence-band complex, just below the 24-electron gap, consists of a narrow (~ 40 mRy) doubly degenerate band derived from the $e_g(x^2 - y^2)$, the $t_{1u}(xz, yz)$, the $t_{2g}(xz, yz)$, and the $t_{2u}(x^2 - y^2)$ molecular orbitals. The e_g character dominates but the t_{1u} and t_{2g} characters increase with decreasing intercluster distance and with increasing elongation of the octahedron. The covalent mixing with the X p-orbitals is very important, whereas the overlap of the $e_g(x^2 - y^2)$ orbital with the site of the third component M is extremely small. The calculated values of the density of states at the Fermi level $N(E_F)$ correlate with the measured superconducting transition temperatures. But the shapes, N(E), of the densities of states differ significantly from compound to compound, and a rigid e_g-band model is therefore only a first approximation.

A structural phase transition which lowers the symmetry of the octahedron was recently found in the 22-electron compounds $EuMo_6S_8$ and $BaMo_6S_8$, and it was suggested that this phase transition is responsible for the unexpected absence of superconductivity in these compounds. We shall see that this deformation of the octahedron creates a gap in the doubly degenerate conduction band and thus may turn a high-T_c superconductor into a semiconductor or an insulator.

The calculations for the anisotropic chain compounds $(MMo_3X_3)_2$ with 26 electrons show that there is usually just one, singly degenerate band at the Fermi level, which suggests why superconductivity has so far been found only in one compound, $TlMo_3Se_3$. In other cases a Peierls distortion may have taken place.

Long before the discovery of the Chevrel phases the Me_6X_8 cluster had been known as an ion (e.g., $Mo_6Cl_8^{4+}$) in watery solution and as the building block in a number of crystals. At the beginning of the 1950s DUFFEY [6.1] and ten years later CROSSMAN et al. [6.2], COTTON and HAAS [6.3], and KETTLE [6.4] studied the metal-metal bonding in this cluster using simple molecular orbital methods. These early investigators found the symmetries of the 12 bonding states, but since the results were either expressed in terms of undetermined parameters [6.1,2,4] or neglected the hybridization with the ligands (X) [6.3], the ordering of the levels could not be ascertained. This was done for $Mo_6Br_8^{4+}$ by GUGGENBERGER and SLEIGHT [6.5] who used the extended Hückel method and found a_{1g}, t_{1u}, t_{2g}, t_{2u}, and e_g in order of increasing energy. This order of the Mo d-like levels was recently confirmed for all the molybdenum halogen ionic clusters by COTTON and STANLEY [6.6] and by SEIFERT et al. [6.7] using the self-consistent-field scattered-wave Xα method. However, these authors did find a number of nonbonding X p-like levels mixed in with the Mo d-like levels.

The first calculations for the building block of the Chevrel phases, the molybdenum chalcogenide cluster, were performed by ANDERSEN et al. [6.8] who used the linear muffin-tin orbital (LMTO) method in the atomic-sphere approximation (ASA) and by MATTHEISS and FONG [6.9] who used the tight-binding method with parameters determined from an augmented-plane-wave calculation for a hypothetical Mo_3S crystal. The issues were the level ordering, in particular the nature of the levels at the Fermi energy, and the influence of the departure from the ideal, cubic cluster geometry. For Mo_6S_8, Mo_6Se_8, Mo_6Te_8, $PbMo_6S_8$, and $PbMo_6Se_8$ crystals, ANDERSEN et al. furthermore calculated the band structure of the e_g band, believed to be the essential conduction band, neglecting the hybridization with all other subbands, and estimated various physical properties, comparing them with experiments. The first full band structure calculations for the above mentioned crystals were performed by BULLETT [6.10]. He used a localized-orbital method and could prove that for a given number of valence electrons the function of the third element, Pb, is primarily a passive one of distorting the Mo_6X_8 cluster. For the level ordering and the densities of states at the Fermi level, he obtained results which, except for a somewhat smaller Mo d bandwidth, seemed to correlate well with those of ANDERSEN et al. Self-consistent LMTO-ASA band-structure calculations for $SnMo_6S_8$, $SnMo_6Se_8$, $GdMo_6S_8$, and $EuMo_6S_8$, neglecting noncubic distortions, were thereafter performed by JARLBORG and FREEMAN [6.11,12] who, in addition to obtaining results roughly similar to those of the previous investigators, studied the observed magnetic isolation of the rare-earth component. One problem left unsolved by the full band-structure calculations [6.10-12] was the detailed investigation of the nature of the conduction band which is so important for understanding the difference in chemical and physical properties between various Chevrel-phase compounds. NOHL and ANDERSEN [6.13] therefore extended their previous study [6.8] by performing full, although not self-consistent, LMTO-ASA band-structure calculations for a number of compounds including $BaMo_6S_8$ and $EuMo_6S_8$ in the newly discovered low-symmetry phases. KELLY and ANDERSEN [6.14] finally calculated the band structures of the chain compounds and discussed their physical properties as well as the bonding in these condensed clusters.

In this chapter we shall essentially follow the exposition given in the most recent works [6.13,14] and then, at the appropriate places, compare and augment the results with those of the previous investigators. The question about the magnetic interactions in those Chevrel phases where the third element is a rare earth will be treated by FREEMAN and JARLBORG in [Ref.6.15, Chap.6].

6.2 Input to the Band-Structure Calculations

In this section we shall specify the input to the calculations, that is, the crystal structure and the stoichiometry or, more specifically, the atomic positions and the one-electron potentials of the atomic constituents. In order to discuss trends we consider the 20-electron binary Chevrel phases — Mo_6S_8, Mo_6Se_8, and Mo_6Te_8 — the 22-electron ternaries — $PbMo_6S_8$, $PbMo_6Se_8$, and $BaMo_6S_8$ — and the 23-electron ternary — $GdMo_6S_8$. In the ternaries chosen, the third element is a large cation with a well-defined position (Chap.3). We furthermore consider the In and Tl selenides and tellurides with the chain structure. Finally, we shall state the rudiments of the LMTO-ASA formalism necessary to define the potential parameters presented in this section and necessary to provide a semiquantitative understanding of the results of the full computer calculations to be presented in the following sections.

6.2.1 Crystal Structure

The crystal structures of the MMo_6X_8 compounds were dealt with in Chaps.2 and 3, and here we shall repeat and discuss some of this information in a form suitable for a subsequent interpretation of the calculated band structures.

The primitive cell of the Chevrel phases and the top layer of a neighboring cell is shown in Fig.6.1. The Mo atoms (black spheres) form an elongated octahedron whose corners are approximately at the face centers of a distorted cube, with edge length b, formed by the X atoms (large, white spheres). This Mo_6X_8 cluster is located inside a larger, distorted cube, a rhombohedral unit cell with lattice constant a and rhombohedral angle $\alpha \approx 90^o$. In the large-cation ternaries the cations M occupy the corners of the rhombohedron (small, white spheres). The space group is $\bar{R}3$ and the major deviation from cubic symmetry is the turning by an angle $\phi \approx 26^o$ of the Mo_6X_8 cube with respect to the rhombohedron about their common body diagonal, the threefold axis.

Crystallographic work employs hexagonal or rhombohedral coordinates with axes along the translation vectors (ijk-axes in the figure), but since the electronic interactions are strongest *within* the Mo_6X_8 cluster, we prefer using a cartesian system with axes defined by the Mo octahedron. This xyz system has an inversion center at its origin and the threefold axis along its 111 direction; moreover, it is turned in such a way that the corners of a regular Mo octahedron would be at the coordinate axes, while those of the real, elongated octahedron are slightly displaced along the face diagonals of the cube, towards the threefold axis. In other words, we choose the xyz system in such a way that the Mo atoms are at the positions $\pm(b,c,c)/2$, $\pm(c,b,c)/2$, and $\pm(c,c,b)/2$, where $0 < c \ll b$.

In Table 6.1 we give atomic positions for the Chevrel phases considered and for a so-called regular structure to be defined in (6.2,3). The xyz coordinates of

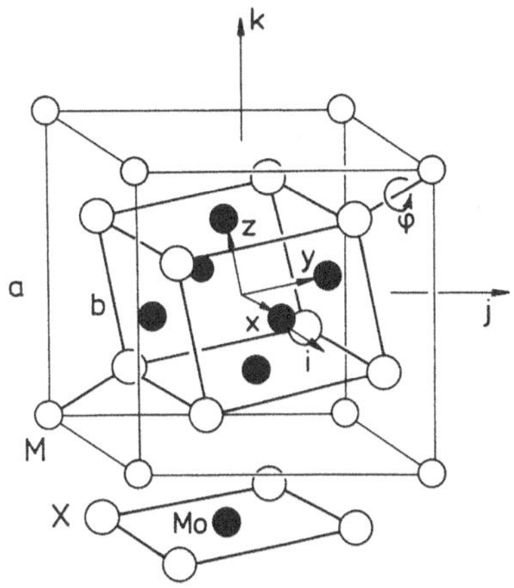

Fig.6.1. Primitive cell of MMo_6X_8 and top layer of neighboring cell

Table 6.1. Atomic positions (Å) in MMo_6X_8 Chevrel phases

Atomic positions [Å] in MMo_6X_8 Chevrel phases		Regular structure	Mo_6S_8 n=20	Mo_6Se_8 20	Mo_6Te_8 20	$PbMo_6S_8$ 22	$PbMo_6Se_8$ 22	$Ba^a Mo_6S_8$ 22	$GdMo_6S_8$ 23
M	x=y=z	$(1+1/\sqrt{2})b/2$	(3.140)	(3.236)	(3.354)	3.314	3.450	3.405	3.276
Mo	x=y	0	0.057	0.053	0.026	0.021	0.013	0.012	0.019
	z	b/2	1.965	1.951	1.927	1.913	1.920	1.898	1.900
X	x	b/2	1.635	1.736	1.872	1.671	1.767	1.696	1.647
	y	-b/2	-1.769	-1.882	-2.030	-1.783	-1.889	-1.761	-1.793
	z	b/2	1.671	1.756	1.889	1.757	1.845	1.764	1.775
X	x=y=z	b/2	1.777	1.853	1.959	1.707	1.779	1.689	1.713
T	x	-b/2	-1.475	-1.492	-1.657	-1.381	-1.418	-1.385	-1.303
	y	b/2	1.741	1.702	1.765	1.889	1.950	2.010	1.784
	z	$(1+1/\sqrt{2})b$	6.014	6.261	6.600	6.119	6.369	6.184	6.072

[a] [6.16,17]. T = 300 K.

the atoms not included may be found by cyclic permutations of x, y, and z and by a change of sign. From the Mo z coordinate (b/2) we realize that the intracluster Mo-Mo distance ($\approx b/\sqrt{2}$) is approximately equal to the nearest-neighbor distance in elemental Mo metal, and it may be concluded that metal-metal bonds play an important role for the cohesion of the octahedron. The octahedron contracts slightly, about 3%, as the number of electrons is increased from 20 to 23, and more importantly, the elongation, as measured by the Mo x = y-coordinate (c/2), decreases significantly. Of the eight X atoms, two occupy special positions on the threefold axis while the remaining six occupy general positions. In the regular structure all eight X atoms would be at the corners of the cube circumscribing the Mo octahedron, but

Table 6.2. Interatomic distances and angles

Interatomic distances and angles			Regular structure	Mo_6S_8 n=20	Mo_6Se_8 20	Mo_6Te_8 20	Pb Mo_6S_8 22	Pb Mo_6Se_8 22	Ba Mo_6S_8 22	Gd Mo_6S_8 23
Mo-Mo intra	d	[Å]	$b/\sqrt{2}$	2.780	2.760	2.725	2.705	2.715	2.684	2.687
			±0	±.082	±.076	±.037	±.030	±.018	±.017	±.027
Mo-Mo inter	d	[Å]	b	3.085	3.266	3.648	3.270	3.490	3.411	3.163
	l		-1/2	-0.515	-0.489	-0.461	-0.435	-0.414	-0.413	-0.424
	m		1/2	0.527	0.489	0.470	0.565	0.551	0.582	0.552
	n		$1/\sqrt{2}$	0.676	0.722	0.753	0.701	0.724	0.700	0.718
Mo-Mo $\frac{\text{inter}}{\text{intra}}$	$\frac{d}{d}$		$\sqrt{2}$	1.110	1.183	1.339	1.209	1.285	1.271	1.171
Mo-X intra	d	[Å]	$b/\sqrt{2}$	2.439	2.563	2.743	2.454	2.577	2.450	2.457
			±0	±.020	±.024	±.045	±.061	±.076	±.069	±.079
Mo-X inter	d	[Å]	$b/\sqrt{2}$	2.426	2.598	2.827	2.566	2.722	2.626	2.568
	l		0	0.098	0.130	0.123	0.148	0.168	0.139	0.183
	m		0	0.005	-0.041	-0.053	0.043	0.034	0.089	-0.004
	n		1	0.995	0.991	0.991	0.988	0.985	0.986	0.983
Mo-X $\frac{\text{inter}}{\text{intra}}$	$\frac{d}{d}$		1	0.995	1.014	1.031	1.046	1.056	1.072	1.045
$\frac{\text{Mo-X}}{\text{Mo-Mo}}$ intra	$\frac{d}{d}$		1	0.877	0.929	1.007	0.907	0.949	0.913	0.914
T	a	[Å]	$b(2+\sqrt{2})^{1/2}$	6.432	6.658	7.030	6.551	6.810	6.648	6.461
	α	[°]	94.2	91.3	91.6	92.6	89.3	89.2	88.6	89.2
	φ	[°]	26.9	25.3	24.2	24.4	25.8	25.5	26.6	24.6

in the real structures, the X-cube is distorted and smaller than the Mo cube. The X cube expands considerably going from the sulfides to the tellurides, where it reaches the same size as the Mo cube. The presence of a large M atom on the threefold axis pushes the two nearest-neighbor X atoms, also on the threefold axis, towards the center of the octahedron whereby the X cube becomes more regular. T is the translation vector in the k direction.

In Table 6.2 we give numbers derived from those in Table 6.1. These include the two Mo-Mo intracluster distances, the shorter being that of the six bonds perpendicular to the threefold axis and the longer being that of the remaining six bonds. There are four different Mo-X intracluster distances, the table giving their average and maximum fluctuation. Whereas the Mo octahedron contracts by about 2% when going from Mo_6S_8 to Mo_6Te_8, the average X-cube and Mo-X intracluster distances increase by over 10%. For a given chalcogen these distances depend little on the number of valence electrons. Although the Mo octahedron and the X cube individually are most distorted in Mo_6S_8 and Mo_6Se_8, and least in $BaMo_6S_8$, the four Mo-X intracluster distances scatter by only 1% in the former compounds but by 3% in the ternaries. Similarly, for Mo_6S_8 and Mo_6Se_8 the Mo-X inter- and intracluster distances are almost equal, whereas for the ternaries, the intercluster distance exceeds the intracluster distance by about 5%. For Mo_6Te_8 the numbers are intermediate between those mentioned for the two other binaries and those mentioned for the ternaries. These trends indicate that the role of the Mo-X covalent interaction in determining the structure is most important for Mo_6S_8 and Mo_6Se_8 and least for the ternaries considered.

The numbers in Table 6.2 include the translation vector expressed in terms of its length [the rhombohedral lattice constant (a)], the rhombohedral angle (α) and the turn angle (ϕ). The relations between these and the xyz coordinates of the translation vector are

$$a = (x^2 + y^2 + z^2)^{1/2}, \quad \cos\alpha = (xy + yz + zx)a^{-2} \quad ,$$

and

$$\sin\phi = (\sqrt{3/2})(y - x)(1 - \cos\alpha)^{-1/2}a^{-1} \quad . \tag{6.1}$$

The volume of the primitive cell is $V = x^3 + y^3 + z^3 - 3xyz \approx a^3$.

Table 6.2 furthermore gives the lengths (d) and directional cosines (l, m, and n) of the shortest Mo-X and Mo-Mo intercluster vectors. It can be seen that the Mo-X intercluster vector is almost in the z direction, and as mentioned above, its length is remarkably close to those of the four Mo-X intracluster vectors. It is therefore plausible that the translation vector is determined by the requirement that the strength of the Mo-X intercluster bond be optimized [6.9]. For our regular structure we consequently define that the position of the nearest X atom in the neighboring cell is $(0,0,1 + \sqrt{2})b/2$, whereby

$$d^{\text{inter}}_{\text{Mo-X}} = d^{\text{intra}}_{\text{Mo-X}} = b/\sqrt{2} \quad , \quad \text{and} \quad d^{\text{inter}}_{\text{X-X}} = d^{\text{intra}}_{\text{X-X}} = d^{\text{inter}}_{\text{Mo-Mo}} = b \quad . \tag{6.2}$$

The translation vector of the regular structure thus becomes $(-1,1,2 + \sqrt{2})b/2$, and for this structure, the lattice constant, the rhombohedral angle, and the turn angle are

$$a = b(2 + \sqrt{2})^{1/2}$$

$$\alpha = \text{arc}\cos[-4(2 + \sqrt{2})]^{-1} = 94.2^\circ \quad ,$$

and

$$\phi = \text{arc}\sin[\sqrt{3}(1 + 2\sqrt{2})^{-1}] = 26.9^\circ \quad . \tag{6.3}$$

The agreement with the corresponding numbers for the real materials is quite good considering the fact that for the regular structure, the Mo and X cubes are chosen to be identical. Except for Mo_6Te_8 this is an oversimplification. For the large-cation ternaries the rhombohedral angle is smaller than 94.2°. This is presumably a mismatch imposed by the large size of these cations, because for the small-cation ternaries and the pseudo-binary compounds of type $Mo_6(S_6I_2)$ or $(Mo_4Ru_2)Te_8$ with n = 22-24 and correspondingly small distortion of the octahedron, the rhombohedral angle lies between 93° and 95.5° (Chaps.2 and 3). The superconducting transition temperature within subgroups of materials has often been quoted to correlate with the rhombohedral angle and the cell volume such that high transition temperatures usually occur in small-angle, high-volume materials. Exceptions are $Mo_6(S_6Br_2)$ and $Mo_6(S_6I_2)$, which have $T_c \approx 14$ K and $a \approx 6.5$ Å, similar to $PbMo_6S_8$, but $\alpha \approx 94°$ [6.16,17]. As we shall see, this is in accord with the band-structure calculations which indicate that within a subgroup of materials the cell volume, rather than the rhombohedral angle, is of importance for the shape of the conduction band.

The structural parameter of most obvious importance for the bandwidth is the Mo-Mo intercluster distance. It increases from being 11% larger than the Mo-Mo intracluster distance in Mo_6S_8 to 34% larger in Mo_6Te_8 and to 27% larger in $BaMo_6S_8$. The factor $\sqrt{2}$, valid for the regular structure, is never reached primarily because the Mo atoms lie outside the X cube.

It was recently observed that $EuMo_6S_8$ appears to be semiconducting at normal pressure [6.20] but superconducting with $T_c \approx 10$ K at pressures above 7 Kbar [6.20, 21]. Shortly thereafter BAILLIF et al. [6.16] found that $EuMo_6S_8$ and $BaMo_6S_8$ at normal pressure transform into a triclinic distorted low-temperature modification, this, they inferred, being the reason for the absence of superconductivity. The distortion involves primarily the octahedron and is illustrated in Fig.6.2. In the high-temperature rhombohedral phase the octahedron is elongated along the $\bar{3}$-axis as mentioned above, and in the low-temperature phase, further distortions ranging from -0.012 to +0.037 Å occur such that only the center of inversion ($\bar{1}$) remains as symmetry element. The distortion may essentially be viewed as a compression of a regular octahedron along one of its $\bar{4}$-axes, the z-axis, followed by a tilt of this axis towards one of the perpendicular $\bar{2}$-axes. The x=y-plane is thus approximately a mirror plane for the distorted octahedron but not for the entire crystal structure.

In the attempt to synthesize Chevrel phases with In, Tl, or an alkali-metal atom as the ternary element, new phases containing Mo_6X_8 units condensed along the $\bar{3}$-axis to form face-sharing bioctahedra, e.g., Mo_9Se_{11}, trioctahedra, e.g., $Mo_{12}S_{14}$, or infinite columns, $(Mo_6X_6)_\infty$ with X = Se or Te, were found (Chaps.2 and 3). We shall only consider the latter member of this $Mo_{3n}X_{3n+2}$ family, specifically the four chain compounds with In and Tl [6.22]. The hexagonal crystal structure illustrated in Fig.6.3 is characterized by the presence of infinite chains $(M_\infty)_2$ and columns $(Mo_6X_6)_\infty$ running in the c direction. The center of each column is a $\bar{3}$-axis as well as a twofold screw axis with translation c/2. Each column consists of two, coaxial tubes of staggered, equilateral triangles of which the inner is formed by the Mo atoms and the outer by the X atoms. The Mo and X triangles lie in the same plane perpendicular to the c-axis. The lattice constant c is such that for the inner

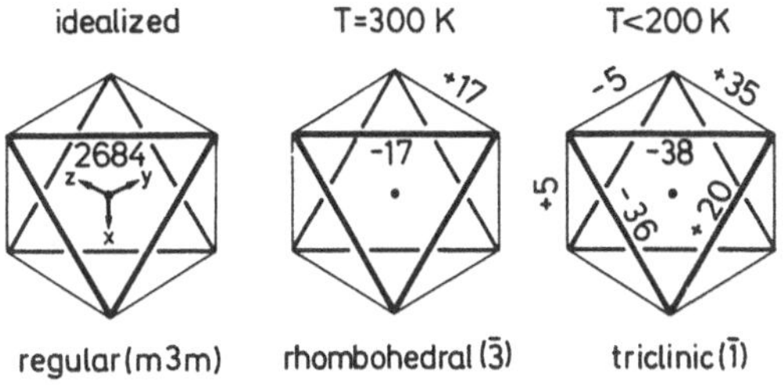

Fig.6.2. Mo_6 octahedron in the high- and low-temperature phases of $BaMo_6S_8$. The interatomic distances (in 10^{-3} Å after [6.16]) are given with respect to the idealized, regular octahedron shown in the first figure

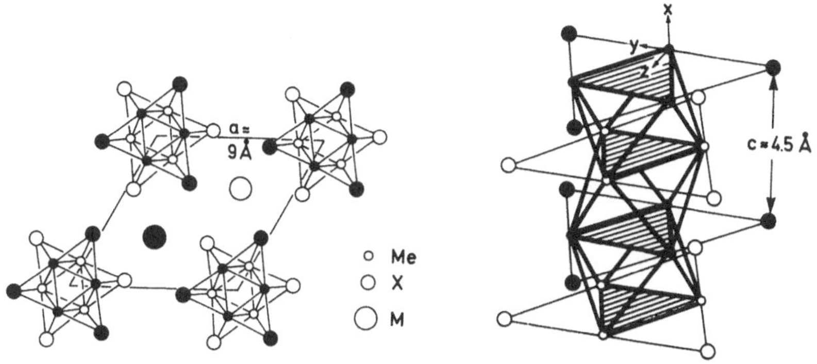

Fig.6.3. Hexagonal crystal structure of the $M_2Me_6X_6$ chain compounds. Black-colored and white-colored atoms lie in different planes separated by c/2 perpendicular to the c-axis (= x-axis)

tube a pair of staggered triangles, at the distance c/2, form a slightly elongated octahedron. The shortest intracluster Mo-Mo distances are 2.69 ± 0.03 Å in both the In and Tl selenides, and they are 2.69 ± 0.06 Å in both tellurides. The octahedra of these 26-electron chain compounds are slightly more contracted than those of the 20-electron Chevrel phases Mo_6Se_8 and Mo_6Te_8 whose data were given in Table 6.2. The X tube is slightly turned with respect to the Mo tube. The distances from a Mo atom to the two nearest X atoms in the plane perpendicular to the c-axis are 2.61 ± 0.01 Å in the selenides, and they are 2.76 ∓ 0.03 in $In_2Mo_6Te_6$ and 2.78 ∓ 0.05 in $Tl_2Mo_6Te_6$. The turn which breaks the vertical mirror symmetry of a single column is thus largest and of opposite sign in the tellurides, and the Mo-X intra-cluster distances are slightly larger in the chain compounds than in the Chevrel phases, particularly in the selenides. The distance from a Mo atom to the two nearest X atoms in the neighboring planes is 2.69 Å in the selenides and 2.83 and 2.85 Å in the In and Tl telluride, respectively. Between columns the shortest Mo-Mo distance is about 2 1/2 times the intracluster distance. Thus the direct coupling of the Mo d-electrons is negligible. The strongest coupling between columns is via two X atoms, but the distance between these is about 3.80 Å in the selenides and 3.97 Å in the tellurides so that the coupling must be extremely weak. Like in the Chevrel phases the M atoms are far away from the Mo atoms and their shortest distance to an X atom is 3.35 and 3.39 Å in the In and Tl selenides and 3.48 and 3.52 Å in the In and Tl tellurides.

6.2.2 Atomic-Sphere Potentials

The band-structure calculations reported in [6.8,11-14] were performed with the LMTO-ASA method [6.23]. Those of JARLBORG and FREEMAN were self-consistent, but since evidence later accumulated that energy gaps obtained with the self-consistent local density-functional method might be inaccurate, self-consistency was not attempted in the more recent calculations [6.13-14], which instead employed potentials constructed by superposition of neutral-atom charge densities and use of full Slater exchange. The relativistic shifts but not the spin-orbit coupling were included. As far as can be judged, for the cluster compounds, this simple, well-tried potential construction yielded the same results as obtained with fully self-consistent potentials.

In the LMTO-ASA formalism [6.8,23] the one-electron potential is written as the sum of (slightly overlapping) atomic-sphere potentials and the one-electron Hamiltonian may be expressed as

$$H = C + \Delta^{1/2} S(1-\gamma S)^{-1} \Delta^{1/2} \qquad (6.4)$$

in a basis of localized, orthogonal orbitals. Here, all quantities are matrices in site (R) and angular momentum (ℓm) indices and are all diagonal in the Bloch vector ($\underline{k}$). Moreover, the parameters C, Δ, and γ, which give the energy, width, and shape of the resonance in the scattering from the potential at site R, are diagonal in $R\ell m$ and are independent of m. The Hamiltonian (6.4) is of the two-center form with C being the (diagonal) energies of the muffin-tin orbitals and $\Delta^{1/2} S(1-\gamma S)^{-1} \Delta^{1/2}$ being the hopping or transfer matrix. The latter is expressed in terms of the potential parameters Δ and γ and a so-called canonical structure matrix S which contains all the structural information but is independent of the potential.

The muffin-tin orbital energies, C, are listed in the upper halves of Tables 6.3 and 4. Although s, p, and d orbitals on all the atoms, plus on the empty M-site in the binaries, were included in the calculations [6.13-14], we list only those orbital energies which lie less than 1 Ry from the energy of the Mo 4d orbital. Moreover, in Table 6.4 we have only listed the M- and X-orbital energies because the Mo-orbital energies are like those in Table 6.3. Furthermore, only the compounds $TlMo_3Se_3$ and $InMo_3Te_3$ have been included, because for $InMo_3Se_3$ and $TlMo_3Te_3$ the In-orbital energies were essentially the same as in $InMo_3Te_3$, the Se-orbital energies were essentially the same as in $TlMo_3Se_3$, and so on.

From the values given in the tables and the fact that covalent hybridization with the X p orbitals pushes the M s bands to energies higher than C_{Ms}, it is plausible that Pb donates its two 6p electrons, Ba its two 6s electrons, Gd its

Table 6.3. Potential parameters in the Chevrel phases

Potential Parameters in Chevrel phases			Mo_6S_8	Mo_6Se_8	Mo_6Te_8	$PbMo_6S_8$	$PbMo_6Se_8$	$BaMo_6S_8$	$GdMo_6S_8$
C [Ry]	M	s	0.43	0.47	0.56	-0.58	-0.59	0.05	-0.03
		p				0.14	0.15	0.61	0.72
		d						0.28	0.24
	Mo	s	-0.15	-0.13	-0.12	-0.15	-0.13	-0.15	-0.15
		p	0.85	0.86	0.88	0.85	0.87	0.85	0.85
		d	0	0	0	0	0	0	0
	X6	p	-0.29	-0.23	-0.14	-0.28	-0.23	-0.25	-0.28
	X2	p	-0.25	-0.20	-0.12	-0.29	-0.24	-0.27	-0.29
$\hat{C}$ [mRy]	Mo	d	0	0	0	0	0	0	0
	X6	p	-340	-260	-156	-330	-269	-282	-318
	X2	p	-286	-226	-130	-338	-288	-306	-337
$\hat{\Delta}$ [mRy]	Mo	d	1.59	1.60	1.68	1.81	1.73	1.88	1.88
	X6	p	13.6	15.9	19.4	14.7	17.1	14.2	14.8
	X2	p	13.0	15.4	18.6	14.9	17.4	14.5	15.0
s [Å]	M		1.48	1.39	1.22	2.00	1.96	2.22	1.89
	Mo		1.55	1.55	1.55	1.55	1.55	1.55	1.55
	X6=X2		1.21	1.35	1.57	1.23	1.40	1.15	1.20

Table 6.4. Potential parameters in the chain compounds

Potential Parameters in Chain Compounds			$TlMo_3Se_3$	$InMo_3Te_3$
C [Ry]	M	s	-0.45	-0.29
		p	0.30	0.27
		d	-0.84	-1.16
	X	p	-0.20	-0.14
s [Å]	M		2.08	2.07
	X		1.40	1.62

three 5d and 6s electrons, and In and Tl their single 5p or 6p electron to the Mo 4d band. Since, moreover, the X p orbital energies lie below the Mo 4d energy and covalent hybridization will push the Mo 5s bands above the 4d bands, we do arrive at the electron counts mentioned previously and listed at the top of Table 6.1. These counts are confirmed by the band-structure calculations (Sect.6.4.1).

For the Chevrel phases the six X atoms at general positions are named X6 and the two at the $\bar{3}$-axis, X2. It may be seen that the presence of an M atom lowers the potential in the X spheres. The Te 5p energies are considerably higher than the S 3p and the Se 4p energies. But as we shall see, this does not necessarily imply that the covalent hybridization between the X p and Mo d orbitals is strongest in the tellurides since the Mo-X distances (Table 6.2) also increase considerably when going from the sulfides to the tellurides.

The band-structure calculations [6.11-14] were performed with the Hamiltonian (6.4), although in a nonorthogonal representation. For the interpretation of the computed energy bands in a narrow range around the Fermi level it is, however, simpler to use the equivalent Hamiltonian

$$\hat{H} = \hat{C} + \hat{\Delta}^{1/2} S \hat{\Delta}^{1/2} \quad , \qquad (6.5)$$

which avoids a matrix inversion at the expense of having energy-dependent potential parameters, $\hat{C}$ and $\hat{\Delta}$. These Mo d and X p potential parameters, valid in a region of about 100 mRy around the Fermi level, have been given in the central part of Table 6.3. We have found it convenient to include the contraction of the Mo octahedron, as given by the average Mo-Mo intracluster distance (first row of Table 6.2), in the definition of the width parameter $\hat{\Delta}$, rather than in the definition of the structure matrix S. This involves a $(d^{\text{intra}}_{\text{Mo-Mo}})^{-2\ell-1}$-dependence of $\hat{\Delta}_\ell$, and this is the reason why $\hat{\Delta}_{\text{Mod}}$ increases from the sulfides to the tellurides and why it increases for increasing numbers of valence electrons. The width parameters for the X p-orbitals increases from sulfur to tellurium.

At the bottom of Tables 6.3,4 are listed the radii of the atomic spheres used to generate the potential parameters given above. The Mo sphere was chosen equal to the Wigner-Seitz sphere in elemental (bcc) Mo metal, and the X2, X6, and M spheres were chosen such that the overlap with the nearest-neighbor sphere is always about 14%. With these sphere sizes the space filling is about 65%. In the remaining space the potential is assumed to be constant and equal to the energy in question, or said in another way, the kinetic energy is assumed to be zero. This is a good approximation for the compounds and energy range considered. The results of the band-structure calculations were not sensitive to the choice of sphere radii as long as the space filling was above 60% and no two spheres were allowed to overlap by more than 20%. Comparison with standard covalent, atomic, and ionic radii shows that our radii correlate well with the *atomic* radii. This is consistent with the fact that although the band-structure calculations, for instance, showed that the X p band is full and hence that X is formally in the state X^{2-}, they also showed that within half an electron the spheres are neutral.

The expressions (6.4,5) show that the potential- and structure-dependent part of the transfer or hopping matrix essentially factorize and that the structural part is given by the "screened" structure matrix $S(1-\gamma S)^{-1}$ or, in a narrow energy

Table 6.5. Canonical structure matrix

		$S(3z^2-1,3z^2-1)$	$= (-3/4)(35n^4-30n^2+3)\ h$
$S(x,y)$	$= 3lm\ f$	$S(x^2-y^2,x^2-y^2)$	$= [-(35/4)(l^2-m^2)^2-5n^2+4]\ h$
$S(x,x)$	$= (3l^2-1)\ f$	$S(xy,xy)$	$= (-35l^2m^2-5n^2+4)\ h$
$S(x,yz)$	$= -5lmn\ g$	$S(xy,x^2-y^2)$	$= (-35/2)lm(l^2-m^2)\ h$
$S(x,xy)$	$= (1-5l^2)m\ g$	$S(xz,x^2-y^2)$	$= -5[(7/2)(l^2-m^2)-1]ln\ h$
$S(z,x^2-y^2)$	$= (-5/2)(l^2-m^2)n\ g$	$S(xz,3z^2-1)$	$= (-5\sqrt{3}/2)(7n^2-3)ln\ h$
$S(x,x^2-y^2)$	$= [1-(5/2)(l^2-m^2)]l\ g$	$S(x^2-y^2,3z^2-1)$	$= (-5\sqrt{3}/4)(7n^2-1)(l^2-m^2)\ h$
$S(x,3z^2-1)$	$= (\sqrt{3}/2)(1-5n^2)l\ g$	$S(xy,3z^2-1)$	$= (-5\sqrt{3}/2)(7n^2-1)lm\ h$
$S(z,3z^2-1)$	$= (\sqrt{3}/2)(3-5n^2)n\ g$	$S(yz,xz)$	$= -5(7n^2-1)lm\ h$

$f \equiv 6(d^{intra}_{Mo-Mo}/d)^3 = -pp\pi$, $g \equiv 6\sqrt{5}(d^{intra}_{Mo-Mo}/d)^4 = pd\pi$, $h \equiv 10(d^{intra}_{Mo-Mo}/d)^5 = -dd\delta$

range, simply by the structure matrix itself. The pp, pd, and dd elements of the structure matrix have been listed in Table 6.5 as a function of the length d and directional cosines l, m, and n of the vector from the first to the second orbital. The entries not given in the table may be found by cyclically permuting (x,y,z) and (l,m,n). Expressed as two-center integrals with the z-axis along the interatomic vector [6.24] the elements are simply

$$pp(\sigma,\pi) = (12,-6)(d^{intra}_{Mo-Mo}/d)^3\ , \quad pd(\sigma,\pi) = (-6\sqrt{15},6\sqrt{5})(d^{intra}_{Mo-Mo}/d)^4\ ,$$

$$dd(\sigma,\pi,\delta) = (-60,40,-10)(d^{intra}_{Mo-Mo}/d)^5\ . \qquad (6.6)$$

6.3 One-Electron States of a Single Mo_6X_8 Cluster

The Chevrel phases contain fairly well-separated Mo_6X_8 units and it is useful to describe their band structure in terms of the molecular-orbital states of a single cluster in the same way as one describes the band structure of elemental Mo metal in terms of the d orbitals of a single Mo atom. These single-cluster states will now be derived, and we shall see that they may be described in terms of orbitals bonding and antibonding along the edges of the Mo octahedron, of orbitals bonding and antibonding along the edges of the X cube, and of Mo-X bonding and antibonding orbitals.

In the chain compounds the clusters are condensed and there is no unique way of dividing the structure up into individual Mo_6X_8 clusters. In this situation the useful orbitals are not those of a single cluster but rather some of the bond orbitals and certain directed orbitals. We shall return to this topic in a later section.

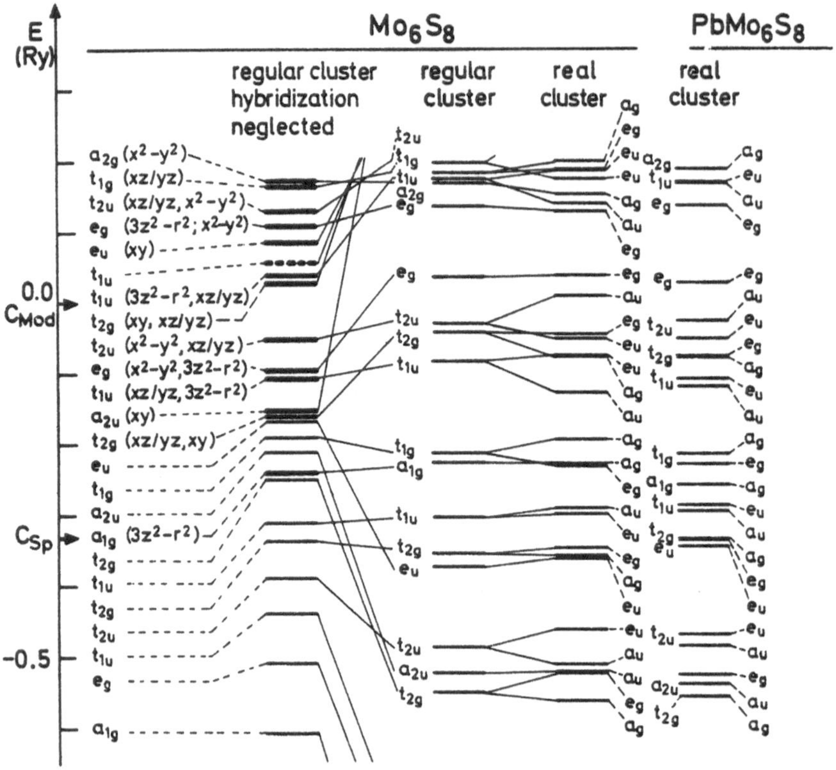

Fig.6.4. Level schemes for the Mo_6S_8 clusters in Mo_6S_8 and $PbMo_6S_8$ Chevrel phases. The zero of energy is at C_{Mod}. The solid, broken, and weak lines in the leftmost scheme indicate, respectively, Mo d, Mo s, and S p levels

The one-electron levels computed for single Mo_6S_8 clusters and lying within half a Rydberg from the center of the Mo d band are shown in Fig.6.4. The first three level schemes were obtained using the atomic positions and atomic-sphere potentials of crystalline Mo_6S_8 and the last was obtained using the data from crystalline $PbMo_6S_8$. In the first level scheme the deviation of the Mo_6S_8 cluster from cubic symmetry is neglected as well as the hybridization between orbitals of different $\underline{R}$ and ℓ. The hybridization is turned on in the second scheme and the distortion in the third. The e_g orbital with energy close to 0 Ry ≡ C_{Mod} is the lowest unoccupied level in Mo_6S_8, and in $PbMo_6S_8$ it is half occupied.

In order to understand the orbital symmetries, and eventually the energy band structure, it is useful first to consider the unhybridized X p and Mo d levels distinguished in the first level scheme by, respectively, weak and solid lines. The broken solid line indicates the lowest of the Mo s levels. (The Mo s a_{1g} level, which in [6.8] occurred in the lower part of the Mo d band, lies above the top of the d band in the recent calculation [6.13] which included the partial waves outside a large sphere circumscribing the cluster.)

6.3.1 The X p States of a Regular Cube

The 12 X p states of lowest energy transform according to the irreducible representations a_{1g}, e_g, t_{1u}, t_{2u}, and t_{2g}, and these are exactly the representations generated by σ bonds directed along the 12 edges of a cube. The symmetry of these states may therefore be illustrated as in Fig.6.5 where bond orbitals of positive and negative signs are indicated by, respectively, full and broken solid lines. In the present case the bond orbitals are made up of the X x-, y-, and z-atomic orbitals which, of course, have lobes of opposite sign pointing away from the cube. These parts of the bond orbitals have not been indicated in the figure although they will provide coupling between the clusters as we shall see in Sect.6.4.1.

If we assume that the bonds are simple two-center bonds of energy $-pp\sigma$, then the average of the 12 bonding states is $C_{Xp} - pp\sigma$ and the deviation of the individual levels from this average is due to the nearest-neighbor $pp\pi$ interaction and to second-nearest-neighbor interactions. Neglecting the latter, the energies relative to the average are easily seen to be $2pp\pi$ (a_{1g}, e_g), 0 (t_{1u}, t_{2u}), and $-2pp\pi$ (t_{2g}); and using the canonical values (6.6) of the two-center integrals, these energies become -4 (a_{1g}, e_g), -2 (t_{1u}, t_{2u}), and 0 (t_{2g}) in units of $6(d^{intra}_{Mo\text{-}Mo}/d^{intra}_{X\text{-}X})^3\,\hat{\Delta}_{Xp}$ relative to C_{Xp}. Using the structural and potential parameters given in Tables 6.1-3 ($\hat{\Delta}_{Xp}$, strictly speaking, applies to levels near C_{Mod} rather than near C_{Xp}), the results of these crude estimates are in reasonably good agreement with the computed levels shown in the first scheme of Fig.6.4.

The 12 X p states of highest energy transform according to the irreducible representations t_{1u}, t_{2g}, t_{1g}, a_{2u}, and e_u which are, in fact, spanned by σ^* antibonds directed along the 12 edges of the cube. These states may therefore be illustrated as in Fig.6.6. The computer calculation showed that in those cases (t_{1u} and t_{2g}) where there are bonding and antibonding states transforming according to the same irreducible representation the uppermost level had always over 99% antibonding character. Using the same approximations as for the bonding states, the energies of the antibonding states are $2pp\pi$ (t_{1u}), 0 (t_{2g}, t_{1g}), and $-2pp\pi$ (a_{2u}, e_u) relative to $C_{Xp} + pp\sigma$ or, using the canonical values 0 (t_{1u}), 2 (t_{2g}, t_{1g}), and 4 (a_{2u}, e_u), relative to C_{Xp} and in the units mentioned above. For the t_{1u} state the π bonds thus cancel the σ^* antibonds and the state is nonbonding. We shall later see that this state plays an important role for the intercluster coupling.

From Fig.6.4 it may be seen that when hybridization is included, the t_{1g} level will mark the top of the X p band. The energy of this level is approximately $12(d^{intra}_{Mo\text{-}Mo}/d^{intra}_{X\text{-}X})^3\hat{\Delta}_{Xp} + \hat{C}_{Xp}$ with respect to $\hat{C}_{Mod}$. Using Tables 6.1,3, we find for $Mo_6S_8 \rightarrow GdMo_6S_8$ the following values: -0.26, -0.18, -0.07, -0.25, -0.19, -0.20, and -0.24 Ry, which essentially reflects the increasing trend in $\hat{C}_{Xp}$ when going from a sulfide towards a telluride.

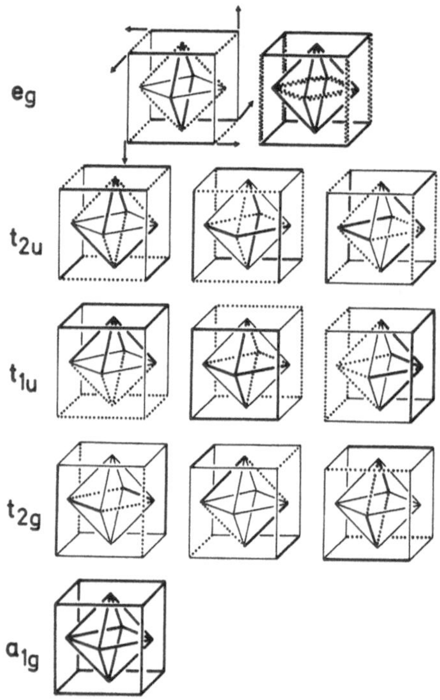

Fig.6.5

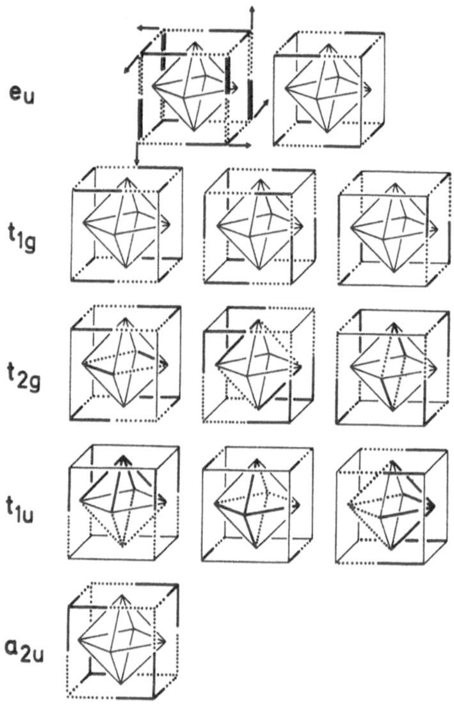

Fig.6.6

Fig.6.5. Irreducible representations of the cubic group generated by 1) σ bonds along the edges of a cube and 2) σ-like bonds along the edges of an octahedron. These illustrate the symmetries of 1) the bonding X p states and 2) the bonding Mo d states of a regular Mo_6X_8 cluster. The signs of the bonds are indicated by strong full and broken lines, respectively. The sign of the cube states relative to those of the octahedron states has been chosen as bonding. The normalizations of the various states may easily be evaluated and have therefore not been included. The arrows indicate for the Chevrel phases the direction from each of the X6 atoms towards the Mo atom in the neighboring cluster. The two corners without arrows are the sites of the X2 atoms and the cube diagonal between them is the $\bar{3}$ axis

Fig.6.6. Irreducible representations of the cubic group generated by 1) σ^* antibonds along the edges of a cube and 2) σ-like bonds along the edges of an octahedron

6.3.2 The Mo d States of a Regular Octahedron

The pure d states of the Mo octahedron are seen in Fig.6.4 to be grouped into 13 low-lying and 17 high-lying states separated by a gap. Of the low-lying states the a_{2u} state has xy lobes pointing towards the X atoms. This state is therefore going to hybridize strongly with the X p state of the same symmetry and thus pushed above the gap. The 12 remaining low-lying states transform according to the irreducible representations a_{1g}, t_{1u}, t_{2g}, t_{2u}, and e_g, exactly the representations generated by bonds directed along the 12 edges of an octahedron [6.4]. This is the reason for the gap at 24 electrons in the Chevrel phases and all other transition-metal compounds containing separated Me_6X_8 units. These bond representations of the octahedron happen to be the same as the bond representations of the cube. We have therefore included the symmetries of the 12 bonding Mo d states in Fig.6.5. The symmetries of the t_{1u} and t_{2g} states have been included together with the antibonding states of the cube in Fig.6.6 too. In these figures the relative signs of the X and Mo states have been chosen as bonding.

The 17+1 d-like states above the gap include the 12 antibonding states of the octahedron as well as the 5+1 Mo-X antibonding states based on the Mo xy orbitals. These unoccupied states are of less interest to us.

The atomic d orbitals compatible with the various irreducible representations have been indicated in Fig.6.4. Here, and in the following, the x-, y-, and z-axes are chosen as in Fig.6.1 and as described in Sect.6.2.1. The orbital symmetry refers to the Mo atom at the positive z-axis, i.e., the one with the coordinates given in Table 6.1. The symmetries and signs of the orbitals at the remaining five Mo atoms may, for the bonding states, readily be obtained from Fig.6.5 if it is kept in mind that the back lobes, (which for d orbitals have the same sign as the front lobes) are not shown in the figure, or in other words, that the solid lines representing the bonds should be extended beyond the two terminal atoms.

The energies and wave functions of the 30 unhybridized Mo d states may even be calculated by hand, as was done in [6.8], because no irreducible representation occurs more than twice and because the hopping matrix is easily obtained from (6.5) with the use of Tables 6.1, 3, and 5. Including nearest-neighbor interactions only, the energies are -143, -111, -65, -59, and $-34\hat{\Delta}_{Mod}$ for the a_{1g}, t_{2g}, t_{1u}, e_g, and t_{2u} states, respectively. The orbital characters have been given in the second panel of Table 6.6, and it may be seen that the xz/yz character dominates the t_{1u} and t_{2g} states, whereas the (x^2-y^2) character dominates the t_{2u} and, in particular, the e_g states. For the series of Chevrel phases considered the scale of the d band is $\hat{\Delta}_{Mod}$ as given in Table 6.3. The trend is that of $(d^{intra}_{Mo-Mo})^{-5}$ as explained in Sect.6.2.2.

6.3.3 The d Bond Orbital of a Regular Octahedron

It is, of course, possible to perform a unitary transformation of the 12 a_{1g}, t_{1u}, t_{2g}, t_{2u}, and e_g states into 12 degenerate bond orbitals, each one being directed (essentially) along an edge of the octahedron. The average of the 12 levels is the diagonal element of the one-electron Hamiltonian in the representation of these bond orbitals, i.e., the bond energy, and the spread around the average is due to the interactions between bond orbitals directed along different edges, as discussed for the example of the X p states of the cube in the previous section. The shape of the bond orbital depends on the characters of the 12 original states, for instance on the amount of $(3z^2-1)$ character in the t_{1u} and e_g states. As can be seen in the third column of Table 6.6, these characters change

Table 6.6. Orbital characters of the bonding states of isolated Mo_6S_8 clusters

	Regular clusters										Real clusters											
Population [%]	2-center d-bond orbital model					Mo_6S_8 d orbitals only					Mo_6S_8						$PbMo_6S_8$					
											Mo_6 d					S_8	Mo_6 d					S_8
	$3z^2-1$	x^2-y^2	xy	xz	yz	$3z^2-1$	x^2-y^2	xy	xz	yz	$3z^2-1$	x^2-y^2	xy	xz	yz	p	$3z^2-1$	x^2-y^2	xy	xz	yz	p
e_g	25	75				4	96				8	74	1	3	2	13	3	80				16
t_{2u} a_u		50		25	25		68		16	16	0	40		28	28	4		49		24	24	3
t_{2u} e_u											9	41		21	24	5		48		25	24	3
t_{2g} e_g				50	50			14	43	43	1	4	6	37	38	14			2	42	43	14
t_{2g} a_g											0	0	6	39	39	17			8	39	39	15
t_{1u} e_u	50			25	25	14			43	43	35	6		23	24	9	37			25	25	10
t_{1u} a_u											48	0		20	20	9	45			21	21	9
a_{1g}	100					100					96					2	97					2

somewhat when the hybridization with the Mo s and the X p atomic orbitals is included. As a consequence the shape of the bond orbital depends on the compound considered. We have therefore not projected the bond orbital out of the fully hybridized states for each of the various Mo_6X_8 clusters, but instead present the simplest possible analytical d bond orbital for an octahedron: if we insist that the bond orbital be constructed from d atomic orbitals centered *only* on the *two* atoms directly involved in the bond, it then turns out that the bond orbital is uniquely determined. The corresponding characters are given in the first panel of Table 6.6. The t_{1u} and t_{2u} states generated from this two-center d bond orbital are seen to agree better with the fully hybridized states than the eigenfunctions of the dd block of the Hamiltonian given in the second panel. However, for the e_g state, the bond orbital is less accurate than the eigenfunction of the dd block, but for the remaining states the accuracy is about the same.

The two-center d bond orbital is most simply expressed as $B = (D^{I} + D^{II})/\sqrt{2}$, where D^{I} and D^{II} are orbitals directed along the bond and centered on the two terminal atoms. These directed orbitals are given in Table 6.7 for the Mo atom at the z-axis in terms of the usual d orbitals excluding the xy orbital. The directed orbitals are orthogonal and approximately directed along [011],[0$\bar{1}$1], [101], and [$\bar{1}$01]. The angular variation of D_{011} in the yz and xζ planes is illustrated in Fig.6.7, where ζ is the direction of the maximum of the orbital which lies in the yz mirror plane 38° from the z-axis. The angle between D_{011} and $D_{0\bar{1}1}$ is thus 76° rather than 90°, and the angle between D_{001} and D_{101} is 51° rather than 60°. Nevertheless, the bond energy of $-57\hat{\Delta}_{Mod}$, as obtained from Tables 6.5,7, is only marginally smaller than the maximum value possible, which is $dd\sigma = -60\hat{\Delta}_{Mod}$. Of the off diagonal elements of the Hamiltonian in the bond-orbital representation, the interaction between two edges of a triangle as well as the interaction between two opposite edges of a square are $-15\hat{\Delta}_{Mod}$. The interaction between two

Table 6.7. d^4-Directed Orbitals

D	$3z^2-1$	x^2-y^2	yz	xz
011	1/2	-1/2	$1/\sqrt{2}$	0
$0\bar{1}1$	1/2	-1/2	$-1/\sqrt{2}$	0
101	1/2	1/2	0	$1/\sqrt{2}$
$\bar{1}01$	1/2	1/2	0	$-1/\sqrt{2}$

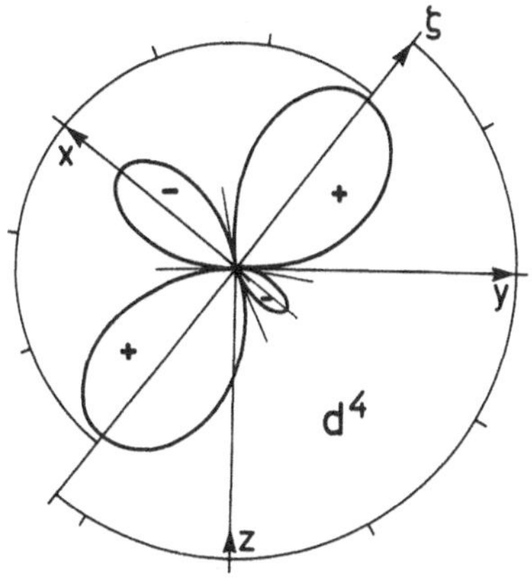

Fig.6.7. The angular variation of the D_{011}-directed orbital in the yz and xζ planes. The unitary transformation from the normalized cubic harmonics to the d^4-directed orbitals is given in Table 6.7

neighboring edges of a square is $+12\hat{\Delta}_{Mod}$. Finally, the interaction between two edges of the octahedron which neither belong to the same triangle nor to the same square is $-9\hat{\Delta}_{Mod}$.

With these matrix elements and Fig.6.5 the energies of the 12 bonding Mo d states, without hybridization, are easy to understand. In the a_{1g} state there are 24 corners of triangles, 12 corners of squares, 6 pairs of opposite edges of squares and 24 pairs of edges neither belonging to the same triangle nor to the same square. The energy is consequently $[24(-15) + 12(+12) + 6(-15) + 24(-9)](2/12) - 57 = -144$ in units of $\hat{\Delta}_{Mod}$. The normalization factor 12/2 is the number of pairs of bonds in the a_{1g} state. In a similar way we find that the energy of the t_{2g} state is $[-4(+12) + 2(-15)](2/4) - 57 = -96$, that the energy of the t_{1u} state is $[8(-15) - 4(-15) - 8(-9)](2/8) - 57 = -54$, that the energy of the e_g state is $[-8(-15) + 8(+12) + 4(-15) - 8(-9)]\ (2/8) - 57 = 0$, and that the energy of the t_{2u} state is $[-8(-15) - 4(-15) + 8(-9)](2/8) - 57 = -30$ in units of $\hat{\Delta}_{Mod}$. According to the variational principle, all these energies are higher than the eigenvalues of the dd block of the Hamiltonian, but only for the e_g state is the difference significant.

6.3.4 Mo-X Hybridization

We now let the pure states, discussed in Sects.6.3.1,2 and illustrated in Figs. 6.5,6, hybridize and thereby proceed from the first to the second level scheme of Fig.6.4.

The six occupied, lowest-lying a_{1g}, e_g, and t_{1u} bonding X p states hybridize strongly with the six unoccupied Mo s states, which transform according to the same irreducible representations, as well as with the (partly) unoccupied bonding e_g Mo d state and the unoccupied antibonding t_{1u} Mo d state. In this way strong Mo-X covalent bonds are formed. The resulting bonding X p-like levels lie below and the antibonding Mo s-like levels lie above the energy range of Fig.6.4. There is, in addition, a weak hybridization with the occupied bonding t_{1u} Mo d state. Furthermore, the t_{2g} antibonding X p state, which hybridizes strongly with the unoccupied Mo p and antibonding Mo xy states of symmetry t_{2g}, and the a_{2u} and e_u antibonding X p states, which hybridize strongly with the unoccupied a_{2u} and e_u Mo xy states as mentioned previously contribute to the covalent bonding. The t_{2u} bonding X p state hybridizes weakly with the occupied bonding and unoccupied antibonding Mo t_{2u} states and with the unoccupied t_{2u} Mo p state. The weak coupling between the antibonding X p and the Mo xz/yz t_{1g} states adds slightly to the strength of the covalent bond.

Of the remaining X p states the bonding t_{2g} state and the nonbonding t_{1u} state hardly hybridize with Mo states, because as seen in Fig.6.5, the d orbital on the Mo atom has the opposite parity of the p orbitals on the four neighboring X atoms with respect to reflection in the plane containing all 5 atoms. This argument, of course, only holds for the regular structure (Sect.6.2.1), but the calculations illustrated in the last two level schemes of Fig.6.4 show that the hybridization of these levels is negligible also in properly distorted Mo_6S_8 clusters. In the following section we shall see that in the *crystalline* Chevrel phases use is made of the t_{1u} states in the bonding.

We now consider the rearrangement of the bonding Mo d levels. The lowest-lying Mo d level, the a_{1g} $(3z^2-1)$ level, hardly hybridizes with the Mo s and X p a_{1g} levels because these, due to their strong mutual interaction, have been pushed far away in energy. The a_{1g} level thus remains below the Mo d-X p hybridization gap which therefore occurs at a nominal occupancy of 2 Mo d electrons per cluster rather than at 0, as one might naively have expected.

The Mo bonding t_{1u} state does not hybridize with the X p antibonding state as explained above, and it only hybridizes weakly with the X p bonding state and with the high-lying Mo s state. The Mo bonding t_{2u} state hardly hybridizes either. The reason follows from Fig.6.6. The pdπ interaction between the (x^2-y^2) orbitals of the Mo atoms on the z-axis and the (x-y) orbitals of the nearest-neighbor X atoms is nearly cancelled by the interaction between the xz orbitals of the Mo atoms on the x-axis and the x orbitals of the X atoms and by the interaction between the yz orbitals of the Mo atoms on the y-axis and the X atom y orbitals. The t_{2g} and e_g states do hybridize with the X p states. From Fig.6.5 we see that the hybridization matrix element for the t_{2g} state is $\sqrt{2}pd\pi$, while that for the e_g state, from Fig.6.6, is $2pd\pi$. The bonding X p e_g state, however, lies much lower than the antibonding X p t_{2g} state, and as a consequence, the upwards shifts of the Mo d e_g and t_{2g} states are about equally large.

As a result, the e_g level is the highest of the Mo d-like bonding levels. This is true for all the clusters considered. The conduction band in the Chevrel phases will therefore essentially be the doubly degenerate band formed from the Bloch sums of the e_g molecular orbitals. As mentioned in the Introduction, the same ordering of the Mo d-like bonding levels as seen in Fig.6.4 has been found in a number of recent calculations [6.5-8,10]. An exception is the Mo_6S_8-cluster calculation of MATTHEISS and FONG [6.9], where the t_{2u} level was found to lie marginally higher than the e_g level.

The trend followed by the X p-Mo d hybridization matrix element is given by, for instance, $pd\pi = 6\sqrt{5}\,(d^{intra}_{Mo-X}/d^{intra}_{Mo-Mo})^{-4}(\hat{\Delta}_{Xp}\hat{\Delta}_{Mod})^{1/2}$, which from Tables 6.2,3 for $Mo_6S_8 \rightarrow GdMo_6S_8$ takes the values 105, 91, 74, 102, 90, 100, and 102 mRy. This trend, which is dominated by the behavior of $(d^{intra}_{Mo-X})^{-4}$, is the one followed by the energy shifts in the limit where the hybridizing p and d levels are degenerate. For levels originally separated by typically $\hat{C}_{Mod}-\hat{C}_{Xp}$, the shifts, however, follow the trend given by $(pd\pi)^2/(\hat{C}_{Mod}-\hat{C}_{Xp})$, which for $Mo_6S_8 \rightarrow GdMo_6S_8$ is nearly constant and equal to 34 ± 4 mRy. For each two levels separated by $\hat{C}_{Mod}-\hat{C}_{Xp}$, the X p character mixed into the Mo d state, or vice versa, follows the trend given by $(pd\pi)^2/(\hat{C}_{Mod}-\hat{C}_{Xp})^2$, which is 0.11 for the sulfides, 0.14 for the selenides, and 0.27 for the telluride. In conclusion, when proceeding from the sulfides to the tellurides, the energy shifts caused by X p hybridization of the Mo d levels, on the average, decrease slightly whereas the wave-function-mixing increases, on the average.

6.3.5 Real Clusters

The Mo_6X_8 clusters in the Chevrel phases do not have cubic symmetry but rather $\bar{3}$ as explained in Sect.6.2.1. For the splittings of the one-electron levels, the most essential distortion seems to be the elongation of the octahedron, and this is specified in the second row of Table 6.2. By comparison of the level splitting in Mo_6S_8 with those in $PbMo_6S_8$ shown in the third- and fourth-level schemes of Fig.6.4, it seems as if the distortion of the X cube and the displacement of the Mo octahedron with respect to the X cube are of less importance.

The lowering of the symmetry splits the triply degenerate t levels into singly degenerate a and doubly degenerate e levels. In Fig.6.8 we have shown these linear combinations of the partner functions for the t_{2g} and t_{1u} Mo bonding and X antibonding states and for the t_{2u} Mo bonding and X bonding states. Since, to first

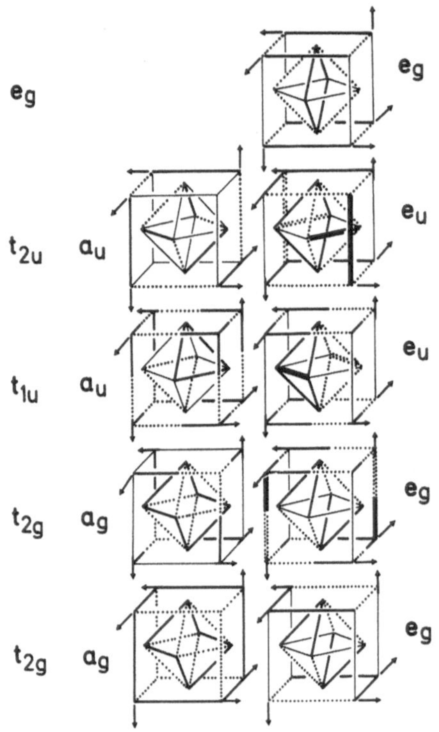

Fig.6.8. Irreducible representations of the cubic group when the partner functions are chosen compatible with $\bar{3}$ symmetry. Only one of the e partners is shown

order in the elongation, the center of gravity of a triplet is constant, we only need to consider the displacement of the a level; the e level will then be displaced in the opposite direction by half the amount and the splitting will be 1.5 times the displacement of the a level. In the $t_{2g} \rightarrow a_g$ state all twelve bonds of the octahedron are equally occupied, and therefore the level does not split to first order. In the $t_{1u} \rightarrow a_u$ state only the triangles perpendicular to the $\bar{3}$ axis are occupied. Since the edges of these triangles become shorter by the relative amount $1-\partial$, the a_u level drops. Using the bond-orbital model presented in Sect. 6.3.3, we estimate the shift of the a_u level to be $[6(-15)5\partial - 3(-15)(-5\partial) -6(-9)(-5\partial)](2/6) - 57(5\partial) = -600\partial$ in units of $\hat{\Delta}_{Mod}$. For Mo_6S_8 we thus estimate the a_u shift to be -28 mR_y and hence the splitting to be 42 mRy as compared with 50 mRy obtained from the computation. For $PbMo_6S_8$ we estimate a splitting of 18 mRy and compute 10 mRy. In the $t_{2u} \rightarrow a_u$ state all edges but the triangles are occupied such that the a_u level rises by the amount $[-6(-15)5\partial - 3(-15)5\partial](2/6) -57(-5\partial) = 560\partial$ in units of $\hat{\Delta}_{Mod}$. For Mo_6S_8 and $PbMo_6S_8$ we thus estimate splittings of, respectively, 39 and 17 mRy as compared with the computed values 57 and 25 mRy, which include the effects from the distortion of the X cube and the displacement of the cube with respect to the octahedron.

The orbital characters of the real clusters are listed in Table 6.6, and the agreement with earlier results for regular Mo_6S_8 is fair [6.9] or good [6.10].

6.4 Energy Bands of Chevrel Phases

In the Chevrel phases the Mo_6X_8 clusters are built into crystals as indicated in Figs.6.1,9, and we now consider their energy bands. We first discuss the gross features of the densities of states and then examine how the energy bands arise from the molecular-orbital states of an individual cluster. The dispersion and wave-function character of the narrow conduction band, the so-called e_g-band, will be of particular interest. We shall try to understand the variation of this band from compound to compound.

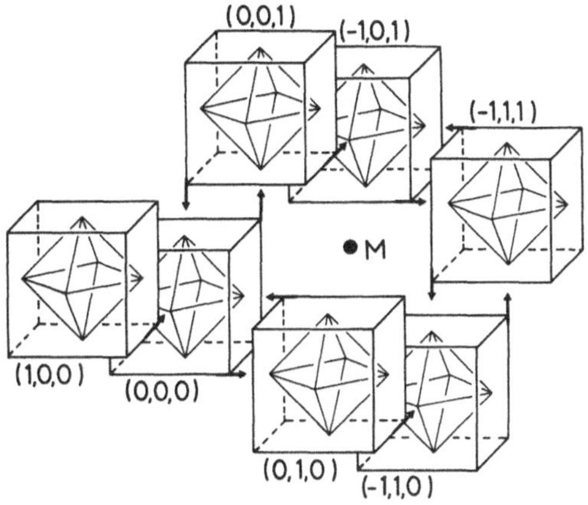

Fig.6.9. Stacking of the Mo_6X_8 clusters in the Chevrel phases. For clarity only some of the clusters have been shown

6.4.1 Gross Features of the Densities of States

The total density of states N(E), the site- and orbital-projected densities of states $N_{t\ell}(E)$, and numbers of states $n_{t\ell}(E) \equiv \int^E N_{t\ell}(E')dE'$ are shown in Fig.6.10. These results for $PbMo_6S_8$ are rather similar to those published for $EuMo_6S_8$ by JARLBORG and FREEMAN [6.11] and for $PbMo_6S_8$ by BULLETT [6.10]. The latter calculation did, however, give d bands of only half the width obtained in [6.11-13]. The S s bands lie below the range shown. The lowest set of bands, extending from -1.21 to -0.82 Ry, are the 24 S p bands into which are mixed the Pb s band and the Mo $3z^2$-1 a_{1g} band. On the energy scale shown C_{Ss} = -2.07, C_{Sp} = -0.96, C_{Pbs} = -1.26, and C_{Mos} = -0.83 Ry. It may be seen that the S p bands have significant Mo s, p, and d character due to the covalent hybridization.

The X p-Mo d gap at -0.82 Ry exists in the sulfides but not in the selenides and telluride considered because C_{Xp} increases as explained in the final paragraph of Sect.6.3.1.

The next set of 11 bands extending from -0.82 to -0.67 Ry are the 12 Mo d bonding bands minus the a_{1g} band. These bands contain the Fermi level and are almost pure d like because they, so to say, lie in a gap created by the strong hybridization between the X p and the M s, p, and xy orbitals. We shall return to the details of these bands in the following sections.

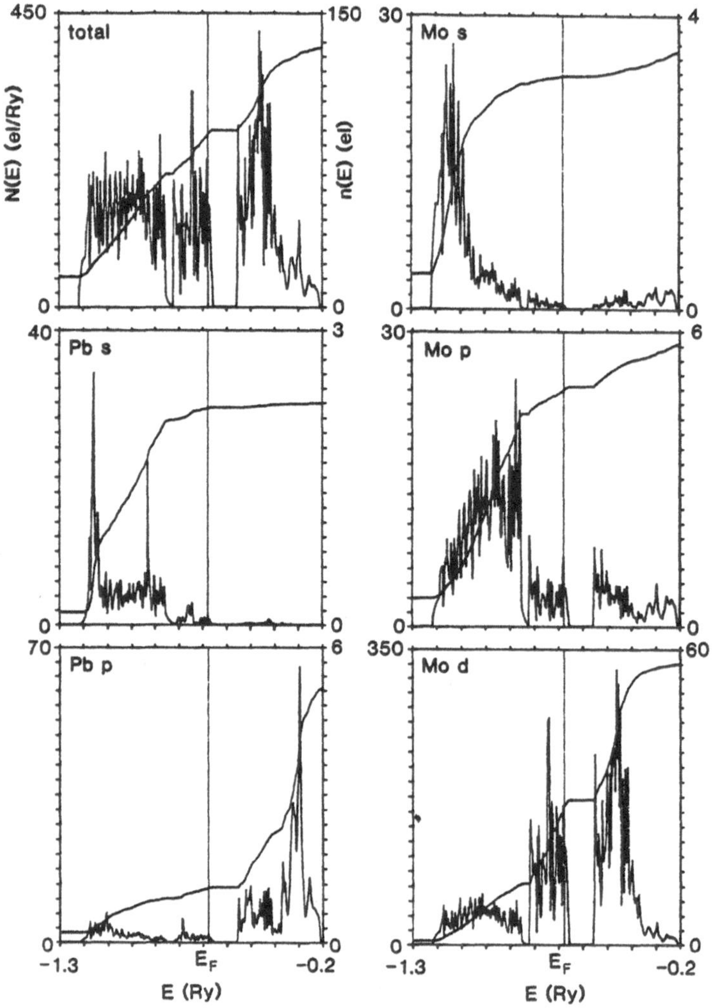

Fig.6.10a

Fig.6.10a,b. Total density of states and site- and orbital-projected densities of states and numbers of states for $PbMo_6S_8$. On the energy scale shown C_{Mod} = -0.68 Ry

On the energy scale shown, C_{Mod} = -0.68 Ry, and the gap between -0.67 and -0.57 Ry is essentially the Mo d bonding-antibonding gap at 24 Mo d electrons discussed in Sects.6.3.2,3. The size of this gap is 53, 67, 38, 97, 83, 110, and 105 mRy in the series $Mo_6S_8 \rightarrow GdMo_6S_8$. As pointed out by BULLETT [6.10] its small size in Mo_6S_8 and Mo_6Te_8 seems to correlate well with the experimental facts that these compounds are relatively difficult to synthesize and that ternary tellurides with large cations so far do not exist. The reason why this gap is smaller in some compounds than in others will be discussed in the following sections.

The highest set of bands extending upwards from the dd gap are the 12 Mo d antibonding bands, the 6 Mo xy-X p antibonding bands and the 3 Pb p bands; C_{Pbp} = -0.54

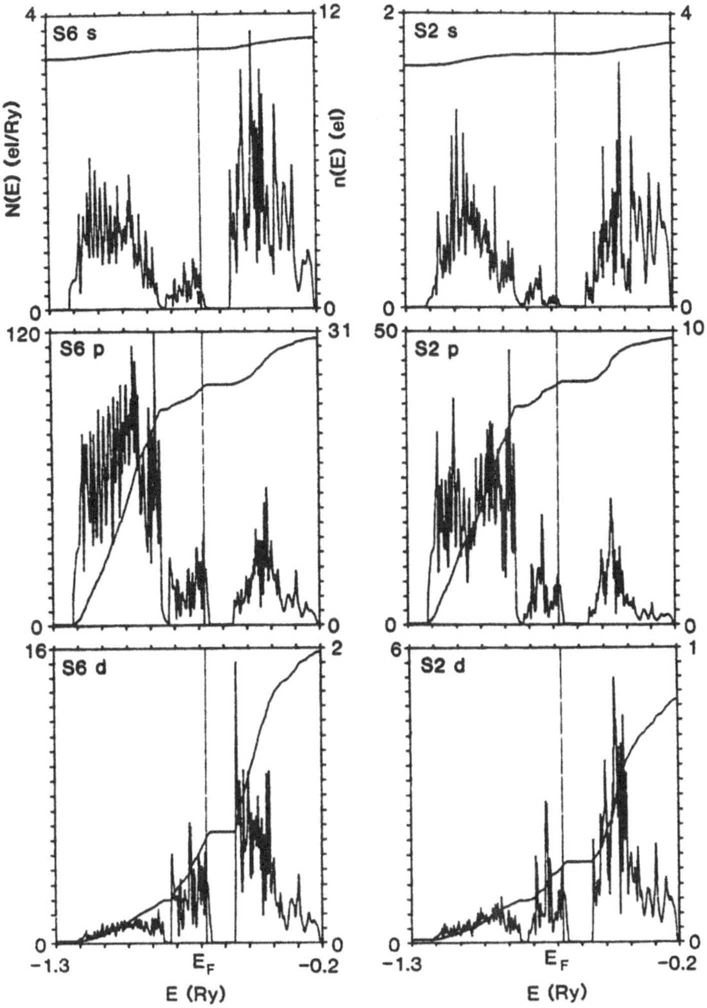

Fig.6.10b
Legend see p.188

Ry. These bands have some X d character, in particular in the selenides and the telluride. Above the range of Fig.6.10 lie the 6 Mo s bands.

The partial densities of states integrated up to the Fermi level, i.e., the numbers $n_{t\ell}(E_F)$ of ℓ electrons in spheres of type t, reveal trends in the covalent mixing for the various compounds. In all compounds considered the number of Mo s electrons is 0.53 per Mo sphere. The number of Mo p electrons is about 0.83 in the sulfides, about 0.75 in the selenides, and about 0.73 in the telluride. This points to a decreasing Mo p-X p mixing which comes about because the trend caused by the increase of the Mo-X distances dominates the trend caused by the rather small relative decrease of the large energy difference $C_{Mop} - C_{Xp}$. For the Mo s electrons these two trends apparently balance, and as was discussed at the end of Sect.6.3.4, for the Mo d electrons the increasing trend caused by $(C_{Mod} - C_{Xp})^{-2}$

dominates. Therefore the number of Mo xy electrons mixed into the X p-band increases when going from Mo_6S_8 over Mo_6Se_8 to Mo_6Te_8, the number of d electrons per Mo sphere being, respectively, 4.55, 4.64, and 4.96. For the same reason the number of X p electrons mixed into the unoccupied Mo d bands increase, while the much smaller number of X p electrons mixed into the Mo s and p bands are constant or decrease. As a result, the number of X p electrons in the occupied bands decrease from 4.06 in Mo_6S_8, over 4.03 in Mo_6Se_8, and to 3.78 in Mo_6Te_8. These numbers apply to the X6 spheres. For the X2 spheres lying at a larger distance from the Mo atoms in the sulfide and at a smaller distance in the telluride, this trend is enhanced and the numbers are 4.16, 4.01, and 3.64.

When we go from Mo_6S_8 to $PbMo_6S_8$, $BaMo_6S_8$, and $GdMo_6S_8$, the number of d electrons in the Mo sphere increases from 4.55 to 4.66, 4.75, and 4.76. This increase is much smaller than the number of electrons added, i.e., 2/6 or 3/6 per Mo atom times the 74% Mo d character found in the e_g-conduction band (Table 6.8). The reason is that nearly all the added electrons remain in the M sphere or, specifically, that some of the Mo d character mixed into the occupied X p band is exchanged with Pb p or with Ba or Gd s, p, and d character. The number of p electrons in an X6 sphere increases from 4.06 and 4.03 in Mo_6S_8 and Mo_6Se_8 to 4.17, 4.16, 4.07, and 4.13 in $PbMo_6S_8 \rightarrow GdMo_6S_8$. This increase is somewhat larger than expected from the number of electrons added times the 15% X6 p character of the conduction band. The reason is that a rearrangement of the electrons between the X6 and X2 atoms takes place. While in the binaries the X2 atoms are farthest away from the Mo atoms, they are closest in the ternaries. Therefore the Mo d-X2 p hybridization increases such that the number of occupied X2 p electrons falls despite the number of electrons added. The number of p electrons in an X2 sphere is thus 4.09, 4.06, 3.99, and 4.14 in $PbMo_6S_8 \rightarrow GdMo_6S_8$.

By addition of the numbers of Mo s, p, and d electrons one realizes that, with the sphere sizes chosen on the basis of the actual crystal structure and given in Table 6.3, the Mo spheres are neutral within ±0.2 electrons. Similarly, the X spheres are nearly neutral. The empty M sphere contains 0.56, 0.43, and 0.29 electrons in Mo_6S_8, Mo_6Se_8, and Mo_6Te_8, respectively. Taking the decreasing sphere volume into account, this corresponds to a constant electron density. The number of electrons in the Pb spheres are 3.87 and 3.58 in the sulfide and the selenide, respectively. The number of electrons in the Ba and Gd spheres are 2.67 and 2.37.

Although the partial densities of states computed for $PbMo_6S_8$ resemble the distribution of the cluster levels shown in Fig.6.4 for $PbMo_6S_8$, the X-X interactions are much stronger in the crystal than in the single cluster, the X p-Mo d interactions are somewhat stronger, and only the Mo d-Mo d interactions have about the same strength. This is essentially due to the increased number of nearest neighbors in the crystal.

For an X6 atom there are three near X atoms in the same cluster, but in the crystal there are 4 + 2 + 1 = 7 X atoms at similar distances in the three neighboring clusters plus one near X atom at a 20% larger distance (if the structure is regular) in a next-neighbor cluster. This may be understood from Figs.6.1,9. For an X2 atom there are three nearest X atoms in the same cluster but 2 + 2 + 2 = 6 near X atoms in the three neighboring clusters. The number of short X-X distances thus increases by a factor of 3.25 when going from the single cluster to the crystal. This means that the width of the pure X p-band complex increases by a factor of $\sqrt{3.25} = 1.8$, but it also means that the X s-X p interaction increases by the same factor. This pushes the bottom of the X p-band complex up in energy. Moreover, the increased hybridization with the Mo atoms, which we shall consider next, pushes the top of the X p band down in energy. As a result, the width of the hybridized X p-band complex is almost the same in the $PbMo_6S_8$ crystal as in the single cluster. However, the details of the energy levels are changed considerably.

For a Mo atom there are four near X atoms in the same cluster and one additional X atom at nearly the same distance (Table 6.2) in the neighboring cluster. This means that the Mo-X hybridization, on the average, increases by a factor of about $\sqrt{(5/4)}$ or by roughly 10%. More important than this increase of the average hybridization matrix element is the fact that the strongly increased X-X interactions cause the topmost X p levels to approach, or even overlap, the Mo d levels. In this way strong Mo-X hybridizations not present in the isolated cluster occur in the crystal.

For a Mo atom there are four near Mo atoms in the same cluster, the distance to the one additional Mo-atom in the neighboring cluster being 10-30% larger. Since the dd interactions decrease strongly with distance, the width of the Mo d-band complex is essentially the same in the crystal as in the cluster. The broadening of the individual Mo d levels into (sub)bands in the crystal will be considered in subsequent sections. We shall see that the dpd intercluster coupling via the intermediate X atom is often more important than the direct dd intercluster coupling between Mo atoms.

In order to substantiate these remarks we shall make use of the following formula for the second moment of the structure matrix defined in Sect.6.2.2:

$$|S^{t\ell}_{t'\ell'}|^2 \equiv \sum_{\underline{R}\neq\underline{R}'}\sum \sum_{m}^{2\ell+1} \sum_{m'}^{2\ell'+1} |S_{R\ell m;R'\ell'm'}|^2$$

$$= \frac{4(2\ell+1)(2\ell'+1)(2\ell+2\ell')!}{(2\ell)!(2\ell')!} \sum_{\underline{R}\neq\underline{R}'}\sum \left(|\underline{R}-\underline{R}'|/d^{\text{intra}}_{\text{Mo-Mo}}\right)^{-2(\ell+\ell'+1)} \quad . \qquad (6.7)$$

Here, $\underline{R}$ runs over all n_t atoms of type t and $\underline{R}'$ runs over all $n_{t'}$ atoms of type t'. If we assume that the eigenvalues of the $t\ell$-$t\ell$ diagonal block of the structure matrix are uniformly distributed, the *width of the unhybridized tℓ-band complex* becomes

$$W_{t\ell} = [12|S_{t\ell}^{t\ell}|^2/(2\ell+1)n_t]^{1/2} \quad ,$$

$$= 12\sqrt{6}[\Sigma(d_{X-X}/d_{Mo-Mo}^{intra})^{-6}]^{1/2} \quad \text{for the X p band} \quad , \qquad (6.8)$$

$$= 20\sqrt{42}[\Sigma(d_{Mo-Mo}/d_{Mo-Mo}^{intra})^{-10}]^{1/2} \quad \text{for the Mo d band} \quad , \qquad (6.9)$$

in units of $\hat{\Delta}_{t\ell}$. In the expression for the width of the X p-band complex the sum runs over all X neighbors of a given X atom and in the expression for the width of the Mo d-band complex the sum is over all Mo neighbors of a given Mo atom. For the latter sum the Mo-Mo intercluster distance is negligible and the estimate of the d bandwidth is therefore $20\sqrt{42}\sqrt{4}\hat{\Delta}_{Mod} = 259\hat{\Delta}_{Mod} = 410$ and 470 mRy for Mo_6S_8 and $PbMo_6S_8$, respectively. This is in good agreement with the widths found in Figs. 6.4,10 and with the energy $-143\hat{\Delta}_{Mod}$ of the lowest Mo d-level found in Sects.6.3.2 and 3.

In Sect.6.3.1 we saw that the width of the unhybridized X p-band complex of a single cluster is 48 in units of $(d_{Mo-Mo}^{intra}/d_{X-X}^{intra})^3\hat{\Delta}_{Xp}$, and this is in good agreement with the value $12\sqrt{6}\sqrt{3} = 51$ obtained from the simple formula (6.8). For a crystal the width of the X p-band complex is about 1.8 times larger due to the above mentioned increase in the number of near X neighbors. By actually performing the sum (6.8) over the X-X distances, we find for the seven crystals Mo_6S_8, Mo_6Se_8, Mo_6Te_8, $PbMo_6S_8$, $PbMo_6Se_8$, $BaMo_6S_8$, and $GdMo_6S_8$ the following X p-band widths: 639, 663, 659, 615, 642, 558, and 630 mRy. The trend expressed by the ratios between these numbers, i.e., 1.04, 1.08, 1.07, 1, 1.04, 0.91, and 1.02, is the product of the trend expressed by the ratios between the potential parameters $(d_{Mo-Mo}^{intra})^3\hat{\Delta}_{Xp}$, i.e., 1.00, 1.14, 1.34, 1, 1.18, 0.94, and 0.99, and the trend expressed by the ratios between the lattice sums $(\Sigma d_{X-X}^{-6})^{1/2}$, i.e., 1.04, 0.95, 0.80, 1, 0.88, 0.97, and 1.03. If we now use the values for the relative position of the unhybridized X p and Mo d bands, i.e., $\hat{C}_{Xp} - \hat{C}_{Mod}$, given in Table 6.3, we find that the position of the top of the unhybridized X p band with respect to the center of the unhybridized Mo d-band complex is, respectively, -8, 80, 180, -25, 47, -9, and -8 mRy for the seven crystals considered. This means that pd-mixing effects will be strong, particularly in Mo_6Te_8. Most of the high-lying X p bands hybridize with the Mo xy states and thereby shift even further away from the Fermi level than in the single cluster. In the crystal there are, however, high-lying X p bands other than those already hybridizing strongly within the cluster. Of these, we shall be particularly interested in the t_{1u} band near the center of the Brillouin zone because, here, it shifts the Mo d-like t_{1u} band upwards and into the conduction band.

From (6.7) we may express the root-mean-square of the hybridization matrix elements for a Mo d state hybridizing with all the X p states as

$$(|S_{Xp}^{Mod}|^2/30)^{1/2} = 6\sqrt{5}[\Sigma(d_{Mo-X}/d_{Mo-Mo}^{intra})^{-8}]^{1/2} \tag{6.10}$$

in units of $(\hat{\Delta}_{Mod}\hat{\Delta}_{Xp})^{1/2}$. The sum is here over all X neighbors of a given Mo atom. If all these neighbors are at the same distance, the hybridization constant is seen to equal pdπ times the square root of the number of neighbors. The trend in the values of pdπ for single clusters of the seven compounds considered was discussed in Sect.6.3.4. The average, of course, includes the strong intracluster hybridization with the Mo xy states, the average of the hybridization matrix element for the 12 Mo d-bonding states being therefore considerably smaller.

6.4.2 Energy Bands Along the $\bar{3}$-Axis from Γ to R

The dispersion of the energy bands in the Chevrel phases is complicated and varies from compound to compound. Although we shall limit ourselves to considering the conduction band only, it is necessary to consider some of the more remote bands too because these interact with parts of the conduction band and hence cause variations in its shape. As a simplification in the present and following two sections, we shall constrain the Bloch vector to lie along the only symmetry line in the rhombohedral structure, the $\bar{3}$-axis, and furthermore focus on the high-symmetry points Γ and R.

The energy bands calculated for Mo_6S_8 and $PbMo_6S_8$ are shown in Fig.6.11 together with the single-cluster levels discussed in Sect.6.3 and shown in Fig.6.4. At the center, Γ, and the corner, R, of the Brillouin zone (Fig.6.16) there are four (six) irreducible representations — e_g, a_g^{*}, e_u, and a_u — and in between, at general points of the $\bar{3}$-axis, there are only two (three) — doubly degenerate e states ($m = \pm 1$) denoted by dark lines in Fig.6.11 and singly degenerate a states ($m = 0$) denoted by light lines. The energy scale is the same as used in Fig.6.10 such that the gaps around -0.82 and -0.6 Ry are, respectively, the pd and dd gaps. Below the pd gap are the 24 X p-like bands, the a_{1g} Mo $(3z^2-1)$-like band and for $PbMo_6S_8$ the Pb s-like band. These 25 (26) bands, of which the six lowest are below the frame of the figure, hybridize strongly. The Mo $(3z^2-1)$ character is distributed over a number of a bands. At Γ the a_g level at -1.06 Ry has about 40% Mo $(3z^2-1)$ character. In $PbMo_6S_8$ the Pb s character at Γ is mainly distributed between the three a_g levels at -1.16 (46%), -0.99 (32%) and -0.82 Ry (14%). Between the pd and dd gaps are the remaining 11 of the 12 bonding Mo d-like bands. In $PbMo_6S_8$ the Γa_u level at -0.8 Ry has 10% Pb p character. Above the dd gap are the antibonding Mo d-like bands of which only the few lowest lie within the frame of the figure. In $PbMo_6S_8$ there is a narrow Γe_u - Re_g band at -0.30 Ry for which the Pb p character dominates. The remaining Pb p a band is strongly hybridized with the topmost Mo d bands.

In our description of the energy bands we shall imagine that the wave functions are expressed as linear combinations of Bloch sums of the molecular-orbital states

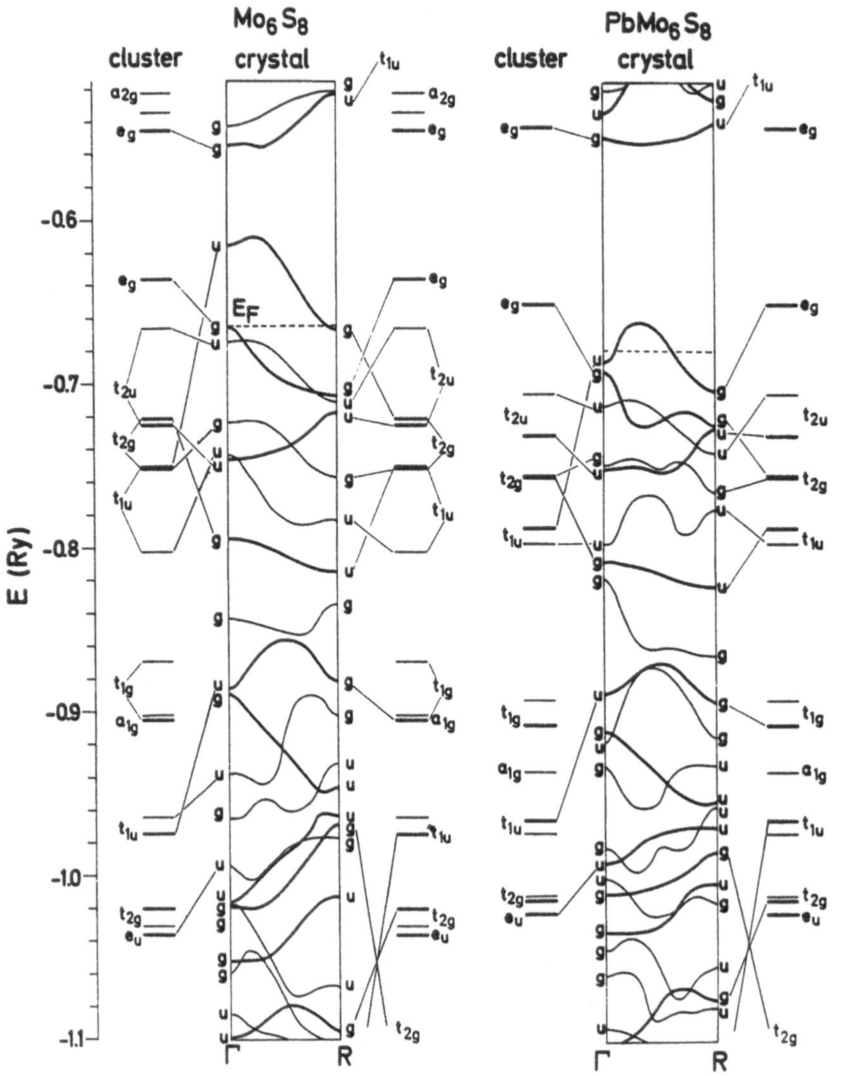

Fig.6.11. Cluster levels and energy bands along the $\bar{3}$-axis for Mo_6S_8 and $PbMo_6S_8$. Doubly degenerate e bands are indicated by dark lines and singly generate a bands are indicated by light lines. On the energy scale shown C_{Mod} = -676 mRy

ψ_j of the single cluster. The basis functions are thus $\sum_{\underline{T}} \exp(i\underline{k}\cdot\underline{T})\psi_j(\underline{r}-\underline{T})$, where $\underline{T}$ are the translation vectors. The molecular-orbital states may be taken as the eigenfunctions of the single cluster or as the unhybridized functions shown in Figs.6.5,6 of the octahedron, $\psi_j^d = \sum_{\underline{O}} x_{jO}^d$, plus those of the cube $\psi_j^p = \sum_{\underline{C}} x_{jC}^p$. Here χ are the atomic X p or Mo d orbitals, and $\underline{O}$ and $\underline{C}$ run over, respectively, the sites of the octahedron and the sites of the cube. In this representation the Hamiltonian becomes

$$H_{ij}(\underline{k}) = N^{-1}\langle \sum_{\underline{T}'} \exp(i\underline{k}\cdot\underline{T}')\psi_i(\underline{r}-\underline{T}')|H| \sum_{\underline{T}} \exp(i\underline{k}\cdot\underline{T})\psi_j(\underline{r}-\underline{T})\rangle$$

$$= \sum_{\underline{T}} \exp(i\underline{k}\cdot\underline{T})\langle\psi_i(\underline{r}-\underline{T})|H|\psi_j(\underline{r})\rangle = \langle\psi_i(\underline{r})|H|\psi_j(\underline{r})\rangle \quad (6.11)$$

$$+ \sum_{\underline{T}\neq 0} \exp(i\underline{k}\cdot\underline{T})\langle\psi_i(\underline{r}-\underline{T})|H|\psi_j(\underline{r})\rangle \equiv H_{ij}^{intra} + \sum_{\underline{T}\neq 0} \exp(i\underline{k}\cdot\underline{T})H_{ijT}^{inter} \ .$$

In order to interpret the computed energy bands in this language, we have projected the X p and Mo d parts of the computed wave functions inside the primitive cell onto the molecular-orbital states of the cube and the octahedron, respectively.

Within the X p-band complex the hybridization between the different X p states of the cube is very strong because of the large increase in the number of X neighbors taking place when going from the cluster to the crystal. Nevertheless, in Γ and R a particular cube state often dominates. For a few important levels this dominance has been indicated in the figure by a line joining onto the corresponding cluster level. The fact that for instance, the t_{1u} levels at Γ and R have very different energies and are not even approximately connected by one band, is then a manifestation of the strong intercluster coupling within the X p band complex.

The direct Mo d-Mo d intercluster coupling, on the other hand, is very weak; $H_{ii}^{inter} \equiv \sum_{\underline{T}\neq 0} H_{iiT}^{inter}$ is at the order of $dd\pi^{inter}/2 = 20(d_{Mo-Mo}^{inter}/d_{Mo-Mo}^{intra})^{-5}\hat{\Delta}_{Mod} = 19$, 14, 8, 14, 10, 11, and 17 mRy for the compounds considered (Tables 6.2,5). The indirect Mo d-X6 p-Mo d intercluster coupling is, on the average, somewhat larger but still small compared with the intracluster coupling. In the crystal-field approximation it is at the order of $pd\pi^{intra}\cdot pd\pi^{inter}/(\hat{C}_{Mod}-\hat{C}_{Xp}) = 34$, 31, 33, 26, 24, 26, and 26 mRy for the compounds considered. Here, the values of $pd\pi^{intra}$ were given in Sect.6.3.4 and the values pf $pd\pi^{inter}$ will be given below. This weak-hybridization, or crystal-field, approximation, according to which the Mo d-band is assumed to hybridize with an X p-band complex whose width is small compared with its position, does not always hold. For the Γt_u level just below the dd gap it is necessary to consider the Mo d-X p hybridization explicitly at each $\underline{k}$ point. In such a case the upper limit to the shift of a Mo d-like level due to Mo d-X6 p intercluster coupling will be at the order of $pd\pi^{inter} = 6\sqrt{5}\,(d_{Mo-X}^{inter}/d_{Mo-Mo}^{intra})^{-4}$ $(\hat{\Delta}_{Mod}\hat{\Delta}_{Xp})^{1/2} = pd\pi^{intra}\,(d_{Mo-X}^{inter}/d_{Mo-X}^{intra})^{-4} = 107$, 85, 66, 85, 72, 76, and 85 mRy. Shifts that large will occur when the hybridizing p and d levels lie closer than about 100 mRy before hybridization. With the exception of the last mentioned case it is thus a reasonable approximation to describe the Mo d bands as Bloch sums formed from the molecular d-orbital states of the Mo_6 octahedron.

The computed energy bands and their projections onto the molecular-orbital states at the points Γ and R (Fig.6.11) show that the Bloch sums of the e_g^d and $t_{2u}^d \rightarrow a_u^d + e_u^d$ orbitals give rise to narrow bands whose widths do not exceed 50 mRy and that these are crossed by and hybridize with rather broad bands derived from

Table 6.8. Orbital characters of conduction-band states

		E			Mo_6			S_6			S_2			Mo d					
		s	p	d	s	p	d	s	p	d	s	p	d	$3z^2-1$	x^2-y^2	xy	xz	yz	
Γ	u	0	0	0	0	5	56	0	34	2	0	3	0	6	1	6	16	28	t_{1u}
R	g	0	0	0	0	1	81	0	12	4	0	1	0	2	9	14	29	28	t_{2g}
M	g	0	0	0	1	3	71	0	21	3	0	2	0	0	2	13	17	38	t_{2g}
	u	0	0	0	0	2	80	1	11	4	0	2	0	13	6	0	42	20	$t_{1u}(t_{2u})$
X	u	0	0	0	0	5	69	1	20	3	0	1	0	1	9	2	40	18	$t_{1u}(t_{2u})$
	g	0	0	0	0	3	79	0	13	2	0	2	1	3	62	1	4	9	e_g
CB		0.04	0.09	0.07	0	3	74	0	17	3	0	2	0						
E_F		0.05	0.09	0.06	0	2	79	1	13	3	0	2	0						
		E			Mo_6			Se_6			Se_2								
Γ	u	0	0	0	0	5	44	0	44	1	0	5	0	1	0	4	18	20	t_{1u}
R	g	0	0	0	0	0	79	0	9	3	0	8	0	12	38	9	9	11	e_g
M	g	0	0	0	0	4	68	0	21	2	0	3	0	2	3	12	13	38	t_{2g}
	g	0	0	0	0	0	80	0	8	2	0	9	0	1	72	4	2	1	e_g
X	g	0	0	0	0	3	78	0	11	1	0	6	1	3	72	0	2	0	e_g
	u	0	0	0	0	4	64	1	25	2	0	3	0	0	14	2	35	12	t_{1u},t_{2u}
CB		0.03	0.07	0.05	0	3	71	0	18	2	0	5	0						
E_F		0.10	0.13	0.06	0	2	75	0	12	2	0	7	0						
		E			Mo_6			Te_6			Te_2								
Γ	u	0	0	0	0	6	33	0	49	2	0	9	1	7	0	9	11	6	t_{1u}
R	g	0	0	0	0	1	65	0	15	3	0	15	1	24	33	5	1	2	e_g
M	g	0	0	0	1	2	58	1	20	2	0	15	0	16	33	3	0	6	e_g
	g	0	0	0	0	2	71	0	18	2	0	7	1	4	62	4	0	0	e_g
X	g	0	0	0	0	3	67	1	12	2	0	14	1	4	58	3	1	0	e_g
	u	0	0	0	1	6	35	1	42	2	0	12	1	1	6	3	25	0	t_{1u}
CB		0.03	0.05	0.03	0	3	57	1	24	2	0	12	1						
E_F		0.14	0.11	0.07	0	4	52	1	29	2	0	11	1						
		Pb			Mo_6			S_6			S_2								
Γ	u	0	4	0	1	5	52	0	35	2	0	2	0	4	1	3	13	31	t_{1u}
R	g	0	4	0	0	0	78	0	8	3	0	6	1	11	49	6	5	6	e_g
M	g	1	0	1	0	2	73	0	16	2	0	4	1	1	47	5	9	11	e_g
	g	0	0	1	0	2	78	0	10	2	0	5	1	0	67	3	4	3	e_g
X	g	0	0	0	0	3	78	0	11	2	0	4	1	3	72	0	1	2	e_g
	u	0	0	1	0	4	71	0	20	2	0	1	0	2	24	1	28	16	t_{1u},t_{2u}
CB		0.34	0.84	0.35	0	3	74	0	15	2	0	4	1						
E_F		0.36	0.84	0.44	0	3	73	0	15	2	0	4	1						

Table 6.8. (continued)

		E			Mo_6			S_6			S_2			Mo d					
		s	p	d	s	p	d	s	p	d	s	p	d	$3z^2-1$	x^2-y^2	xy	xz	yz	
			Pb			Mo_6			Se_6			Se_2							
Γ	u	0	4	0	1	4	41	0	43	1	0	6	0	2	1	3	11	24	t_{1u}
R	g	0	4	0	0	0	73	0	10	3	0	8	1	13	49	7	2	2	e_g
M	g	1	0	0	0	1	75	0	12	2	0	8	1	1	72	0	0	1	e_g
	g	0	0	1	0	4	72	0	15	2	0	5	1	13	38	4	7	10	$e_g(t_{2g})$
X	g	0	1	0	0	3	75	0	13	2	0	6	1	3	70	0	0	1	e_g
	u	0	0	1	0	3	67	0	23	2	0	3	1	1	30	2	24	10	t_{2u}
CB		0.23	1.0	0.31	0	3	70	0	16	2	0	6	1						
E_F		0.21	1.4	0.41	0	3	69	0	17	2	0	7	1						
			Ba			Mo_6			S_6			S_2							
Γ	u	0	1	0	1	5	46	0	43	1	0	4	0	4	1	1	12	28	t_{1u}
R	g	0	2	0	0	0	76	0	11	2	0	8	1	16	58	1	1	1	e_g
M	g	2	0	2	0	1	75	0	10	2	0	6	1	2	71	1	1	1	e_g
	g	0	0	3	0	3	74	0	12	2	0	5	1	10	57	1	1	4	e_g
X	g	0	0	0	0	3	78	0	12	1	0	5	1	4	73	0	0	0	e_g
	g	0	0	0	0	4	73	0	14	2	0	5	1	0	68	4	0	1	e_g
CB		0.27	0.38	0.90	0	2	74	0	15	1	0	5	1						
E_F		0.42	0.49	1.5	0	2	74	0	14	2	0	6	1						
			Gd			Mo_6			S_6			S_2							
Γ	u	0	1	0	1	5	50	0	39	1	0	3	0	4	1	4	12	29	t_{1u}
R	g	0	1	0	0	0	80	0	9	2	0	6	1	13	51	6	5	5	e_g
M	g	0	0	3	0	0	79	0	10	2	0	5	1	1	66	10	2	1	e_g
	g	0	0	4	0	4	69	0	13	2	0	4	1	9	9	3	5	43	$t_{2g}(e_g)$
X	g	0	0	0	0	3	77	0	12	1	0	5	1	3	71	0	0	2	e_g
	g	0	0	0	0	4	73	0	12	3	0	6	1	1	60	3	0	9	e_g
CB		0.10	0.27	1.5	0	3	73	0	14	2	0	5	1						
E_F		0.08	0.26	1.1	0	3	74	0	15	2	0	4	1						

the $t_{1u}^d \to a_u^d + e_u^d$ and the $t_{2g}^d \to a_g^d + e_g^d$ orbitals. The conduction band is the doubly degenerate e band just below the dd gap. In Mo_6S_8 it is composed of the top of the $t_{1u}^d \to e_u^d$ band at Γ and of the top of the $t_{2g}^d \to e_g^d$ band at R. At Γ the conduction band has t_{1u} character for all the compounds considered. In R the character is t_{2g}^d-like only in Mo_6S_8, and in all other compounds it is e_g^d-like. The e_g^d character, in fact, dominates the conduction band throughout most of the Brillouin zone, except in Mo_6S_8. This can be seen from the wave-function characters listed in Table 6.8 and from the conduction bands shown in Fig.6.17.

The reason why the e_g^d and t_{2u}^d bands are narrow is that for these bands the (x^2-y^2) character dominates. Since the (x^2-y^2) function has angular momentum ±2 about the z axis, it cannot couple to any p function centered at the z axis. Therefore, if the e_g^d and t_{2u}^d bands had pure (x^2-y^2) character, if the crystal structure were regular, and if the pd interaction were limited to nearest neighbors, then these bands could not be broadened by intercluster pd coupling. The lack of $(3z^2-1)$ character in the e_g^d band is due to the pd interaction itself. The hybridization with the low-lying X p-band complex shifts the $(3z^2-1)$ character from the bonding to the antibonding d bands. This lack of $(3z^2-1)$ character is demonstrated in Table 6.9 for the Γ point where the bonding and antibonding e_g^d levels lie particularly close and the effect therefore is pronounced. The intercluster Mo-X interaction proceeds via the X6 atoms. The suspicion that the e_g^d band lacks this interaction is furthermore supported in Tables 6.8,9 by the fact that the e_g^d states, in contrast to all t_{1u}^d and t_{2g}^d states, have less than three times as much p-charge density on the six X6 atoms as on the two X2 atoms.

Table 6.9. Orbital character of the bonding e_g-state at Γ

	M			Mo_6			X_6			X_2			Mo d				
	s	p	d	s	p	d	s	p	d	s	p	d	$3z^2-1$	x^2-y^2	xy	xz	yz
Mo_6S_8	0	0	0	1	4	77	1	8	3	0	5	1	0	74	2	1	0
Mo_6Se_8	0	0	0	0	4	76	1	8	3	0	7	1	1	67	7	0	1
Mo_6Te_8	0	0	0	1	4	63	2	9	3	0	17	1	1	42	17	0	3
$PbMo_6S_8$	0	0	1	0	4	75	1	9	3	0	5	1	1	71	2	1	0
$PbMo_6Se_8$	0	0	1	0	4	74	1	9	3	0	6	1	2	67	4	0	1
$BaMo_6S_8$	0	0	2	0	4	74	1	9	2	0	6	1	1	70	2	1	0
$GdMo_6S_8$	0	0	3	0	4	74	1	8	3	0	5	1	2	69	2	0	0

The mixing of t_{1u}^d and t_{2g}^d character into the e_g conduction band was not anticipated in the early crystal-field treatment of ANDERSEN et al. [6.8]. This mixing is partly responsible for the variation in shape of the conduction band between the binary and the ternary compounds and hence for some of the variation in their physical and structural properties. Before considering the conduction band as a whole, we shall try to explain the origins of this mixing.

6.4.3 The t_{1u} and e_g^d Levels at Γ

The energy difference between the $t_{1u}^d \rightarrow e_u^d$ levels in Γ and R is as large as 0.2 Ry in Mo_6S_8 and 0.15 Ry in $PbMo_6S_8$. As judged from the position of the $t_{1u}^d \rightarrow e_u^d$ cluster level, the upwards shift in Γ is larger than the downwards shift at R (Fig.6.11). As a consequence of these two facts the reason for the high-lying t_{1u}^d level at Γ can only be intercluster coupling with a high-lying X p-like e_u level at Γ. This is confirmed by the conduction-band characters listed in Table 6.8. The t_{1u}^d level in Γ has only about 50% Mo d character, as much as 40% X6 p character, and considerably less than 40/3% X2 p character. This is in distinct contrast to the only 9% X6 + X2 p character listed in Table 6.6 for single clusters. By projection of the X p part of the wave function onto the molecular p orbitals of the cube (Figs. 6.5,6) it turns out that the hybridizing level has dominant t_{1u}^p character too, more specifically antibonding p character, i.e., t_{1u}^{Ap} character. By projection of the X p-like levels onto the molecular orbitals of the octahedron and the cube, the Γe_u level at about -0.89 Ry in Mo_6S_8 and $PbMo_6S_8$ (Fig.6.11) is found to be responsible for the strong hybridization. For a single cluster the corresponding t_{1u}^{Ap} level is nonbonding not only with respect to pp bonding in the cube but also with respect to pd bonding between the cube and the octahedron. This was discussed in Sects.6.3.1,4 and was illustrated in Fig.6.4. The questions are therefore: why does the X p t_{1u}^{Ap} band of the crystal have such a high energy in Γ, and what is the strength of the pd intercluster coupling?

From Fig.6.6 it is seen that the t_{1u}^{Ap} molecular orbital is the only one for which the p orbitals have the same direction on all X atoms. This means that the dispersion of the t_{1u}^{Ap} band near Γ is dominated by long-range interactions. In the atomic sphere approximation the pp interaction is simply a dipole-dipole interaction [see (6.6) or Table 6.5]. This means that for a given direction of the Bloch vector there will be an e band with transverse polarization of the p orbitals and an a band with longitudinal polarization of the p orbitals. Near Γ the former, which is of interest to us, does not hybridize with any s orbitals, and as shown in [6.23] its energy with respect to $\hat{C}_{Xp}$ is simply $6s_{WS}^{-3}$ in units of $(d_{Mo-Mo}^{intra})^3\hat{\Delta}_{Xp}$. Here $(4/3)\pi s_{WS}^3$ is the volume per X atom. In the present case of 8 X atoms per cell, we may conveniently express the energy of this Γe_u level in terms of the rhombohedral cell volume V as

$$E(\Gamma t_{1u}^{Ap} \rightarrow e_u) = \hat{C}_{Xp} + (64\pi/V)(d_{Mo-Mo}^{intra})^3\hat{\Delta}_{Xp} \quad . \tag{6.12}$$

(We should, of course, remember that there is no explicit dependence on d_{Mo-Mo}^{intra} which is simply our chosen unit of length; $(d_{Mo-Mo}^{intra})^3\hat{\Delta}_{Xp}$ is independent of d_{Mo-Mo}^{intra}.) The $t_{1u}^{Ap} \rightarrow e_u$ level at Γ does not therefore depend on the detailed arrangement of the X atoms but merely on the volume per atom. The pp interaction between an X2 atom and its 9 *nearest* neighbors or between an X6 atom and its 11 *nearest* neighbors is in fact very close to zero for all seven compounds. Since V equals a^3 to almost three significant figures for the range of rhombohedral angles considered, we find from Tables 6.2,3 that the unhybridized Γt_{1u}^{Ap} level lies 220, 228, 228, 208, 218, 188, and 214 mRy above $\hat{C}_{Xp}$ or at the energies -107, -24, +78, -124, -56, -100, and -109 mRy with respect to C_{Mod} for our seven compounds. The Γt_{1u}^{Ap} level thus lies about 100 mRy lower than the top of the unhybridized X p-band complex estimated in the previous section. (The topmost e_u^p level seen in Fig.6.4 gets pushed down below the center of the p band by the strong intracluster coupling with the Mo e_u^{xy} band.)

We must now estimate the position of the unhybridized bonding Mo $t_{1u}^d \to e_u$ level at Γ. According to the results of the full band-structure computations in Table 6.8, the d-orbital character of this level is almost pure xz/yz and the strong $3z^2-1$ character found for single clusters (Table 6.4) is suppressed. For our estimates in the following it is therefore more accurate, and simpler as well, to use just the xz and yz orbitals, rather than the bond orbitals, combined into t_{1u}^d molecular orbitals as indicated in Figs.6.6,8. The intr*a*cluster interaction for a regular octahedron is seen, for instance, from the first t_{1u}^d partner function in Fig.6.6 and Table 6.5 to be -50 in units of $\hat{\Delta}_{Mod}$. (The single-cluster result from Sect.6.3.2 was -65 and the result of the bond-orbital model in Sect.6.3.3 was -54.) The upwards shift of the e_u^d level due to the elongation of the octahedron was previously estimated to be 300∂ = 9, 8, 4, 3, 2, 2, and 3, the center of the e_u^d band therefore being $(-50 + 300\partial)\hat{\Delta}_{Mod}$ = -65, -67, -77, -85, -83, -90, and -88 mRy below C_{Mod}. The dd intercluster interaction is $S(\chi^*, -\chi; \underline{R}_{Mo-Mo}^{inter})$, where χ is the normalized atomic d orbital on site (0,0,b/2) obtained from the three partner functions in Fig.6.6 by taking the first function, plus $\exp(m\cdot 2\pi i/3)$ times the second function, plus $\exp(-m\cdot 2\pi i/3)$ times the third function. For an e band m = +1 or -1, and the partner function shown in the second column of Fig.6.8 is the imaginary part of the linear combination just mentioned. Hence

$$\chi^m(t_{1u}^d) = [xz\,\exp(2\pi im/3) + yz\,\exp(-2\pi im/3)]/\sqrt{2} \quad \text{and} \tag{6.13}$$

$$H_{dd}^{inter}(t_{1u}^d) = \frac{1}{2}\,[-S(xz,xz) - S(yz,yz) + S(yz,xz)]\hat{\Delta}_{Mod} = 17,\ 11,\ 7,\ 13,$$

9, 13, and 15 mRy, as obtained from Tables 6.2,5. These small values mainly reflect the trend of $(d_{Mo-Mo}^{inter})^{-5}$. The level at Γ now lies H^{inter} above the center of the e_u^d band, and its energy is therefore -48, -56, -70, -72, -74, -77, and -73 mRy for the seven compounds.

Comparison between the above estimated positions of the unhybridized X p and Mo d $t_{1u} \to e_u$ levels in Γ reveals that the levels are nearly degenerate and that for the selenides and the telluride the X p level lies above the Mo d level. This is in accord with the computed wave-function characters given in Table 6.8. The matrix element for X p-Mo d intercluster hybridization is $\sqrt{(6/8)}S(\chi_d^*,\chi_p;\underline{R}_{Mo-X}^{inter})$, where χ_p is the normalized atomic p orbital on site (1,-1,-1)b/2 obtained from the partner functions in Fig.6.6 as described above. The factor $\sqrt{(6/8)}$ appears

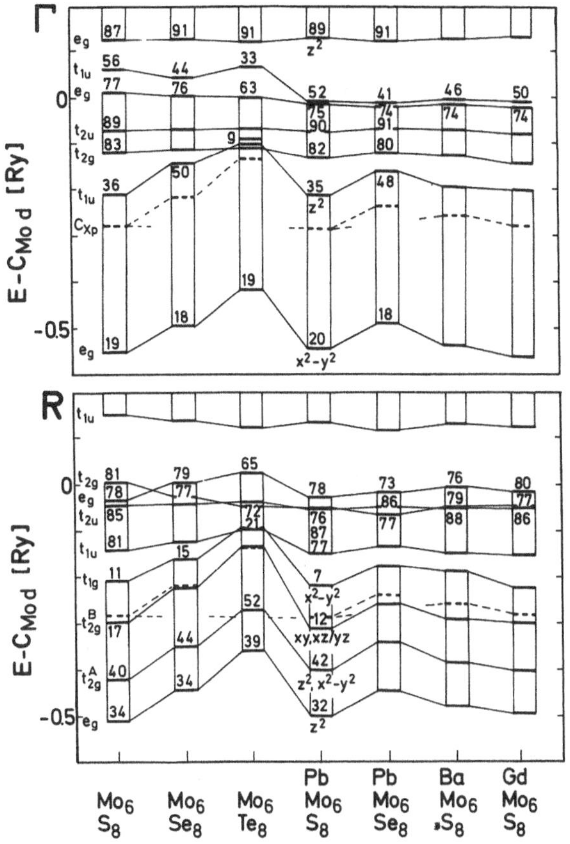

Fig.6.12. Some of the doubly degenerate e levels at the points Γ and R in the Brillouin zone

because only 6 out of the 8 X atoms take part in the intercluster coupling with the Mo atoms. In the present case

$$\chi_p = \chi^m(t_{1u}^{Ap}) = [z + x\,\exp(2\pi i m/3) + y\,\exp(-2\pi i m/3)]/\sqrt{3} \tag{6.14}$$

and using Table 6.5 the matrix element is seen to equal $H_{pd}^{inter}(t_{1u}) = [1 + l_{Mo\text{-}X}^{inter} + m_{Mo\text{-}X}^{inter} \pm i\sqrt{3}(l_{Mo\text{-}X}^{inter} - m_{Mo\text{-}X}^{inter})](-pd\pi^{inter})/\sqrt{2}$ for $m = \pm 1$ and to first order in the deviation from the regular structure defined in Sect.6.2.1. With the values of $l_{Mo\text{-}X}^{inter}$ and $m_{Mo\text{-}X}^{inter}$ given in Table 6.2 and the values of $pd\pi^{inter}$ given previously, we find $|H_{pd}^{inter}(t_{1u})|$ = 83, 65, 50, 71, 61, 66, and 71 mRy for the seven compounds. These values, with the exception of the one for Mo_6Te_8, are larger than half the energy difference between the unhybridized levels. Our estimate of the uppermost $t_{1u} \rightarrow e_u$ level at Γ finally becomes:

$$(E_p + E_d)/2 + [|H_{pd}|^2 + |E_p - E_d|^2/4]^{1/2} = 10,\ 27,\ 93,\ -22,\ -3,\ -22,\ \text{and}\ -18\ \text{mRy}.$$

The computed $t_{1u} \rightarrow e_u$ levels at Γ are shown in the upper half of Fig.6.12 together with the remaining bonding Mo d-like e levels and the only other X p-like e level (the e_g level at the bottom of the X p band) which hybridizes significantly with the Mo d-like bonding levels. In addition the antibonding e_g^d level just above the dd gap is shown. The numbers on the levels give the Mo d character in per cent and the dominant d-orbital character has been indicated. The broken lines give the position, C_{Xp}, of the center of the unhybridized X p band. For Mo_6Te_8 the e_g level lying between the t_{1u} and t_{2u} levels and marked g is almost purely X p-like. Our estimate for the position of the uppermost t_{1u} level is seen to reproduce the computed trend from the binaries to the ternaries reasonably well, although the trend from the sulfides to the selenides is not reproduced. This may be due to our neglect of the interactions beyond the Mo-X nearest neighbors and to our neglect of the $(3z^2-1)$- and xy-orbital characters evident from Table 6.8 and hence to an underestimation of the pd-hybridization matrix. Important factors causing the t_{1u} level to lie low with respect to the Mo d-like levels are: a low-lying C_{Xp}, a large cell volume, a large Mo-X intercluster distance, a small value of $l_{Mo-X}^{inter} + m_{Mo-X}^{inter}$, and a small and regular octahedron. The Mo-Mo intercluster distance seems to be of less importance.

An additional factor which can supress the Γt_{1u}^d level and also the Re_g^d-level is direct hybridization with the M p Γe_u-Re_g band. From Table 6.8 it appears that the Pb compounds have as much as 4% Pb p character in the Γt_{1u} and Re_g states but none at the points M and X of the Brillouin zone. Since the Pb p e band lies only 0.4 Ry above the conduction band, the latter is shifted downwards by about 16 mRy when the Bloch vector lies along the $\bar{3}$-axis. This may be seen by comparing the energy bands of $PbMo_6S_8$ shown in Fig.6.11 with those of "Pb"Mo_6S_8 shown in Fig.6.13. The crystal potential for the latter band structure was constructed using the atomic positions of $PbMo_6S_8$ but omitting the Pb atoms. Apart from the above mentioned downwards shift of the conduction band along the $\bar{3}$-axis and from changes in the structure of the $t_{1u}^d \to a_u$ band, which has 10% Pb p character, the bands of $PbMo_6S_8$ are quite similar to those of "Pb"Mo_6S_8. This agrees with the findings of BULLETT [6.11] that the function of Pb is primarily a passive one of distorting the Mo_6X_8 cluster.

The e_g^d level at Γ can be seen from Fig.6.12 and Table 6.9 to be rather purely (x^2-y^2)-like and not to hybridize with any particularly high-lying X p level. In fact, the essential pd hybridization is the intracluster interaction with the low-lying e_g^p level. The center of gravitiy of the pure $e_g(x^2-y^2)$ band is therefore essentially the intracluster dd energy $-59\Delta_{Mod}$ = -94, -94, -99, -107, -102, -111, and -111 mRy obtained in Sect.6.3.2 plus the intracluster dpd shift of $(2pd\pi^{intra})^2/(E_d-E_p) \approx 90$ mRy for all seven compounds. In the latter expression the value of the nominator was obtained in Sect.6.3.4 and the value of the denominator was taken from Fig.6.12 to be 500, 460, and 400 mRy for the sulfides, selenides, and the telluride, respectively. The center of gravity thus follows the slowly decreasing trend caused by the contraction of the octahedron.

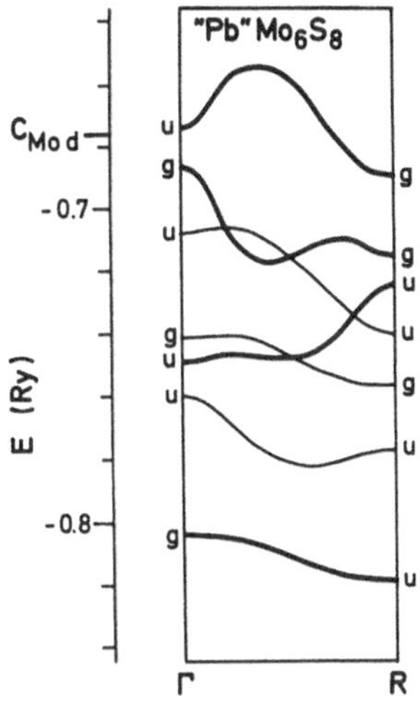

Fig.6.13. Energy bands for $PbMo_6S_8$ without Pb

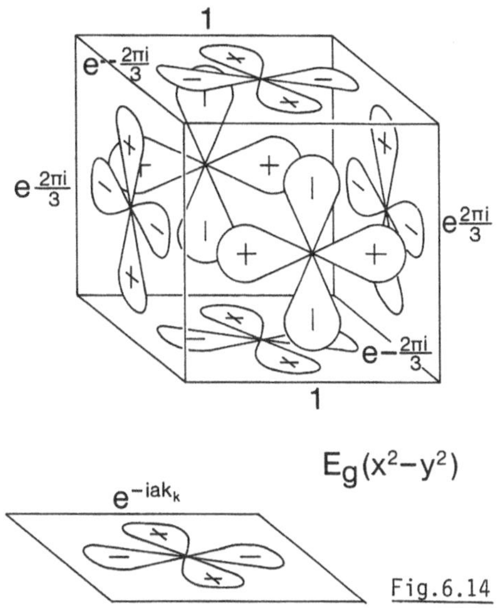

Fig.6.14

Fig.6.14. Symmetry of the $e_g(x^2-y^2)$ state with m = -1

The $e_g(x^2-y^2)$ orbital is shown in Fig.6.14 and its real and imaginary parts, including the $3z^2-1$ contribution according to the bond-orbital model, were shown in Fig.6.5. The e_g^d-bond orbital may be included in Fig.6.14 if the atomic orbital

$$\chi(e_g) = (x^2 - y^2) \tag{6.15}$$

is substituted by

$$\chi(e_g) = [(x^2 - y^2) + (3z^2 - 1)i/\sqrt{3}]/\sqrt{(4/3)} \quad . \tag{6.16}$$

The latter form will only be relevant around the point R where according to Table 6.8 the $(3z^2-1)$ contribution may be important. The intercluster dd interaction may now easily be obtained from Table 6.5 using the values of R^{inter}_{Mo-Mo} given in Table 6.2 and we obtain

$$H^{inter}_{dd}(e_g) = 27,\ 22,\ 20,\ 25,\ 21,\ 23,\ \text{and}\ 24\ \text{mRy} \quad , \tag{6.17}$$

which is dominated by the trend of $(d^{inter}_{Mo-Mo})^{-5}$. The results (6.17) were obtained using the (x^2-y^2) orbital only. The results obtained with the bond orbital (6.16) are 31, 26, 23, 30, 26, 29, and 29 mRy, essentially the same. Since H^{inter} is positive, the e_g^d level at Γ lies *above* the center of gravity by the amounts given in (6.17).

The position of the e_g^d level at Γ may, in addition to the decreasing trends caused by the contraction of the octahedron and the increasing Mo-Mo intercluster distance, be lowered through interaction with (high-lying) d levels on the ternary component. The 2-3% Ba or Gd d character evident from Table 6.9 pushes the e_g^d band down by about 10 mRy at Γ.

6.4.4 The t_{2g}^{d} and e_{g}^{d} Levels at R

In the lower half of Fig.6.12 we show for the point R the following e-levels. The lowest Mo d-like antibonding level t_{1u}^{d}, the Mo d-like bonding levels, and those of the X p levels which hybridize with the e_{g}^{d} and t_{2g}^{d} Mo d-like bonding levels. From the numbers given in the figure the e_{g}^{d} and t_{2g}^{d} levels are all seen to be rather purely d-like. In Γ and R the parity is conserved but all e_g levels (and all e_u levels) may in principle hybridize with each other. The e_{g}^{d} and t_{2g}^{d} levels were far apart in Γ, but in R they are close together and interact. This is one reason for the seemingly strange trend followed by these two levels in the binaries. We now first consider the t_{2g}^{d} and e_{g}^{d} levels in the absence of their interaction.

The t_{2g}^{d} level, in contrast to the t_{1u}^{d} level at Γ, does not hybridize with any high-lying X p level. Its center of gravity is essentially given by the dd intracluster interaction of about $-96\Delta_{Mod}$ obtained in Sect.6.3.3 plus the dpd intracluster interaction with the t_{2g}^{Ap} band of about $(\sqrt{2}pd\pi^{intra})^2/(\hat{C}_{Mod} - \hat{C}_{Xp}) \approx 70$ mRy for all compounds (Sect.6.3.4). There is, in addition, some dpd interaction through the xy contribution to the t_{2g}^{d} orbital evident from Table 6.8, as well as dpd interaction with the t_{2g}^{Bp} band as may be realized from Fig.6.12; this must be due to the part of the interaction reaching beyond nearest neighbors. The center of gravity of the t_{2g}^{d} band thus decreases slightly as the octahedron contracts and becomes more regular. The large width of the t_{2g}^{d} band, which exceeds 100 mRy, is due to intercluster coupling via the X6 atoms. This follows from the facts that the X6 character of the t_{2g}^{d} band is clearly more than three times the X2 character (Table 6.8) and that the direct dd intercluster coupling constant is just the negative (band minimum in Γ, maximum in R) of the one estimated above for the t_{1u}^{d} band and hence of the order of only -15 mRy. The indirect, dpd part of the intercluster coupling (6.11) may be estimated in the crystal-field approximation where the energy denominator $E_d - E_p$ is taken to be the same for all the intermediate X p states and equal to $\hat{C}_{Mod} - \hat{C}_{Xp}$. The nominator, before summing over $\underline{T}$, is therefore

$$\sum_i \langle\psi_d(\underline{r})|H|\psi_{pi}(\underline{r}-\underline{T})\rangle\langle\psi_{pi}(\underline{r}-\underline{T})|H|\psi_d(\underline{r}-\underline{T})\rangle$$

$$+ \sum_i \langle\psi_d(\underline{r})|H|\psi_{pi}(\underline{r})\rangle\langle\psi_{pi}(\underline{r})|H|\psi_d(\underline{r}-\underline{T})\rangle \quad ,$$

which is twice the real part of the first term. In the nearest-neighbor approximation this is the hopping from a Mo atom (Fig.6.1) onto the nearest X6 atom in the neighboring cluster and subsequently from the X6 atom onto the three nearest Mo atoms in the same cluster. Using the atomic orbital

$$\chi^m(t_{2g}^{d}) = -[yz\,\exp(2\pi im/3) + xz\,\exp(-2\pi im/3)]/\sqrt{2} \tag{6.18}$$

on the (0, 0, b/2) site and using the notation $e^i \equiv \exp(2\pi im/3)$ $\pi^e \equiv pd\pi^{inter}$, and $\pi^a \equiv pd\pi^{intra}$, we find for the nominator

$$2\mathrm{Re}(\sqrt{2})^{-1}\{e^{i}\pi^{e}[(\pi^{a}/\sqrt{2}) + (-\pi^{a}/\sqrt{2})e^{-i}]$$

$$+ e^{-i}\pi^{e}[(-\pi^{a}/\sqrt{2}) + (-\pi^{a}/\sqrt{2})e^{i}]\}(\sqrt{2})^{-1} = -\sqrt{2}\cdot \mathrm{pd}\pi^{\mathrm{inter}}\cdot \mathrm{pd}\pi^{\mathrm{intra}} \quad , \qquad (6.19)$$

where for simplicity the regular crystal structure has been assumed. The value of (6.19) is $-\sqrt{2}$ times the average value given in Sect.6.4.2. This means that the intercluster dpd coupling constant takes the values $H^{\mathrm{inter}}_{\mathrm{dpd}}(t^{d}_{2g}) = -48, -44, -47, -37, -34, -37$, and -37 mRy for the seven compounds. The dd and dpd contributions thus add and the width, $2|H^{\mathrm{inter}}_{\mathrm{dd}} + H^{\mathrm{inter}}_{\mathrm{dpd}}|$, of the t^{d}_{2g} band is consequently of the order of 100 mRy. The top of the band is at R and it follows a slowly decreasing trend from Mo_6S_8 to $GdMo_6S_8$, as long as the hybridization with the e^{d}_{g} band has not been accounted for.

The $e_g(x^2-y^2)$ band was considered in the previous two sections, where it was concluded that due to the lack of intercluster pd interaction, the width is only $2H^{\mathrm{inter}}_{\mathrm{dd}}(\mathrm{eg}) \approx 50$ mRy, the maximum is at Γ, and the minimum at R. This is not in accord with Figs.6.12,17, which show the R level to lie above the Γ level in many of the compounds, most notably in Mo_6Te_8. Tables 6.8,9 show that the $(3z^2-1)$ contribution to the e^{d}_{g} orbital is anomalously high at R (and sometimes at M) where the orbital resembles the bond orbital (6.16). This orbital couples strongly with the nearest X atoms in the neighbor cluster. To get an idea of the sign and order of magnitude of this coupling, we can calculate the dpd contribution to the intercluster coupling in the crystal-field approximation, as we did for the t^{d}_{2g} orbital. The interaction thus proceeds from the $(3z^2-1)$ part of the e^{d}_{g} orbital to the z orbital on the X6 atom in the neighbor cluster and from here mainly onto the two of the three neighboring (x^2-y^2) orbitals in the same cluster. The resulting interaction is $-2.4\ \mathrm{pd}\pi^{\mathrm{inter}}\cdot \mathrm{pd}\pi^{\mathrm{intra}}$ and after dividing by $\hat{C}_{\mathrm{Mod}} - \hat{C}_{\mathrm{Xp}}$ we obtain $H^{\mathrm{inter}}_{\mathrm{dpd}}(e^{d}_{g}) = -80, -73, -78, -62, -57, -62$, and -62 mRy. This shows that 25% $(3z^2-1)$ character suffices to shift the e^{d}_{g} level at R upwards such that it lies above the level at Γ by the amounts $-H^{\mathrm{inter}}_{\mathrm{dpd}} - 2H^{\mathrm{inter}}_{\mathrm{dd}} = 26, 29, 38, 12, 15, 16$, and 14 mRy. If we then take the actual $(3z^2-1)$ contributions into account we can obtain the trend observed in Figs.6.12,17. But the question why the $(3z^2-1)$ contribution is so high, particularly at R, has not been answered. Figure 6.12 indicates that the contribution comes from the low-lying e^{p}_{g} and t^{Ap}_{2g} states.

In order to rule out the possibility that deviations from the regular structure, which allow intercluster $p(x^2-y^2)$ coupling, are important for the high-lying e^{d}_{g} level at R, we show in Fig.6.15 the band structure of Mo_6Te_8 with the real and regular crystal structure. While the Re^{d}_{g}-level is only weakly affected by the distortion, we call attention to the sensitivity of the $t^{Ap}_{1u} \to a_u$ level at Γ. We now realize that the structure of Mo_6Te_8 in fact deviates from the regular one in order to lower the a_u band and thereby create a gap at the Fermi level.

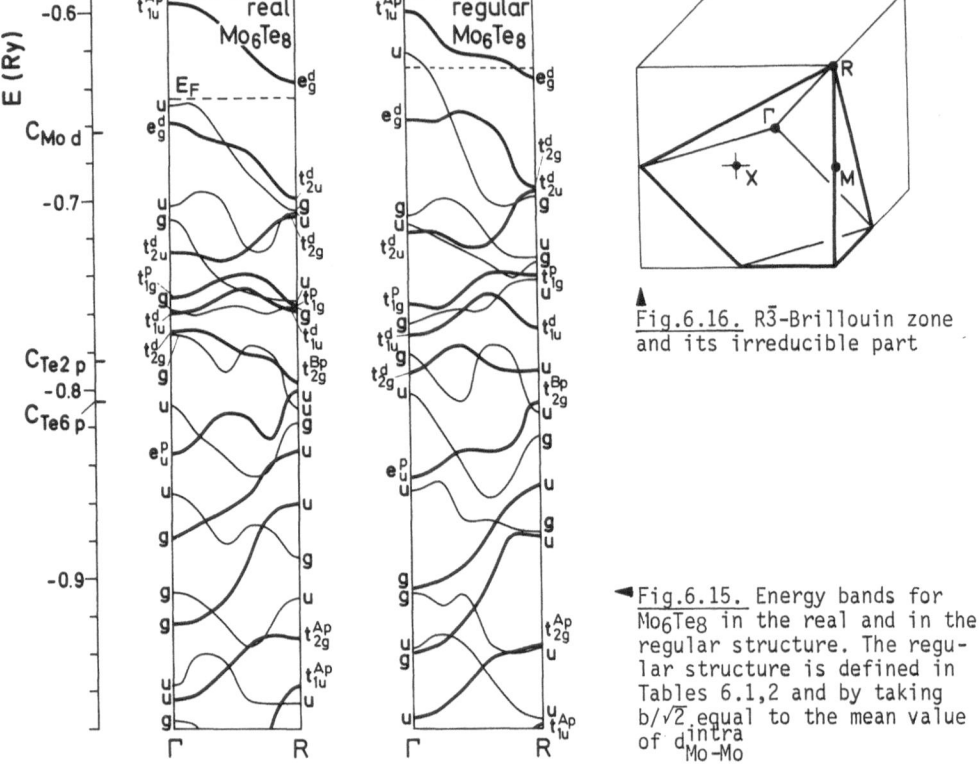

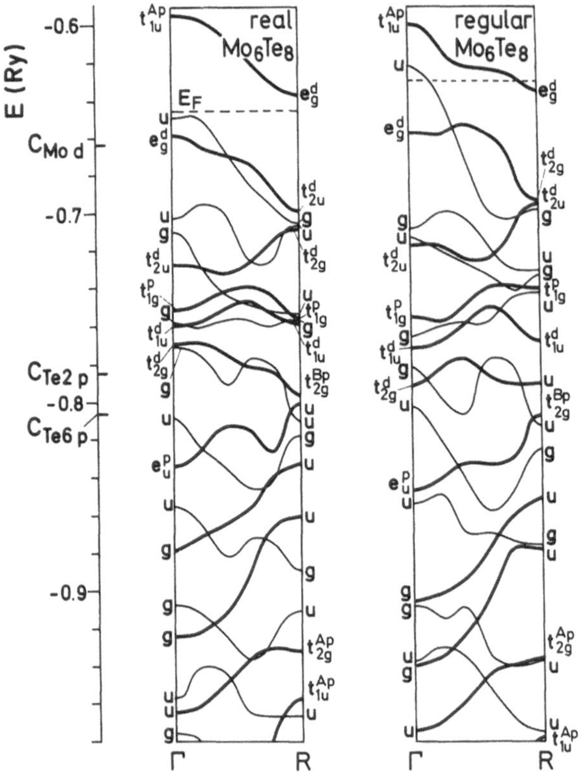

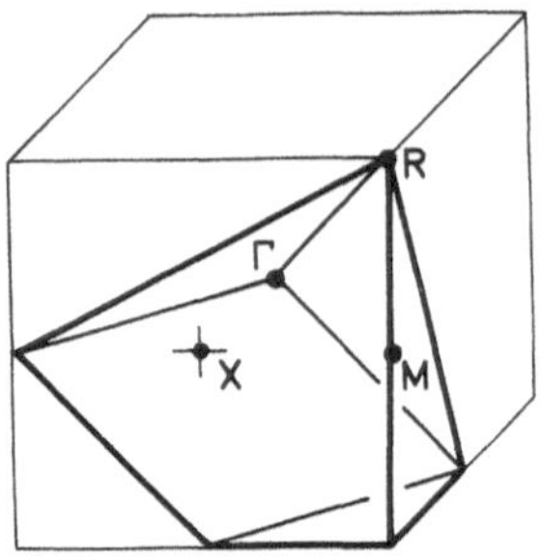

▲ Fig.6.16. R$\bar{3}$-Brillouin zone and its irreducible part

◄ Fig.6.15. Energy bands for Mo_6Te_8 in the real and in the regular structure. The regular structure is defined in Tables 6.1,2 and by taking $b/\sqrt{2}$ equal to the mean value of d^{intra}_{Mo-Mo}

6.4.5 The Conduction Band

The rhombohedral Brillouin zone, with its irreducible part emphasized, is shown in Fig.6.16. In Fig.6.17 we present along a number of lines in the Brillouin zone the conduction bands of our seven Chevrel-phase compounds.

The symmetry points Γ, R, X, and M are at the origin and at the points $(\underline{G}_i + \underline{G}_j + \underline{G}_k)/2$, $\underline{G}_i/2$, and $(\underline{G}_i + \underline{G}_j)/2$, respectively. Here the $\underline{G}$'s are the primitive vectors of the reciprocal lattice. All corners of the zone are R points, all face centers are X points and all edge centers are M points. There are, however, two different types of ΓR lines: the $\bar{3}$ axis (or [111]) along which there are two different representations, a(m = 0) and e(m = ±1); and the lines [11$\bar{1}$] along which there is no symmetry. Similarly, there are two types of ΓM lines, two types of MX lines, two types of RX lines, but only one ΓX and one RM line. Along all of these there is no symmetry. At Γ and R there are e_g, a_g, e_u, and a_u representations, as mentioned previously, and at X and M there are only g and u representations. In the pictures of the bands we have labelled the levels at the symmetry points according to the dominating cluster symmetry as explained in Sect.6.4.2, but it should be kept in mind that only g and u are good quantum numbers plus, at Γ and R, e and a.

In Table 6.8 we have given the orbital characters for the two states of the conduction band at the Γ-, R-, M- and X-points. At Γ and R the two states are degenerate. For the points M and X the orbital character of the state with highest energy appears in the table above that of the state with lowest energy. At the end of the table we have furthermore listed the orbital character averaged over the entire conduction band (CB) and the orbital character averaged over the Fermi surface (E_F). In terms of the total-, N(E), and the site- and orbital-projected densities of states, $N_{t\ell}(E)$, the orbital character averaged over the Fermi surface consists of the numbers $N_{t\ell}(E_F)/N(E_F)$, and the orbital character averaged over the conduction band consists of the numbers $[\int N_{t\ell}(E)dE]/4$, where the integral is over the energy range of the conduction band.

Below the bands of Mo_6Te_8 we have shown the model e band of ANDERSEN et al. [6.8]. We shall now derive it, introduce interactions with other bands, and use these results to explain the band structures computed for the real materials.

Let us first consider the intercluster matrix element, that is, the second term in (6.11), between two molecular-orbitals, ψ_1 and ψ_2, which we choose to have a definite parity (P = ±1) and a definite value of m(= ±1 or 0) about the $\bar{3}$ axis. We then express each molecular orbital as that linear combination of six congruent, normalized atomic orbitals χ^m, each one being associated with a face of the cube, which is given by the parity and the m value. Such atomic orbitals were previously given in (6.13-16,18). The intracluster matrix is diagonal in 1 and 2, and the intercluster matrix becomes

$$H_{12}(\underset{\sim}{k}) = H_{12}^{\text{inter}} \begin{cases} [K(\underset{\sim}{k}) + \omega_{12}^* I(\underset{\sim}{k}) + \omega_{12} J(\underset{\sim}{k})]/3 \ , & \text{for } P_1 = P_2 \\ i[\tilde{K}(\underset{\sim}{k}) + \omega_{12}^* \tilde{I}(\underset{\sim}{k}) + \omega_{12} \tilde{J}(\underset{\sim}{k})]/3 \ , & \text{for } P_1 \neq P_2 \end{cases} \tag{6.20}$$

if we neglect interactions between all other faces besides opposing faces of neighboring cubes (Figs.6.9,14). In the expression above $H_{12}^{\text{inter}} = \langle\chi_1^{m1}|\chi_2^{m2}\rangle$, where χ_1 is centered at the face (0, 0, b/2) and χ_2 is centered at the opposing face at $\underset{\sim}{T}_k$ + (0, 0, -b/2) and is inverted in accordance with the parity of ψ_2. Moreover, $I \equiv \cos(\underset{\sim}{k} \cdot \underset{\sim}{T}_i)$, $\tilde{I} \equiv \sin(\underset{\sim}{k} \cdot \underset{\sim}{T}_i)$, and analogously for J, K, $\tilde{J}$, and $\tilde{K}$. Finally, $\omega_{12} \equiv \exp[(m1-m2)2\pi i/3]$.

From (6.20) the dispersion of a singly degenerate a band becomes $H_{11}^{\text{inter}} \cdot (I + J + K)/3$. By diagonalizing the 2 × 2 matrix $\langle\psi_1^{m1}|H|\psi_1^{m2}\rangle$ the dispersion of a doubly degenerate e band becomes

$$H_{11}^{e}(\underset{\sim}{k}) = H_{11}^{\text{inter}}[I + J + K \pm \sqrt{(I^2 + J^2 + K^2 - JK - IK - IJ)}]/3 \ , \tag{6.21}$$

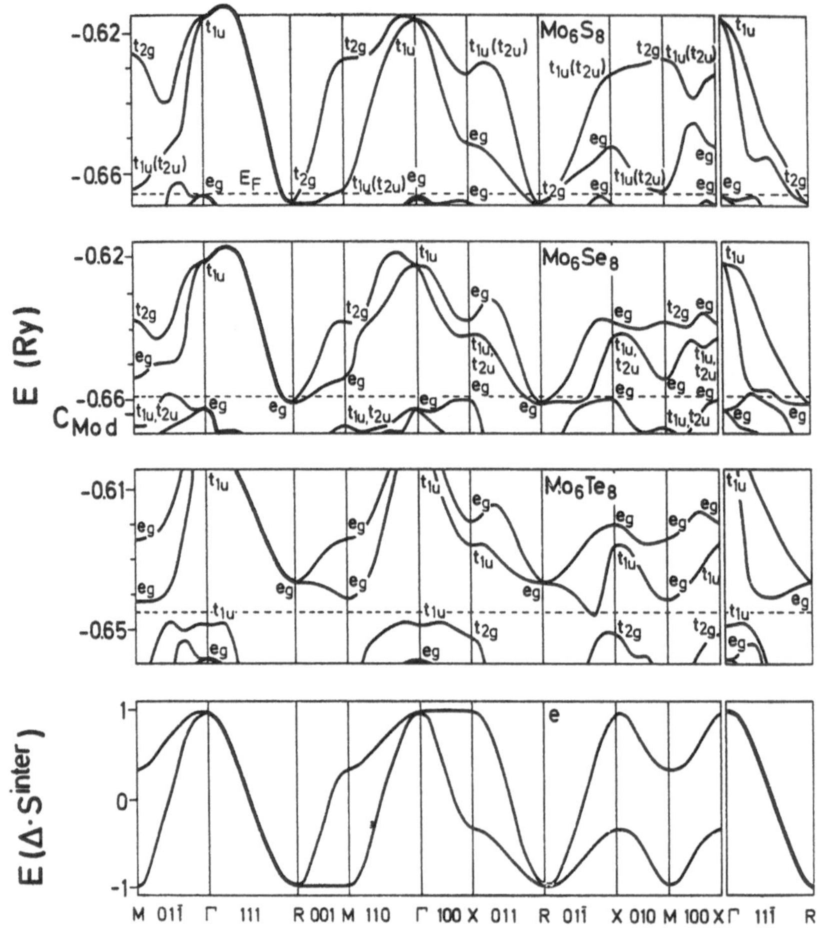

Fig.6.17a

Fig.6.17a,b. Conduction band of $Mo_6S_8 \rightarrow GdMo_6S_8$ Chevrel phases and the model e bands. Doubly degenerate bands are indicated by dark lines

which is our e model shown at the bottom of Fig.6.17 and at the top of Fig.6.18 in units of H_{11}^{inter}.

The unhybridized e band provides a first approximation to the conduction-band structure of the Chevrel phases. It was first derived in [6.8] as a model for the $e_g(x^2-y^2)$ band. It is accidental that its shape most resembles those of the least $e_g(x^2-y^2)$-like bands, i.e., the conduction bands of the binary compounds. The band structure of $BaMo_6S_8$ which is the most $e_g(x^2-y^2)$-like looks different, but after the discussion in the previous sections, we understand that it is essential to let the e_g band hybridize with the top of the broad t_{1u} band near Γ and with the top of the broad t_{2g} band, for which H^{inter} is negative, near R and M.

The matrix elements for hybridization between two e_g (or two e_u) bands are $H_{12}^{inter}(K + I + J)/3$ when m1 - m2 = 0, and $|H_{12}(\underline{k})| = |H_{12}^{inter}|\surd(I^2 + J^2 + K^2 -$

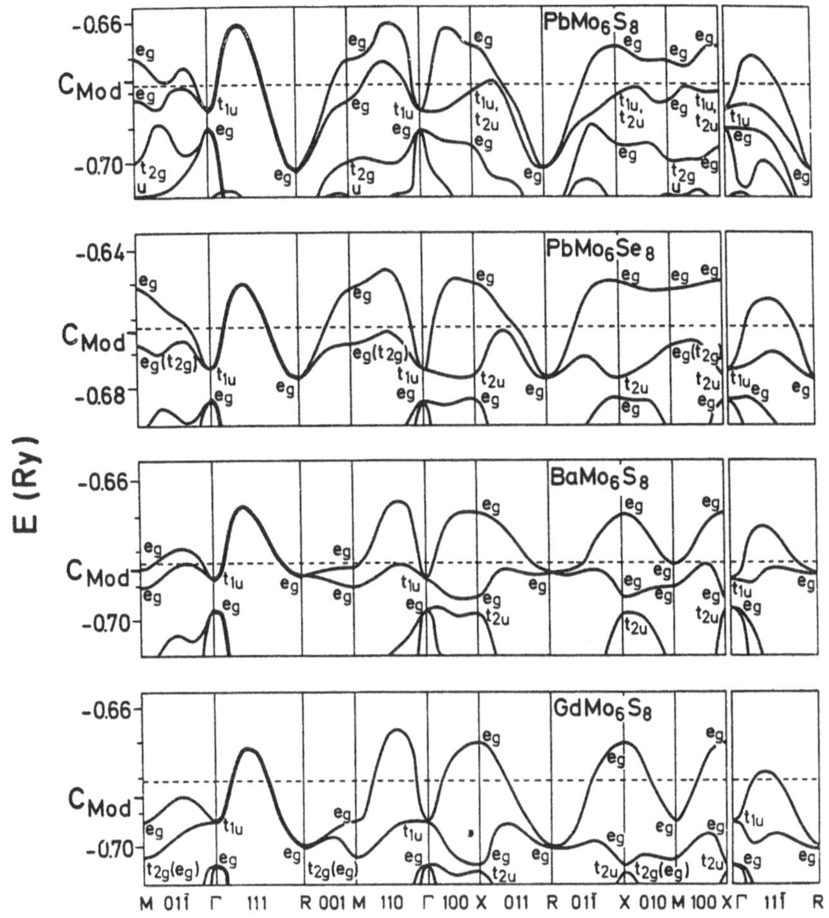

Fig.6.17b

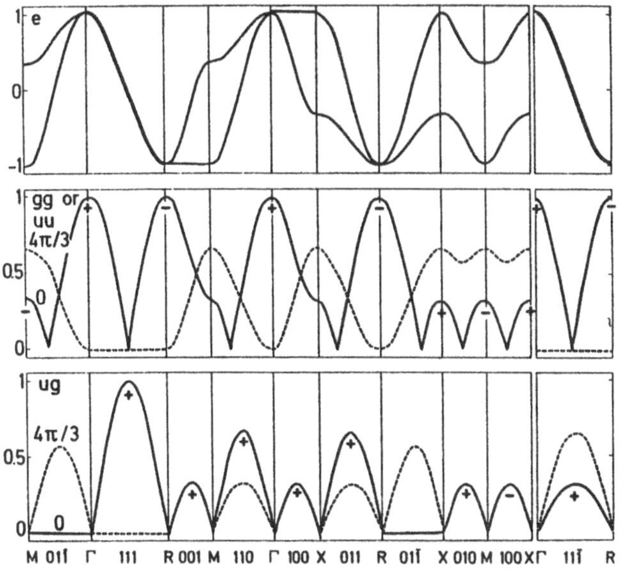

Fig.6.18. Model e band and interband hybridization matrix as defined in the text and in units of H^{inter}. The matrix elements for which m1-m2=0 and $|m1-m2| = 4\pi/3$ are indicated by, respectively, full and broken lines

$JK - IK - IJ)/3$ when $|m1 - m2| = 4\pi/3$. These functions are shown in the central part of Fig.6.18. For $\alpha = 90^{\circ}$, they have cubic symmetry like the unhybridized e bands. The matrix elements for hybridization between an e_g and an e_u band, however, give rise to anisotropy. They are given by the expressions above when I, J, and K are substituted by $\tilde{I}$, $\tilde{J}$, and $\tilde{K}$. These elements are shown in the bottom part of Fig.6.18.

Most important is the hybridization between the t_{1u} and the e_g band. The matrix element reaches its maximum, H_{12}^{inter}, along the $\bar{3}$ axis midway between Γ and R. The resulting behavior is clearly recognized in the calculated band structures. For the dd part alone and neglecting the p contribution to the t_{1u} orbital, we obtain, using (6.13,15), $|H_{dd}^{inter}(t_{1u},e_g)| = 20$ mRy for the binaries and 25 mRy for the ternaries considered. In addition to this comes a dpd = $t_{1u}^{d}\ e_g^{p}\ e_g^{d}$ part, these two together explaining the band maxima occurring for the ternaries between Γ and R. In Mo_6S_8 the top of the broad $t_{1u} \rightarrow e_u$ band, whose maxima are at Γ and X, combine with the top of the broad $t_{2g} \rightarrow e_g$ band, whose somewhat lower-lying maxima are at R and M, into a common e band. The hybridization matrix element along the $\bar{3}$ axis is $H_{dd}^{inter}(t_{1u},t_{2g}) = 15$ mRy. The hybridization between the t_{2g} and the e_g band reaches its maximum at R. The matrix element is rather small and varies between 10 and 20 mRy because it arises as the sum of a negative $t_{2g}^{d}\ e_g^{d}$ part of -20 mRy and a positive $t_{2g}^{d}\ e_g^{p}e_g^{d}$ part which varies between 30 and 40 mRy.

The anisotropy caused by the hybridization between even and odd bands is clearly brought out in the calculated band structures. The Fermi surfaces of Mo_6S_8 and Mo_6Se_8 consist of two electron pockets at R and hole pockets at general points in the zone. Mo_6Te_8 is a semimetal. The Fermi surface of $PbMo_6S_8$ consists of a small electron sphere at Γ, a large electron "octahedron" at R whose axes are turned by 60° around the $\bar{3}$-axis away from the [100]-directions, and two hole pockets lying between the electron spheres and the electron octahedra and touching them along the $\bar{3}$-axis. The hole pockets do not extend to the other six ΓR lines.

The densities of states for Mo_6S_8 through $GdMo_6S_8$ are shown in Fig.6.19 for the 60 mRy range which contains the conduction band. The white part indicates the Mo d and the black part the non-Mo d contribution to the total density of states. At the bottom left of the figure are shown the density of states for the model e band (6.21) as well as the contributions from its two subbands which are symmetrical in $\underline{k}$- and energy space. The shape consists of steep edges arising from the flat bands along RM and ΓX, and two plateaus separated by a valley. This shape, despite individual distortions, can be found in the conduction-band densities of states for the real materials. The most striking difference is the one between the binary and ternary densities of states, this being due to the different positions of the t_{1u}-band as discussed in Sect.6.4.3. The Fermi level of the 20-electron compounds Mo_6S_8 and Mo_6Se_8 falls where the bottom of the $(t_{1u}, t_{2g}, e_g) \rightarrow$ e band overlaps the bottom of the $(e_g, t_{2u}) \rightarrow$ e band. The densities of

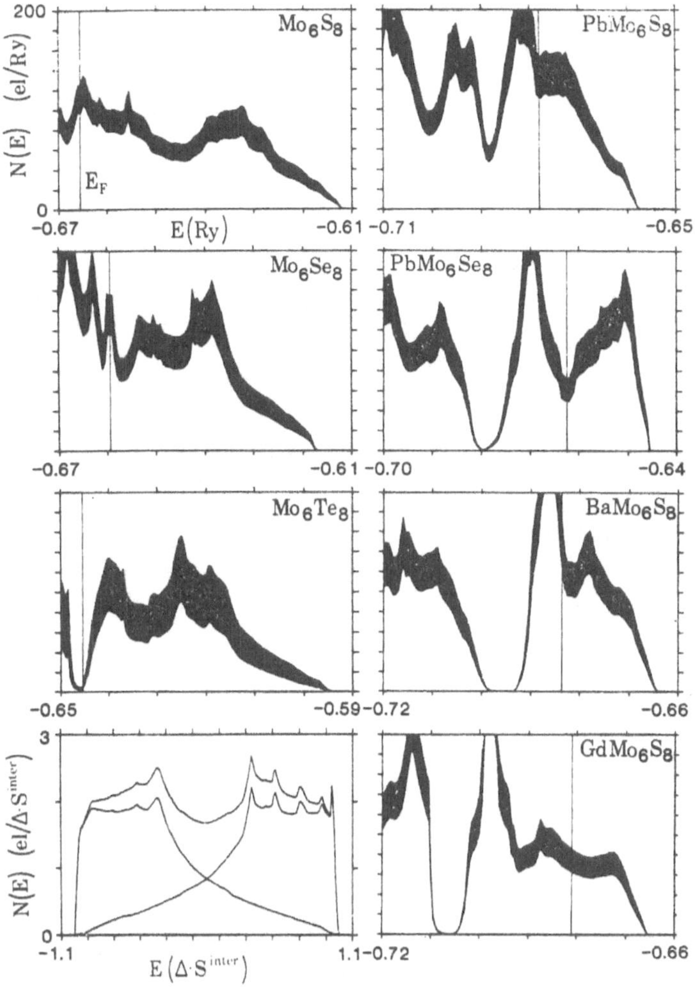

Fig.6.19. Densities of states of the conduction bands of $Mo_6S_8 \rightarrow GdMo_6S_8$ and the model e band. The white part of the curves gives the Mo d contributions and the black part all other contributions to the total state density

Table 6.10. Densities of states at the Fermi level [electrons per Rydberg per cell]

			Pb	Pb	Ba	Gd
Mo_6	Mo_6	Mo_6	Mo_6	Mo_6	Mo_6	Mo_6
S_8	Se_8	Te_8	S_8	Se_8	S_8	S_8
124	154	6	153	73	179	85

states at the Fermi level are high and the Mo e_g^d-contribution is large (Tables 6.8-10). In Mo_6Te_8 the interband interactions are so large, due to the close-lying X p bands, that the $(t_{1u}, e_g) \rightarrow e$ band is nearly separated with the consequence that the density of states at the Fermi level is very low and 44% X-like. The Fermi level in the 22-electron compounds $PbMo_6S_8$, $PbMo_6Se_8$, and (the high-temperature modification of) $BaMo_6S_8$ falls in the central valley of the $(t_{1u}, e_g) \rightarrow e$ band. The density of states is very high, particularly in $BaMo_6S_8$, and it

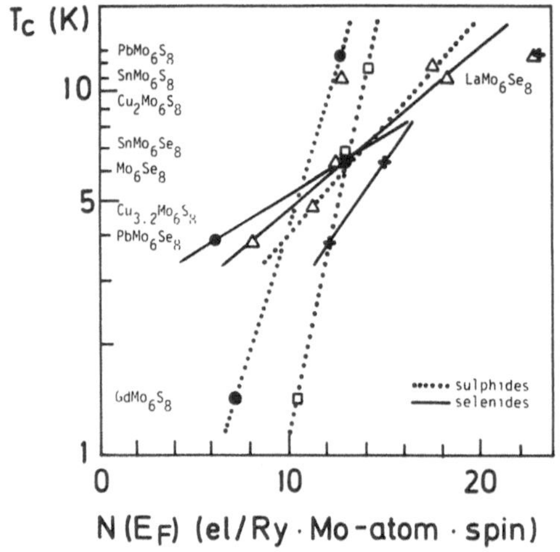

Fig.6.20. Correlation between theoretical densities of states at the Fermi level and the logarithm of the experimental superconducting transition temperature. Crosses: BULLETT [6.10], squares: JARLBORG and FREEMAN [6.11], triangles: FISCHER [6.25], and dots: NOHL and ANDERSEN [6.13] and present work

would be even higher could the Fermi level be decreased without destroying the crystal structure. The densities of states for the ternaries are asymmetric with a flattened-out high-energy side caused by the t_{1u}-e_g hybridization. The Fermi level of the 23-electron compound $GdMo_6S_8$ is situated in this region of relatively low density of states.

The densities of states listed in Table 6.10 agree reasonably well with those found previously [6.8,11,12] in view of the approximations involved, such as the neglect of interband hybridization [6.8] or the use of a $PbMo_6S_8$ crystal structure with $\alpha = 90^{\circ}$ for all compounds [6.11,12]. In this chapter we shall not discuss the relation of band structure to superconductivity but merely show in Fig.6.20 the correlation between the densities of states at the Fermi level, as obtained by various authors, and the experimental transition temperature. The estimates by FISCHER [6.25] were emperical while the remaining ones were the results of band-structure calculations. Of the seven compounds considered in the present chapter, Mo_6S_8 and Mo_6Te_8 seem not to become superconducting. For Mo_6Te_8 this can be understood as a consequence of the very low state density, but for Mo_6S_8 the reason is not so clear. The calculated densities of states are, however, highly uncertain in the region near 20 electrons, as may be gathered from Fig.6.17. For the observed superconductors there seems from Fig.6.20 to be a correlation between T_c and the calculated densities of states, but it also appears that the chalcogen plays a role in the electron-phonon coupling. This is not inconsistent with the 20% X character of the electronic wave functions at E_F (Table 6.8).

6.4.6 Relation to Structural Properties

Certain structural trends as functions of the number of conduction electrons were revealed in Chap.3 and may be found in Table 6.2. We shall now suggest explanations for some of them on the basis of the calculated band structures.

The Mo-X intercluster distance is larger in the ternaries than in the binaries. This is because the $t_{1u} \rightarrow e_u$ band, which has antibonding intercluster Mo-X character, is nearly occupied in the ternaries and therefore does not contribute to the binding. The fact that the t_{1u} level is computed to lie lower in the ternaries than in the binaries (Figs.6.12,17) confirms this explanation.

The Mo octahedron contracts and becomes less elongated as the conduction band is filled. The first fact is caused by the strong e_g Mo d bonding character of the conduction band, and the second fact is connected with the filling of the $t_{1u} \rightarrow e_u$ and $t_{2u} \rightarrow a_u$ band. From Fig.6.8 we realize that the occupied a_u partner of the $t_{1u} \rightarrow e_u$ band binds within the two Mo triangles perpendicular to the $\bar{3}$ axis. The e_u partner must therefore bind the two triangles together such that when both a_u and e_u levels are occupied, the octahedral symmetry is restored. Therefore the octahedron is elongated when only the a_u partner is occupied and the elongation decreases with increasing e_u occupancy. (It might be noted that the t_{2g} states do not have this property.) For the t_{2u} states there is the slight modification in that it is the $t_{2u} \rightarrow a_u$ state which binds the triangles together and the $t_{2u} \rightarrow e_u$ states which bind within the triangles. The filling of the high-lying $t_{2u} \rightarrow a_u$ band near X (Fig.6.17) thus contributes to the decreasing elongation.

It is a well-known effect that clusters distort when the number of valence electrons differs from the number necessary to fill all the bonding and no antibonding levels. Highly distorted Nb_6I_8 clusters with a little as 19 valence electrons are known to exist. They were recently shown to undergo an interesting structural phase transition involving a further distortion of the Nb_6 octahedron [6.26]. Structural phase transitions involving a symmetry lowering of the Mo_6 octahedron at low temperatures in the Chevrel phases $BaMo_6S_8$ and $EuMo_6S_8$ with 22 electrons were reported even more recently by BAILLIF et al. [6.16], as mentioned in Sects.6.1 and 6.2.1. It was speculated that this distortion could create a gap at the Fermi level and hence be the cause for the observed absence of superconductivity in these compounds. (It might be noted, however, that $EuMo_6Se_8$ appears not to be superconducting or semiconducting.)

Band-structure calculations [6.13] for the low-temperature triclinic modification shown in Fig.6.2 yielded the results shown in Fig.6.21. $BaMo_6S_8$ appears to be a semimetal and $EuMo_6S_8$ a semiconductor. For the Eu compound the atomic positions in the cluster were taken as the ones accurately measured for the Ba compound. The distortions of the octahedron in real $EuMo_6S_8$ are known to be even larger [6.16], and the conclusion that the low-temperature phase of $EuMo_6S_8$ is a semiconductor seems to hold.

If we compare the band structures of the three 22-electron Chevrel phases in Fig.6.17 and ask which of these is most susceptible to a breaking of the threefold symmetry and hence the degeneracy along the $\bar{3}$-axis, the answer must be $BaMo_6S_8$, because here the Γt_{1u} and Re_g levels are almost degenerate and the bandwidth along

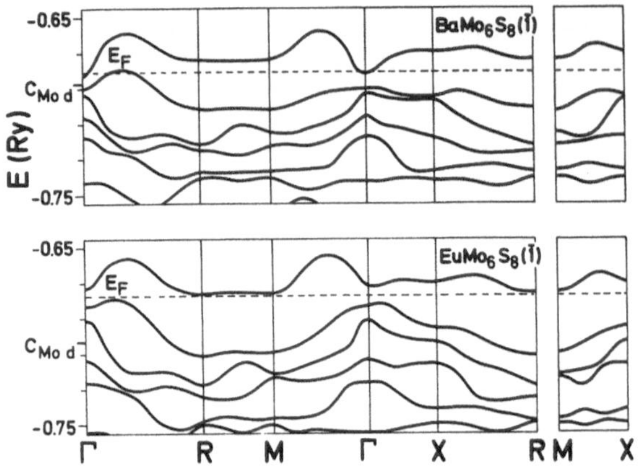

Fig.6.21. Conduction bands of $BaMo_6S_8$ and $EuMo_6S_8$ in the low-temperature phases shown in Fig.6.2. The basis of $EuMo_6S_8$ was taken to be that of $BaMo_6S_8$

the $\bar{3}$ axis is particularly small. From the picture of the two e_g partner functions in Fig.6.5, we can understand the reason for the Jahn-Teller distortion shown in Fig.6.2. With the e_g band half occupied it is energetically profitable to occupy only the state at the left-hand side of Fig.6.5 and then to lower its energy by a distortion involving a compression along the z-axis. (From this mechanism alone the same energy would, of course, be gained by occupation of the state at the right-hand side and a subsequent elongation in the z direction.)

From the above mentioned correlations between the observed structural and the computed electronic properties of the Chevrel phases, we conclude that the proposed energy bands are essentially correct, despite the approximations involved in the construction of the one-electron potential and in the application of the LMTO-ASA method.

6.5 Energy Bands of Chain Compounds

Cluster compounds exist in which the clusters may be described as condensed Me_6X_8 units (Me = Mo, Nb, Ti, etc.) and examples of corner-sharing, edge-sharing, and face-sharing Me_6 octahedra are well known [6.27]. In this section we shall discuss the band structures of the compounds, related to the Chevrel phases, which contain infinite chains of face-sharing Mo_6 octahedra. The structure of these compounds is illustrated in Fig.6.3 and was described in Sect.6.2.1.

The reason why Mo_6 octahedra condense when the appropriate number of valence electrons is available is that the back lobes of the d^4-directed orbitals (Sect. 6.3.3, Fig.6.7) may participate in bonding. At those places where the octahedra condense there are thus directed orbitals which simultaneously participate in two bonds [6.28]. The reason that the directed orbitals have back-lobes is due to the fact that the Mo d bonding levels are rather weakly hybridized and that the directed orbital, obtained from the bond orbital of the Mo_6X_8 unit, has a definite parity, which for d orbitals is even. In this respect the d^4-directed orbital differs from the sp^3 hybrid. The appropriate number of valence electrons is the number, or range of numbers, corresponding to the gap, or low-density-of-states region and this number of course depends on the structure of the condensate.

In the case of the face-condensed columns of octahedra shown in Fig.6.3, each Mo atom has six nearest Mo neighbors; two in the triangle perpendicular to the c-axis, two along the right-hand helix winding along the column, and two along the left-hand helix. It is therefore plausible that the triangle bonds perpendicular to the c-axis are like those in the isolated octahedron but that the back-lobes of the two remaining orbitals, which are directed forwards along the right- and left-hand helices, respectively, are bent slightly (by weak hybridization with X p and Mo p orbitals) and then used for bonding backwards along the two helices. In the approximation where interactions between orbitals not directed along the same bond are neglected, the band structure per Mo_3X_3 unit would thus consist of three flat triangle bonding bands, three flat-triangle antibonding bands, and six broad helix bands, each one being made up of the one-dimensional Bloch states formed from the orbitals directed along the helix. Moreover, there would be 3 Mo xy-X p antibonding bands. Such a band structure would have a broad low-density-of-states region around 12 electrons per Mo_3X_3 unit, i.e., around 24 electrons/Mo_6. In reality, however, the interaction between the helix bands cannot be neglected and the strong hybridization with the X p orbitals cannot be concentrated onto the xy orbital as in the isolated Mo_6X_8 cluster, because in the chain compounds the four X atoms nearest to a given Mo atom do not lie in the same plane. In fact, three of the X atoms are arranged approximately as in the Mo_6X_8 cluster, while the fourth is arranged along the z-axis like the neighbor-cluster X atom in the Chevrel phases. In our imagined "process" of cluster condensation we must therefore allow the orbitals to rearrange themselves into a "final state."

We shall now present the energy band structure of $TlMo_3Se_3$ computed by KELLY and ANDERSEN [6.14] and then explain how they interpreted it in terms of the rearranged orbitals just mentioned. The band structures of the other chain compounds are rather similar. On the basis of their band structures KELLY and ANDERSEN suggested that the physical properties of the chain compounds might be quite interesting, and this will be the topic of the final section.

6.5.1 General Features of the Energy Bands

The band structure of $TlMo_3Se_3$ is shown in Fig.6.22 for $\underline{k}$ along the c direction from $\Gamma(0)$ to $A(2\pi/c)$. In this double-zone representation, corresponding to the $TlMo_3Se_3$ cell, the c component of the Bloch vector labels the irreducible representation of the Abelian subgroup generated by the $(180^\circ, c/2)$ screw axis through the center of a column. Due to the presence of this screw operation, which commutes with all other operations of the space group, the calculated band structures have

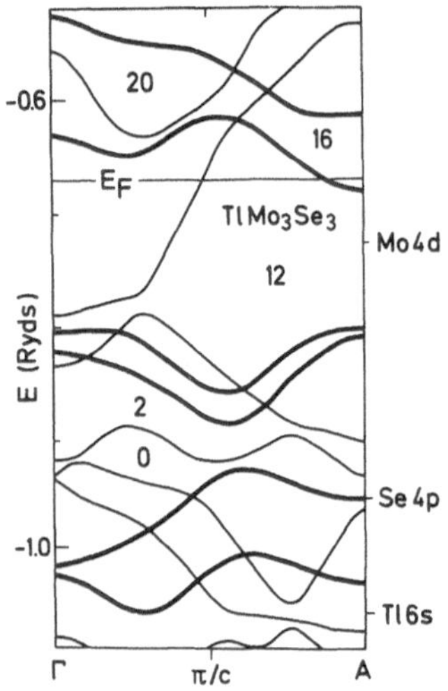

Fig.6.22. Band structure of $TlMo_3Se_3$ in the double-zone representation and for $\underline{k}$ along the chain. C_{Mo} 4d, etc. have been indicated on the right-hand side of the figure and the numbers between the bands give the Mo d-electron count. Doubly degenerate e and singly degenerate a bands are indicated by dark and light lines, respectively

no gaps at π/c and can be unfolded as shown. In addition to this screw axis, and the translations perpendicular to the c-axis, the space group contains a threefold axis through the center of each column. In Fig.6.22 the bands belonging to the singly, a ($m = 0$), and doubly, e ($m = \pm 1$), degenerate representations have been indicated by light and dark lines, respectively. The energy bands depend little on the perpendicular components of $\underline{k}$. We therefore have not shown density-of-states curves but have merely indicated the electron count n between the bands. Zero corresponds to nominally filled Tl s, Se s, and Se p states and the Fermi level is at $n = 26/2 = 13$.

Below $n = 12$ there are two singly degenerate (a) and two doubly degenerate (e) bands of Mo d character, but they are crossed by and hybridize with bands of Se p character such that there is no clear separation of the Se p and the Mo d bands at $n = 0$. The singly degenerate band between $n = 0$ and 2 at Γ, between 10 and 12 at π/c, and between 2 and 4 at A is the Tl 6s band which crosses and hybridizes with the uppermost Se p band lying between $n = 2$ and 4 at Γ, between 0 and 2 at π/c, and between -12 and -10 at A. In the In compounds the In 5s band lies about 60 mRy's higher. All bands between $n = 12$ and 30, and these include the conduction bands, are Mo d-like. Above $n = 30$ are the Tl 6p bands as well as the Mo s and p bands which have been pushed up above the Mo d bands by the covalent hybridization with the Se p bands. The region of low state density in the d-band structure at $12 \leq n \lesssim 13$ is responsible for the stability of the structure. The nature of the d bands at the Fermi level determine the physical properties of these compounds. In the following we shall therefore try to provide an understanding of the d-band structure.

6.5.2 Interpretation of the d-Band Structure

In view of the fact that a Mo_3 column contains six helices, the result of the full band-structure calculation shown in Fig.6.22 is remarkable in yielding only one broad band at the Fermi level plus, in some cases, an electron pocket at A. This has important consequences for the chemical and physical properties of the chain compounds. We shall now try to interpret this band structure. Rather than starting from the arrangement of the orbitals directly following from the arrangement in the isolated Mo_6Se_8 cluster, as described in the introduction to this section, we now give the final arrangement.

It is obvious that two of the Mo d orbitals must be directed approximately along the edges of the triangle (T) perpendicular to the c-axis as in the isolated cluster. Moreover, the band-structure calculations show that the wave function of the singly degenerate broad band is the sum of the Bloch states for the right-hand helices minus the sum for the left-hand helices (H), that is, it has symmetry a_2. We therefore reserve the xy orbital for this band (plus a band of symmetry e). This implies that we use the xy orbital plus the set of d^4-directed orbitals specified in Table 6.7 and Sect.6.3.3, choosing now the x-axis along the column parallel to the c-axis, the y-axis tangential to the column, and the z-axis towards its center. The angle between the T orbitals pointing along the edges of the triangle is now 76° rather than 60°, but weak hybridization with the Mo z orbital will help. Similarly, hybridization with this Mo p orbital will help bend the xy orbital along the edges of the helices. The remaining two orbitals lying in the xz plane have front lobes directed towards the centers of the two condensed octahedra (O), back lobes directed towards the two terminal X atoms (those on the octahedron "z-axes"), and negative lobes in the x direction (Figs.6.3,7) towards the remaining two X atoms. The angle between the "z-axes" of the two condensed octahedra is 71° (when the octahedra are regular) and this fits well with the 76° angle between the directed orbitals.

This orbital coupling scheme has the full symmetry of the column and only the O orbitals hybridize strongly with the X p orbitals. With the approximation that interaction between orbitals not participating in the same bond is neglected, we obtain the schematic band structure shown in Fig.6.23. The bond energy is approximately equal to $dd\sigma = -60\hat{\Delta}_{\mathrm{Mod}}$ and the O levels have been shifted upwards by $pd\sigma/2$ as an approximate way of treating their hybridization with the X p orbitals. We have assumed that the chain has a vertical mirror plane containing the c-axis and can therefore separate the a representations into even (a_1) and odd (a_2) representations.

The orbitals directed along the edges of the triangles of course make bonding-antibonding pairs and therefore give rise to three triangle bonds (T; a_1 + e) and three triangle antibonds (T; a_2 + e) which have no dispersion. In the real calculation for $TlMo_3Se_3$ we can identify these bands by the wave functions. The Ta_1 band is found to be the one between 12 and 14 at Γ and between 0 and 2 at A. The

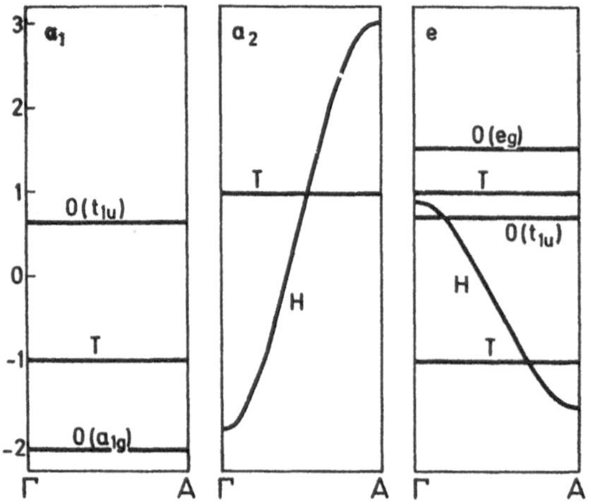

Fig.6.23. The model bands in the triangle-helix-octahedron (THO) orbital coupling scheme for the chain compounds. The unit of energy is $-dd\sigma = 60\hat{\Delta}_{Mod}$

bonding Te band is between 4 and 8 at Γ and between 8 and 12 at A, while the antibonding Te band is between 20 and 24 at Γ and between 16 and 20 at A. The antibonding Ta_2 band can only be seen at A where it lies between 20 and 22.

The xy orbitals on the 3 Mo atoms give rise to 3 Bloch states, running along the helices: an a_2 band and a doubly degenerated e band. The hopping integral along a helix is about $-40\hat{\Delta}_{Mod}$ and the interaction between two xy orbitals on the same triangle is $20\hat{\Delta}_{Mod}$ as obtained from Table 6.5. In the real calculation the Ha_2 band can be recognized at Γ between -2 and 0 and at π/c between 12 and 14, where it forms the conduction band as previously stated. The He band is repelled from the antibonding Ot_{1u} band and therefore appears below the "gap", between 8 and 12 at Γ, between 2 and 6 at π/c, and between 4 and 8 at A.

There are six orbitals pointing into an octahedron and these form a_{1g}, t_{1u}, and e_g states, exactly as the states based on the $(3z^2-1)$ orbital shown in Fig.6.4. The energies of these levels, before they have been adjusted for the X p hybridization, are approximately -140, 10, and $55\hat{\Delta}_{Mod}$. In the real energy bands the $Oa_{1g} \rightarrow a_1$ band is between -14 and -12 at Γ and at π/c, and below the frame at A. The $Oe_g \rightarrow e$ band is between 14 and 18 at Γ, and the $Ot_{1u} \rightarrow e$ band is above the frame at Γ and between 12 and 16 at A where it forms the second part of the conduction band. At A its symmetry is seen from Table 6.7 to be $[(3z^2-1) + (x^2-y^2)]/\sqrt{2}$ where, as usual, $(3z^2-1)$ denotes a *normalized* d orbital.

6.5.3 Physical Properties

What is presently known about the physical properties of the chain compounds is that they are highly one dimensional, that only $TlMo_3Se_3$ is a superconductor ($T_c \approx 3$ K), and that they, in contrast to the Chevrel phases, are stoichiometric. For $TlMo_3Se_3$ ARMICI et al. [6.29] found that the anisotropies of the upper critical field is 26 and the normal-state resistivity is of the order of $1000 \approx 26^2$. They concluded that this is related to the anisotropy of the Fermi surface. Moreover, they realized that the mobility along the columns is very high and that the (phonon enhanced) density of states is correspondingly low, of the order of 3 states/(Ry • Mo atom • spin). Magnetoresistance measurements showed that there may be open orbits perpendicular to the c axis but not along the c axis.

The band-structure calculations show that the Fermi surface comprises two sheets at $\pm k_F$ perpendicular to the c-axis arising from the helix a_2 band and two small, concentric ellipsoids at A arising from the octahedron $t_{1u} \rightarrow e$ band. This agrees with the behavior of the magnetoresistance.

The sheets have a slight wobble proportional to the dispersion in the directions perpendicular to the chain. Using the coordinate axes indicated in Fig.6.3, the dispersion of the helix band near the Fermi level, i.e., near $k_x = \pm\pi/c$, may be parametrized by the following first few terms of its Fourier series:

$$E(\underline{k}) = 2E_x \cos(k_x c/2) + 2E_{yz}\{\cos k_z a + \cos[(k_z+\sqrt{3}k_y)a/2] + \cos[(k_z-\sqrt{3}k_y)a/2]\} ,$$

where c = 4.48 and a = 8.92 Å for $TlMo_3Se_3$. This expression represents the computed bands very well and allows an analytical evaluation of the mass tensor

$$(m^{-1})_{ij} \propto \int_{FS} d^2k \; v_i(\underline{k})v_j(\underline{k})/|\underline{v}(\underline{k})| \equiv \langle v_i v_j \rangle$$

and hence of the parameter

$$\varepsilon = [\langle v_x^2\rangle/\langle v_z^2\rangle]^{1/2} \approx cE_x/(aE_{yz}\sqrt{3})$$

for the sheet, describing the anisotropy of the critical field (in the dirty limit and when the relaxation time due to impurity scattering is isotropic). The density of states per spin contributed by the sheet is $(2\pi|E_x|)^{-1}$, which for $TlMo_3Se_3$, where E_x = -90 mRy, is 0.59 states/(Ry • Mo atom • spin). The corresponding Fermi velocity is almost as high as for free electrons. The anisotropy parameter is very sensitive to the exact value of E_{yz}, which is of the order of 1 mRy, and cannot be calculated with any accuracy from the LMTO method in the atomic-spheres approximation. With the sphere sizes given in Table 6.4 one obtains E_{yz} = -0.5 mRy and ε = 50, while with Se spheres of 15% larger radius, which certainly is the maximum allowed, one finds E_{yz} = -1.4 mRy and ε = 18.

The overlap between the helix band at π/c and the octahedron band at A, upon which the existence of the ellipsoidal pockets relies, cannot be calculated with any accuracy either. In fact, with the sphere sizes given in Table 6.4, the octahedron band lies 9 mRy *higher* than the helix band at π/c, while for the 15% larger Se radius there is an overlap of 14 mRy. Now, if there were no overlap, the structure would be unstable against any distortion which could destroy the two-fold screw axis and hence open up a gap at π/c. Therefore, in $TlMo_3Se_3$ the ellipsoids must exist, and with the large Se spheres one finds an occupancy of 0.1 electrons per $TlMo_3Se_3$ unit. The curvature in the c direction of the octahedron band at A corresponds to a mass of 1.09 while in the perpendicular directions, where this doubly degenerate band splits up, the masses are about 3 and 1, respectively. For the maximum estimated occupancy of 0.1 electrons in the ellipsoids, these will therefore dominate the density of states, which will be 3.2 states/(Ry • Mo atom • spin), as well as the anisotropy parameter ε, which will be reduced from 18 to 7.

By comparison with the experimental values for the density of states and the anisotropy, we conclude that in $TlMo_3Se_3$ there *is* an overlap and that it is at most 10 mRy. The band structures computed for the remaining Tl and In selenides and tellurides are almost identical to that of $TlMo_3Se_3$ and show no clear trend towards larger or smaller overlaps. The effect is, however, subtle. KELLY and ANDERSEN have speculated that the reason why none of the chain compounds apart from $TlMo_3Se_3$ have been found to superconduct is either that the overlap is so small that the density of states is very low or that a Peierls distortion has taken place. So far no structural phase transition of this kind has been observed, but the atomic rearrangement necessary to create a gap could be subtle, for instance, a relative twist of the alternating Se triangles or maybe even the turning of entire columns relative to one another thus leading to a superstructure in the perpendicular directions. The reason why the chain compounds are found to be stoichiometric is also suggested by the band structure. If more than 13 electrons were present, the Fermi level would move up into a high-density-of-states region and this would destabilize the crystal structure. If less than 13 electrons were present, the columns would break up as the result of a Peierls distortion and Chevrel phases with chain fragments, such as bioctahedra, trioctahedra, and so on, would be more stable. It is, however, an interesting thought whether it will not be possible through doping or stress to control within certain limits the superconducting properties of $TlMo_3Se_3$ or other compounds with the chain structure.

Acknowledgements. We are grateful to P. Kelly for many useful discussions and for allowing us to quote results for the chain compounds before publication. The cooperation with K. Yvon, W. Hönle, H.G. von Schnering, and A. Simon is also gratefully acknowledged. We are indebted to V. Heine for his active interest and encouragement. Finally we wish to thank C. Hearne for help in preparing the manuscript.

References

6.1 G.H. Duffey: J. Chem. Physics *19*, 963 (1951)
6.2 C.D. Grossman, D.P. Olsen, G.H. Duffey: J. Chem. Physics *38*, 73 (1963)
6.3 F.A. Cotton, T.E. Haas: Inorg. Chem. *3*, 10 (1964)
6.4 S.F.A. Kettle: Theor. Chim. Acta *3*, 211 (1965)
6.5 L.J. Guggenberger, A.W. Sleight: Inorg. Chem. *8*, 2041 (1969)
6.6 F.A. Cotton, G.G. Stanley: Chem. Phys. Lett. *58*, 450 (1978)
6.7 G. Seifert, G. Grossmann, H. Müller: J. Mol. Struct. *64*, 93 (1980)
6.8 O.K. Andersen, W. Klose, H. Nohl: Phys. Rev. B*17*, 1209 (1978)
6.9 L.F. Mattheiss, C.Y. Fong: Phys. Rev. B*15*, 1760 (1977)

6.10 D.W. Bullett: Phys. Rev. Lett. *39*, 664 (1977)
6.11 T. Jarlborg, A.J. Freeman: J. Magn. Magn. Mater. *15-18*, 1579 (1980)
6.12 T. Jarlborg, A.J. Freeman: Phys. Rev. Lett. *44*, 178 (1980)
6.13 H. Nohl, O.K. Andersen: To be published
6.14 P.J. Kelly, O.K. Andersen: Phys. Rev. B
6.15 Ø. Fischer, M.B. Maple (eds.): *Superconductivity in Ternary Compounds II*, Topics in Current Physics, Vol. 32 (Springer, Berlin, Heidelberg, New York, to be published)
6.16 R. Baillif, A. Dunand, J. Müller, K. Yvon: Phys. Rev. Lett. *47*, 672 (1981); A. Dunand, R. Baillif, K. Yvon: Acta Crystallogr. (1981)
6.17 K. Yvon: Private communication
6.18 M. Sergent, Ø. Fischer, M. Decroux, C. Perrin, R. Chevrel: J. Solid State Chem. *22*, 87 (1977)
6.19 A. Perrin, R. Chevrel, M. Sergent, Ø. Fischer: J. Solid State Chem. *33*, 43 (1980)
6.20 D.W. Harrison, K.C. Lim, J.D. Thompson, C.Y. Huang, P.D. Hambourger, H.L. Luo: Phys. Rev. Lett. *46*, 280 (1981)
6.21 C.W. Chu, S.Z. Huang, C.H. Lin, R.L. Meng, M.K. Wu, P.H. Schmidt: Phys. Rev. Lett. *46*, 276 (1981)
6.22 W. Hönle, H.G. von Schnering, A. Lipka, K. Yvon: J. Less-Common Metals *71*, 135 (1980);
M. Potel, R. Chevrel, M. Sergent: Acta Crystallogr. B*36*, 1545 (1980)
The structure of the chain compounds were first reported by K. Klepp, H. Boller: Acta Crystallogr. A*34*, S160 (1978); Monatsh. Chemie *110*, 677 (1979); M. Potel, R. Chevrel, M. Sergent, M. Decroux, Ø. Fischer: C.R. Acad. Sci. Paris Ser. C*288*, 429 (1979)
6.23 O.K. Andersen: Phys. Rev. B*12*, 3060 (1975); Europhysics News *12*, 5, 4 (1981)
6.24 J.C. Slater, G.F. Koster: Phys. Rev. *94*, 1498 (1954)
6.25 Ø. Fischer: Appl. Phys. *16*, 1 (1978)
6.26 J.J. Finley, H. Nohl, E.E. Vogel, H. Imoto, R.E. Camley, V. Zevin, O.K. Andersen, A. Simon: Phys. Rev. Lett. *46*, 1472 (1981)
6.27 A. Simon: Angew. Chem. *93*, 23 (1981)
6.28 O.K. Andersen, H. Nohl, P.J. Kelly: To be published
6.29 J.C. Armici, M. Decroux, Ø. Fischer, M. Potel, R. Chevrel, M. Sergent: Solid State Commun. *33*, 607 (1980)

Additional References with Titles

J.K. Burdett, Jung-Hui Lin: Structures of Chevrel Phases, Inorg. Chem. *21*, 5 (1982)

B.E. Bursten, F.A. Cotton, G.G. Stanley: Bonding in Metal Atom Cluster Compounds. From the d-Orbital Overlap Model to SCF - Xα - SW Calculations, Israel. J. Chem. *19*, 132 (1980)

J.A. Creighton: Splitting of Degenerate Vibrational Modes due to Symmetry Perturbations in Tetrahedral M_4 and Octahedral M_6 Clusters, Inorg. Chem. *21*, 1 (1982)

T. Hughbanks, R. Hoffmann: The $Mo_3Se_3)_\infty$ chain, J. Am. Chem. Soc. *104* (1982)

7. Phonons in Ternary Molybdenum Chalcogenide Superconductors*

S. D. Bader, S. K. Sinha, B. P. Schweiss, and B. Renker

With 19 Figures

Experimental and theoretical investigations of the phonon properties of Chevrel-phase superconductors are reviewed in historical perspective. The crystallographic considerations and lattice heat capacity data which led to the introduction of the early molecular-crystal model of the lattice dynamics are discussed. The technique of phonon spectrum measurements by inelastic neutron scattering studies on polycrystalline samples is described, and the results for these systems are presented in detail. A more recent Born-von Kārmān force-constant model of the lattice dynamics, that utilizes Lennard-Jones potentials, is described. Physical properties calculated from this model are discussed in the light of experimental results for the heat capacity, neutron-weighted phonon spectra and the Mössbauer effect. The first single-crystal inelastic neutron scattering experiments available are considered briefly. The controversial question of the relationship of the phonon spectra of these materials to the electron-phonon interaction is addressed. Finally, areas of particular interest for future research are identified.

7.1 Preliminary Comments

In this chapter we review the phonon properties of the ternary compounds MMo_6S_8 and MMo_6Se_8 which form the Chevrel-phase (Chap.1 and [7.1]) structure. In these materials, the M site can be occupied by any one of a large class of metal atoms [7.2,3] ranging from Cu, Fe, Pb, Sn, ... to the rare-earth metals. Binary Chevrel-phase compounds (where the M site is empty) can also be prepared. As is evident from the title of this volume, the main interest in these compounds centers on their unique superconducting properties, namely, their generally high superconducting transition temperatures [7.4], high critical fields [7.5-7], and in the case where M is a rare-earth metal, a possible occurrence of both magnetic ordering and superconductivity as the temperature is lowered, either in consecutive order or even simultaneously [7.8-10]. From the above, it should be obvious that the

*Work supported by the U.S. Department of Energy.

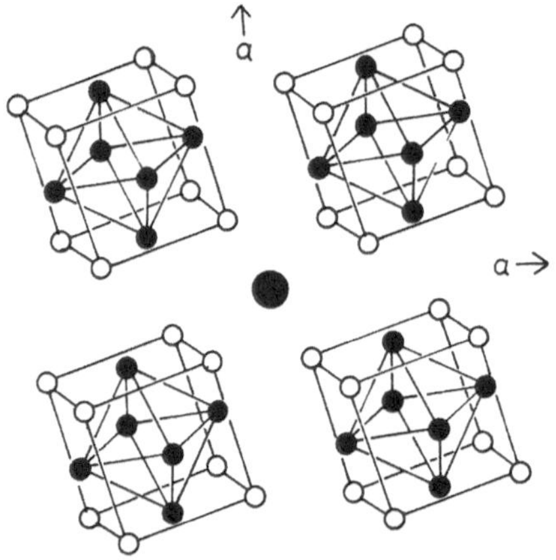

Fig.7.1
Chevrel-phase crystal structure

phonons (and the electron-phonon interaction) in these materials are also of great interest. Unfortunately, a detailed knowledge of the phonon spectra in these compounds has not been possible to attain to date, owing to the pervasive lack (with some notable exceptions discussed below) of single crystals on which inelastic neutron scattering experiments could be performed. If such information were available, we might have as detailed a knowledge of the electron-phonon interactions in these materials as we do for the superconducting 4d transition metals and compounds. Instead, in the present article, we mainly content ourselves with somewhat cruder information as revealed by heat-capacity measurements, Mössbauer measurements, and inelastic neutron scattering measurements on powder samples.

Figure 7.1 shows schematically the crystal structure of the Chevrel-phase compounds [7.1]. It consists essentially of Mo_6S_8 or Mo_6Se_8 octahedra (Mo atoms are small solid circles, chalcogen atoms are open circles in the figure) as the basic units. These units are arranged in a quasicubic structure, with a rhombohedral distortion, with the M atom (large solid circle in Fig.7.1) occupying the interstitial site of the rhombohedral lattice. There is also a distortion and tilting of the octahedra themselves (Chap.2). The lattice constants and rhombohedral angles for the various compounds have been tabulated in [7.1]. The M atom occupies a relatively large volume between the octahedra. Its nearest neighbors are the S or Se atoms in the corners of the octahedra adjacent to the M site. These chalcogen sites are called the $S_2(Se_2)$ sites. The other six equivalent S or Se sites are called $S_6(Se_6)$ sites. As a consequence of the large open volume around the M atom, we expect relatively weak coupling between the M atom and the Mo_6S_8 (Mo_6Se_8) octahedra, so that the dynamics of the M atom may, in the zeroth-order approximation, be thought of as that of an Einstein oscillator. The uniaxial anisotropy along the [111] direction associated with the rhomobohedral distortion might be expected to split the degeneracy between longitudinal and transverse modes leading to the appearance of two sharp peaks in the phonon spectrum at fairly low frequencies. The corresponding electronic property, namely, the relatively weak overlap between electrons associated with orbitals localized on the M atom and electrons associated with Mo and S(Se) orbitals, is believed to be responsible for the fact that magnetic ions at the M site do not seem to perturb strongly the superconducting properties of these compounds. This is borne out by recent band-structure calculations (Chap.6, [7.11,12] and [Ref.7.13, Chap.6]).

Before we develop the lattice-dynamical model further, it is necessary to mention a few additional facts regarding the structures which have some bearing on the dynamics of these crystals. (For detailed up-to-date reviews of the structural aspects, the reader is referred to the first three chapters of this volume.) It was frequently observed [7.1,14] that these compounds appeared not to form stoichiometrically, but with an excess of M atoms. For M atoms from the third row of the periodic table (e.g., Cu, Ni, Fe) the general formula is of the form $M_xMo_6S_8$ where $1 \leq x \leq 4$, whereas for heavy M atoms such as Pb, Sn or the rare-earths, $1 \leq x \leq 1.2$. Recent diffraction studies (Chap.3 and [7.1,15,16]) indicate that the M atoms do *not* occupy the ideal M site described above and in Fig.7.1, but instead there is a *ring* of sites centered symmetrically on the M site and lying in or close to the plane normal to the trigonal axis. For M = Cu, Ni, Fe there are twelve sites, six above the plane, six below. The displacement of these sites (which we may call M_{12} sites) from the central M site is fairly large (~ 1.3 Å), so that these sites can be multiply occupied in a given unit cell, leading to $x \geq 1$. For the heavy M atoms such as Pb, Sn, or the rare-earths, there are only six sites *in* the plane, and the displacement of these M_6 sites from M is much smaller ($\lesssim$ 0.3 Å) (Chap.3 and [7.15,16]), so that only one site can be occupied per cell. Thus, x would appear to be restricted to unity. The apparent values of x greater than 1 (up to 1.2) observed frequently for these compounds are now believed to be due to the presence of phases other than the Chevrel phase [7.16].

The above complication in the structure has potentially important and interesting consequences on the lattice dynamics of these materials. In the first place, if the M_6 or M_{12} sites are randomly occupied, it introduces a degree of *disorder* into what we have considered to be perfectly ordered structures. Secondly, it shows that the potential well in which the M atom sits is probably far from being harmonic. Finally, in principle there is a new degree of freedom in the dynamics, namely, that of *hopping* between M_6 or M_{12} sites and even possibly (at low temperatures) a cooperative ordering of these sites in the crystal. Unfortunately, in the absence of detailed single-crystal neutron scattering data, there is little experimental information regarding such effects, and so we shall treat the lattice dynamics in terms of the more idealized, ordered Chevrel-phase structure pictured in Fig.7.1.

7.2 The Molecular-Crystal Model and Lattice Heat Capacity

Even the idealized Chevrel-phase structure has a complex mode structure, since the 15 atoms per unit cell give rise to 45 branches of the phonon dispersion curves. The stability of the $Mo_6S_8(Mo_6Se_8)$ octahedra suggested to BADER et al. [7.17] a simplified model in which the compounds are regarded as molecular crystals composed of M atoms bound to quasi-rigid Mo_6S_8 units. The lattice dynamics would then simplify since the 45 normal modes of the unit cell would group into 3 acoustic and 3 optic "external" modes, 3 torsional modes associated with the Mo_6S_8 (Mo_6Se_8) clusters, and 36 "internal" vibrational modes of the clusters. A rigorous lattice-dynamical calculation of the mode structure, of course, would show some mixing between these types of modes, as we shall see, but the above may be regarded as the zeroth-order approximation. In this approximation, then, the phonon density of states may be written in the form

$$F(\omega) = 3F_A(\omega) + 3F_O(\omega) + 3F_T(\omega) + 36F_I(\omega) \quad , \tag{7.1}$$

where $F_A(\omega)$, $F_O(\omega)$, $F_T(\omega)$, and $F_I(\omega)$ represent the contributions from the acoustic, optic, torsional, and internal modes, respectively. Equation (7.1) implicitly assumes degeneracy between the longitudinal and transverse external modes.

The most detailed study of the lattice contribution to the heat capacity to date has been reported by BADER et al. [7.17,18] for $PbMo_6S_8$ and $SnMo_6S_8$. [The actual stoichiometry they reported for some samples (e.g., $PbMo_{5.1}S_6$ and $SnMo_5S_6$) differs from that of the Chevrel phase, but it is now believed that the stoichiometry was that of the Chevrel phase with the probable presence of other phases.] Their measured heat capacity of $PbMo_6S_8$ [7.17] is shown in Fig.7.2. As expressed in terms of an effective Debye temperature Θ in Fig.7.3, there is a strikingly large increase of Θ with temperature [7.18], similar to that found in molecular crystals. This indicated that the molecular-crystal approach might not be unreasonable. Figure 7.2 also shows the fit obtained to C/T using (7.1), with the following additional assumptions: $F_A(\omega)$ was taken to be given by a Debye spectrum with a cutoff frequency ω_D, $F_O(\omega)$ and $F_T(\omega)$ were taken to be delta functions centered at the same frequency ω_E, and $F_I(\omega)$ was taken to be a Gaussian distribution function centered at a frequency $\bar{\omega}_I$, with a standard deviation ω_I. The parameters obtained from a fit to C/T were for $PbMo_6S_8$: ω_D = 6.0 meV, ω_E = 12.1 meV, $\bar{\omega}_I$ = 40.1 meV, and ω_I = 0.3 $\bar{\omega}_I$; and for $SnMo_6S_8$: ω_D = 6.4 meV, ω_E = 13.4 meV, $\bar{\omega}_I$ = 40.1 meV, and ω_I = 0.3 $\bar{\omega}_I$. The contributions to C/T from each of the sets of modes is also shown individually in Fig. 7.2, and their sum (curve d) is seen to provide a reasonable fit to the measurements. The deviations could be well accounted for in terms of anharmonic effects and electronic contributions to the heat capacity. The fit for $SnMo_6S_8$ was reported to be of similar quality.

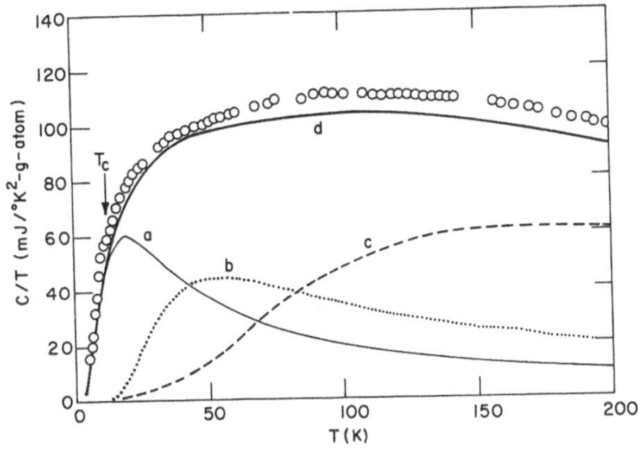

Fig.7.2. Heat-capacity data plotted as C/T vs T for a Pb ternary sulfide. A fit (curve d) using the molecular-crystal model was decomposed into external (curves a and b) and internal (curve c) mode contributions

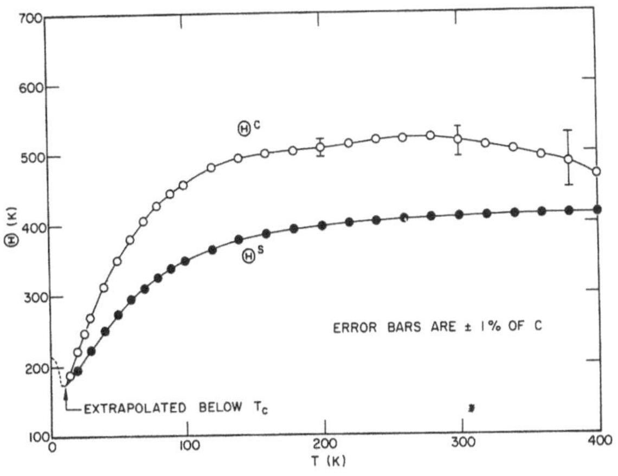

Fig.7.3. Experimental heat capacity (θ^C) and entropy (θ^S) Debye temperatures for a Pb ternary sulfide

While the heat capacity is far too integrated a property to give detailed information about the phonon spectrum, it does provide (together with the assumptions of the molecular-crystal model) a means of setting the scale for the magnitudes of the frequencies of the different types of modes present. As a matter of fact, as we shall see, when compared with the inelastic neutron scattering spectra and a more detailed lattice-dynamical calculation, the values of the obtained parameters cited above seem quite reasonable. Nevertheless, we caution that the molecular-crystal model provides only a qualitative picture which must not be taken too literally.

7.3 Phonon Spectra from Inelastic Neutron Scattering

7.3.1 The Theory of the Technique for Polycrystalline Samples

We now discuss a more detailed microscopic probe of the phonon spectrum, namely, the technique of inelastic neutron scattering (INS). The technique of determining phonon densities of states from polycrystalline samples by coherent inelastic neutron scattering has been discussed extensively in the literature. [7.19-22] The one-phonon coherent partial differential cross section for thermal neutron scattering from a crystal is given by [7.23]

$$\frac{d^2\sigma}{d\Omega dE} = \frac{(2\pi)^3}{V_a}\frac{k_1}{k_0}\sum_{q,j}\left|\sum_{\kappa}(2M_\kappa)^{-1/2}b_\kappa[\underline{Q}\cdot\underline{e}_\kappa(\underline{q}j)]\exp(i\underline{Q}\cdot\underline{r}_\kappa)\right.$$

$$\left.\times\exp(-W_\kappa)\right|^2\frac{1}{\omega}\frac{1}{e^{\beta\omega}-1}\delta(\omega\pm\omega_{qj})\delta(\underline{Q}-\underline{q}-\underline{G}) \quad , \qquad (7.2)$$

where E is the scattered neutron energy; V_a is the unit-cell volume; b_κ is the coherent neutron scattering length for the κth atom in the unit cell; $\underline{e}_\kappa(\underline{q}j)$ denotes the eigenvector for the κth atom (with mass M_κ and basis vector $\underline{r}_\kappa$) associated with phonon mode of wave vector $\underline{q}$ and branch index j having frequency ω_{qj}; $\underline{Q}$ is the neutron momentum transfer vector and is equal to $\underline{k}_0-\underline{k}_1$, where $\underline{k}_0$, $\underline{k}_1$ are, respectively, the incident and scattered neutron wave vectors; $\hbar\omega$ is the neutron energy gain in the scattering process, the ± signs in the first delta function refer to the cases of one-phonon creation and annihilation, respectively; $\underline{G}$ is a reciprocal lattice vector; $\exp(-W_\kappa)$ is the Debye-Waller factor for atom κ; and $\beta = \hbar/k_BT$. For scattering from a polycrystalline sample, the right-hand side of (7.2) must be averaged over all *directions* of $\underline{Q}$ with respect to the reciprocal lattice. In order to ensure that the net result produces something akin to a total phonon spectrum, one must, however, average over all $\underline{q}$ in an individual Brillouin zone. Proper averaging is achieved by using a large number of detectors grouped at various scattering angles (so that the *magnitudes* as well as the orientations of $\underline{Q}$ are averaged over) in such a way that for a given energy transfer, the volume of reciprocal space sampled by $\underline{Q}$ is large compared to the volume of the Brillouin zones. For a complex crystal, the inelastic-structure factor [represented by the sum over κ inside the modulus signs in (7.2)] does not in general repeat with the periodicity of the Brillouin zones, so in addition one must ensure that a large number of zones are sampled. However, here the small dimensions of each Brillouin zone (resulting from the rather large dimensions of the unit cell) are helpful. It follows that the magnitude of $\underline{Q}$ must be reasonably large compared to the size of a Brillouin zone. On the other hand, if $|\underline{Q}|$ is too large, we will obtain significant contributions, which must be corrected for, from n-phonon processes (where $n > 1$), whose cross section goes roughly as $[(\hbar Q^2/2\bar{M}\bar{\omega})^n/n]$, where $\bar{M}$ is an average mass of the types of atoms predominantly involved in the modes doing the scattering and $\bar{\omega}$ is their average frequency. These corrections have been applied when necessary to the data discussed below.

Let us examine (7.2) in more detail. For a given energy transfer $\hbar\omega$, a sufficiently large sampling of $\underline{Q}$ values at points in the reciprocal lattice of the crystal where the energy delta function is satisfied will lead to destructive interference in the factor $\exp[i\underline{Q}\cdot(\underline{r}_\kappa - \underline{r}_{\kappa'})]$ coming from the square of the inelastic structure factor, unless $\kappa = \kappa'$. Thus we have *approximately*

$$\left\langle \left[\frac{k_0\omega}{k_1}\left(e^{\beta\omega} - 1\right)\right] \frac{d^2\sigma}{d\Omega dE} \right\rangle_{av} \propto \sum_j p_j F_j(\omega) \equiv G(\omega) \quad , \tag{7.3}$$

where the left-hand side describes how the neutron counts in a given group of detectors for a given energy transfer are to be averaged over several detectors, and the right-hand side is what we term a neutron-weighted density of states. Here

$$F_j(\omega) = \sum_{\underline{q}} \delta(\omega - \omega_{\underline{q}j}) \tag{7.4}$$

is the partial density of states for phonons in band j, and

$$p_j = \frac{1}{3} \sum_\kappa \frac{b_\kappa^2}{2M_\kappa} \left\langle \exp(-2W_\kappa) |\underline{e}_\kappa(\underline{q}j)|^2 \right\rangle_{av} \quad , \tag{7.5}$$

where the average is over all $\underline{Q}$ sampled by the detectors for energy transfer $\hbar\omega$. Equations (7.3-5) indicate to what extent features in the true phonon spectrum $F(\omega)$ will appear in the neutron-obtained phonon spectrum $G(\omega)$. (Relevant thermal neutron scattering cross-section-to-mass ratios are tabulated in Table 7.1, where $\sigma_\kappa/M_\kappa = 4\pi b_\kappa^2/M_\kappa$). In fact $G(\omega)$ is quite analogous to the "α^2F" spectrum obtained from tunneling measurements on superconductors. In the present case, the electron-phonon matrix elements of the latter are replaced by the inelastic structure factor (or neutron-phonon matrix element). A more realistic model lattice-dynamical calculation for $PbMo_6S_8$ (to be discussed later) showed that the true $G(\omega)$, as calculated by rigorously averaging (7.2), is very similar to the true $F(\omega)$, implying that the weighting factors p_j do not significantly distort the true spectrum.

Table 7.1. Thermal neutron scattering cross-section-to-mass ratios

κ	σ_κ/M_κ [barns/amu]
Pb	0.0560
Mo	0.0771
S	0.0375
Se	0.127
Sn	0.0413
Cu	0.134

7.3.2 Experimental Neutron Scattering Results

We now discuss the experimental results for G(ω) for various Chevrel-phase compounds. Powder phonon spectral data have been obtained for $PbMo_6S_8$, $SnMo_6S_8$, $Cu_2Mo_6S_8$, Mo_6S_8, $PbMo_6Se_8$, and Mo_6Se_8 by the Karlsruhe group [7.17,22,24], for $PbMo_6S_8$, $SnMo_6S_8$, $PbMo_6Se_8$, and Mo_6Se_8 by the Argonne group [7.17,21], and more recently for UMo_6S_8 and UMo_6Se_8 [7.25]. The results for the lead and tin compounds and the pseudobinaries, where they overlap, are essentially in good agreement. We reproduce in Figs.7.4,5 the results of the Karlsruhe group, which are the most complete and have the best statistics and resolution. Figures 7.4,5 show the (neutron-weighted) generalized phonon spectrum G(ω) for the compounds measured, for both 297 K and 5 K. From an inspection of these figures certain qualitative features are already evident.

We first remark on the fairly sharp low-frequency peak (particularly enhanced at low temperatures) apparent in the spectra of the ternaries, which is noticeably absent in the spectra of the pseudobinaries. It is reasonable to ascribe this feature to the vibration of the M atoms against their surrounding octahedra, i.e., the Einstein-like mode referred to earlier. As we shall see later, it is more correct to think of it as arising from the flat optic mode associated with out-of-phase M-atom and octahedral motions polarized *normal* to the trigonal axis. For $SnMo_6S_8$ this mode shows unusual softening as the temperature is lowered (from 5 meV at 297 K to 4 meV at 5 K), but not for $PbMo_6S_8$. For $Cu_2Mo_6S_8$, the low-frequency peak sharpens dramatically at low temperatures and increases slightly in frequency to a value of 8.4 meV. This is consistent with structural data (Chaps.3,4 and [7.15]) indicating that at room temperature, the two copper atoms are randomized over the twelve interstitital sites around the M site, while at low temperatures they are ordered in dumbbell positions. The sharp peak at 8.4 meV at 5 K may then be ascribed to the Einstein oscillation of the rigid copper dumbbell as a whole against the surrounding octahedra as in the case of the Pb and Sn compounds. The low-energy side of this prominent peak seems to show additional structure, distinct from the normal Debye part of the spectrum, which could be due to rotational excitations of the Cu dumbbell.

The higher-frequency structure (above ~18 meV for the sulphides and ~15 meV for the selenides) is ascribed to the internal modes associated with the Mo_6S_8 (Mo_6Se_8) clusters. There is indeed a great similarity in the structure of G(ω) in this region for all the sulphides, although subtle changes do indicate that the internal modes may be somewhat affected by the presence and type of the M atoms. (This is particularly noticeable when we compare $PbMo_6Se_8$ and Mo_6Se_8). The mode structure between roughly 8 meV and 18 meV for the sulphides (and roughly 6 meV and 15 meV for the selenides) consists of the external modes of the octahedral clusters (rotational and translational) mixed in with the optical M atom modes polarized *along* the trigonal axis. Since the latter should be absent in the

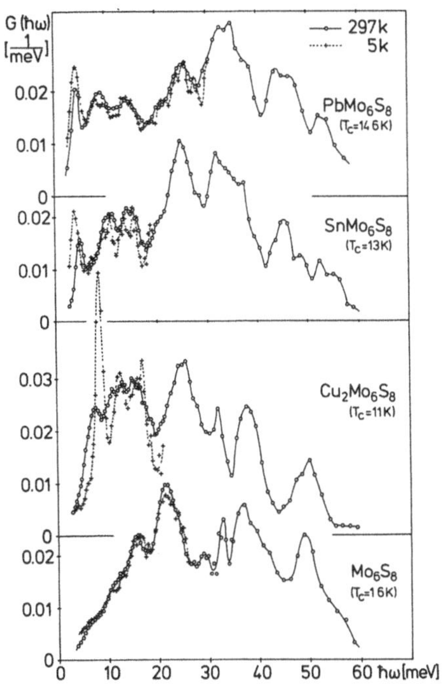

Fig.7.4. Experimental neutron-weighted phonon spectra G(ω) for polycrystalline sulfides

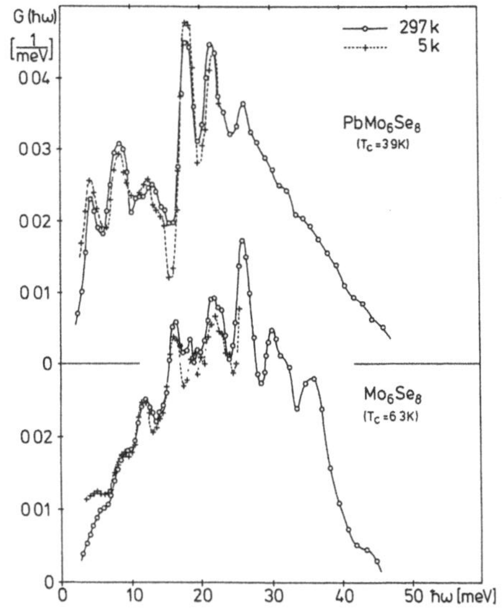

Fig.7.5. Experimental G(ω) spectra for polycrystalline selenides

pseudobinary compounds, we ascribe the peak at ~15 meV seen in both Mo_6S_8 and the MMo_6S_8 compounds to the torsional oscillations of the Mo_6S_8 clusters. Again, slight shifts due to the presence and type of the M atom are discernable. (For instance, in $PbMo_6S_8$ this peak shifts to 13 meV). The remaining maximum around 10 meV is probably due to a combination of optic M-atom modes polarized *along* the trigonal axis mixed with the external modes of the Mo_6S_8 clusters. This in fact is borne out by the more detailed lattice-dynamical model discussed in the following section. The results shown in Figs.7.4,5 were not corrected for instrumental resolution and do not extend below 3 meV because the scattering in the low-frequency region was contaminated by intense elastic or quasi-elastic scattering from the samples. We also see from Figs.7.4,5 that there is no actual gap separating the internal and external modes in these structures, although there is a sharp dip at ~ 18 meV and 15 meV, for the sulphides and selenides, respectively.

Finally, we discuss some other general features of the spectra summarized in Figs.7.4,5. The sharp peak associated with the vibrations of the M atom occurs in $PbMo_6S_8$ and $SnMo_6S_8$ at frequencies which do not scale inversely as the square root of the M-atom masses (or even the reduced masses of the M atom and the Mo_6S_8 clusters), so that the binding of the M atom depends on the atom itself. However, the binding of the Pb atom in the sulphide and the selenide appears to be the same, since the frequencies of the "Pb-atom peaks" are almost identical. As discussed in

Sect.7.1, we expect considerable anharmonicity to be associated with the vibrations of the M atom, and the softening of the "Sn-atom peak" in $SnMo_6S_8$ appears to indicate that this is indeed the case. Mössbauer measurements of the Sn Debye-Waller factor have been performed on $SnMo_6S_8$ [7.26-28] and indicate a dramatic softening (roughly a factor of two between room temperature and 5 K) of the effective Einstein frequency. This cannot be easily reconciled with the much smaller softening observed in the INS experiments unless one ascribes it to anharmonic effects causing the Debye-Waller factor to depart significantly from the usual (harmonic-type) expression in terms of which the Mössbauer results are interpreted. The absence of any significant corresponding frequency shift in the $PbMo_6S_8$ compound indicates that the Pb atom is much less anharmonic than Sn, as is to be expected from its smaller amplitude vibrations, due to its stronger binding and larger mass. The mode softening of the M atom with decreasing temperature can be caused by *quartic* anharmonicity associated with the M site. There is also a general mode softening of the group of *external* mode frequencies with decreasing temperature, and this is analogous to that found in other high-T_c superconductors where the mode softening is driven by the electron-phonon interaction [7.29]. In the case of Mo_6Se_8 there is a remarkable new shoulder in the $G(\omega)$ spectrum at low temperature which seems to correlate with mode-softening effects seen in single-crystal data to be discussed in Sect.7.5. An interesting feature which has been noted previously [7.21] is the considerable sharpening of the peaks in $G(\omega)$ at reduced temperature for all these compounds — another indication of strong anharmonicity.

7.4 A Simple Force-Constant Model for the Lattice Dynamics of Chevrel-Phase Compounds

We have seen how the INS powder data, heat capacity, and Mössbauer measurements have provided us with a qualitative (and in some cases semiquantitative) understanding of the lattice dynamics of the Chevrel-phase compounds, for which the molecular-crystal model has provided a simple, if crude, basis. A more detailed understanding and confirmation of some of the above ideas can only be achieved from an actual calculation of the phonon dispersion curves. There are two aspects to such a calculation. The first problem is to obtain the interatomic force constants, and the second problem is to describe how these yield, within the framework of the rather complex crystal structure of these compounds, the phonon dispersion curves, inelastic structure factors, etc., which can be compared with experiment. We are still a long way from a truly microscopic calculation of the interatomic forces in these materials, although we shall discuss this problem qualitatively in this section. What has been done instead [7.30] is to simulate the interatomic forces with a simple set of potential functions and then to perform a Born-von Kármán

(BvK) lattice-dynamical calculation to yield the normal-mode structure for these crystals. The parameters of the potential functions were adjusted to yield agreement with heat-capacity data, as is described below and in detail in [7.30].

7.4.1 Theoretical Considerations

In a BvK approach [7.31,32] all degrees of freedom are treated together, so that in principle all mixings between internal and external modes are automatically taken into account. The vibrational frequencies $\omega_j(\underline{q})$, where j again denotes a branch index (j = 1,...,3n) and n is the number of atoms in the unit cell, are obtained by solving the secular equations

$$|D^{\kappa\kappa'}_{\alpha\beta}(\underline{q}) - \omega_j^2(\underline{q})\delta_{\alpha\beta}\delta_{\kappa\kappa'}| = 0 \quad . \tag{7.6}$$

$D(\underline{q})$, the dynamical matrix at wave vector $\underline{q}$, is a $(3n \times 3n)$ matrix whose elements $D^{\kappa\kappa'}_{\alpha\beta}(\underline{q})$ are evaluated using the expression

$$D^{\kappa\kappa'}_{\alpha\beta}(\underline{q}) = \frac{1}{(M_\kappa M_{\kappa'})^{1/2}}\left[\sum_\ell \phi^{\kappa\kappa'}_{\alpha\beta}(0,\ell)\exp(i\underline{q}\cdot\underline{r}_\ell) - \delta_{\kappa\kappa'}\sum_{\ell'\kappa''}\phi^{\kappa\kappa''}_{\alpha\beta}(0,\ell'')\right] \quad , \tag{7.7}$$

where $\underline{r}_\ell$ is the vector from the origin to the unit cell ℓ. The force constant $\phi^{\kappa\kappa'}_{\alpha\beta}(0,\ell)$ is the negative of the force exerted in the α direction on atom κ located in the origin (ℓ = 0) unit cell due to a unit displacement in the β direction of atom κ' in the ℓth unit cell. These are derived from a pair-potential function $V_{\kappa\kappa'}(r)$ between atoms of types κ, κ' using the following equation [7.32]:

$$\phi^{\kappa\kappa'}_{\alpha\beta}(0,1) = -\left(\frac{\partial^2 V_{\kappa\kappa'}(r)}{\partial r_\alpha \partial r_\beta}\right)_{\underline{r}=\underline{r}_{\kappa'}-\underline{r}_\kappa} \quad . \tag{7.8}$$

This may be written as

$$\phi^{\kappa\kappa'}_{\alpha\beta}(0,\ell) = X\delta_{\alpha\beta} + Y\frac{r_\alpha r_\beta}{r^2} \quad (\ell \neq 0) \quad , \tag{7.9}$$

where

$$r_\alpha = (\underline{r}_\ell + \underline{r}_{\kappa'} - \underline{r}_\kappa)_\alpha \quad ,$$

$$X = V'_{\kappa\kappa'}(r)/r \quad ,$$

and

$$Y = V''_{\kappa\kappa'}(r) - X \quad .$$

For the model calculation, the potential functions $V_{\kappa\kappa'}(r)$ were taken to be of the Lennard-Jones form:

$$V_{\kappa\kappa'}(r) = A_{\kappa\kappa'}\left[\left(\frac{r}{\sigma_{\kappa\kappa'}}\right)^{-12} - \left(\frac{r}{\sigma_{\kappa\kappa'}}\right)^{-6}\right] \quad , \tag{7.10}$$

where $A_{\kappa\kappa'}$ and $\sigma_{\kappa\kappa'}$ were taken as adjustable parameters to represent the interactions between two Mo atoms, two S atoms, a Mo atom and a S atom, and an M atom and an S atom, respectively, thus yielding four potential functions $V_{Mo-Mo}(r)$, $V_{S-S}(r)$, $V_{Mo-S}(r)$ and $V_{M-S}(r)$. The calculations were carried out for $PbMo_6S_8$. The eight parameters were initially guessed using the intuitive expectation that the $\sigma_{\kappa\kappa'}$ values are closely related to average atomic spacings and that the $A_{\kappa\kappa'}$ values are of the order of the Boltzmann constant k_B times the melting temperature of related materials such as Mo metal, α-S, MoS_2, and PbS.

Using these parameters the dynamical matrix was set up using (7.7) and the unit-cell vectors and atomic basis vectors taken from the X-ray crystallography study by MAREZIO et al. [7.2]. The sum over ℓ in (7.7) included the origin unit cell and atoms in one layer of adjacent unit cells or $3^3 = 27$ unit cells total. Due to the assumed short-range nature of the potentials involved, this proved an adequate range of neighbors to sum over. Increasing the sum to $5^3 = 125$ unit cells changed the resulting Debye temperature by only 0.02%. As usual in such calculations [7.31,32], the "self" force constants $\phi_{\alpha\beta}^{\kappa\kappa'}(0,0)$ were obtained from the translational invariance conditions

$$\phi_{\alpha\beta}^{\kappa\kappa'}(0,0) = -\sum_{\substack{\ell \neq 0 \\ \kappa'}} \phi_{\alpha\beta}^{\kappa\kappa'}(0,1) \quad . \tag{7.11}$$

The procedure used in the calculation [7.30] was first to set up the solutions for the eigenfrequencies at $\underline{q} = 0$ (where 42 nonzero solutions exist) and adjust the eight parameters until reasonable agreement was obtained with heat-capacity-determined moments of the frequency spectrum for $PbMo_6S_8$. The values of $\sigma_{\kappa\kappa'}$ and $A_{\kappa\kappa'}$ are tabulated for $PbMo_6S_8$ in Table 7.2. The values of $\sigma_{\kappa\kappa'}$ were within 8% of the initially guessed values, and the $A_{\kappa\kappa'}$ within a factor of 5 of the initially guessed values.

Table 7.2. Lennard-Jones parameters used in lattice-dynamical calculations

$\kappa\kappa'$	$A_{\kappa\kappa'}$ [erg]	$\sigma_{\kappa\kappa'}$ [Å]
Mo-Mo	3.9×10^{-12}	2.4
S-S	1.8×10^{-13}	3.7
Mo-S	2.0×10^{-12}	2.2
Pb-S	2.15×10^{-12}	2.65

7.4.2 Calculated Dispersion Curves

Let us first comment on the structure of the phonon dispersion curves which such a calculation yields. It is instructive first to compare with the structure of the dispersion curves for "Mo_6S_8" obtained by simply setting the interaction $V_{Pb-S}(r)$ to zero. (The use of quotation marks around Mo_6S_8 is to remind the reader that the

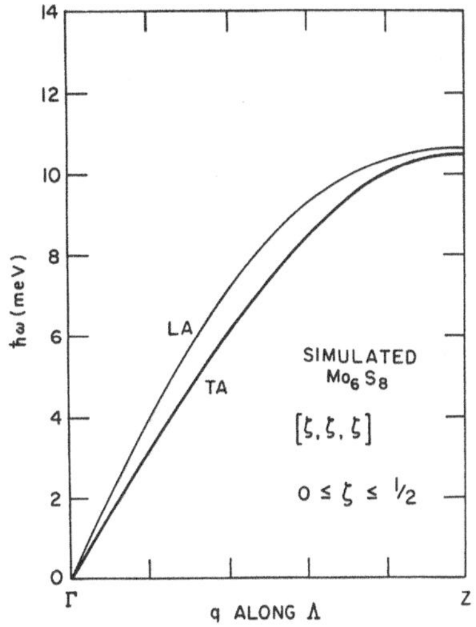

▲ Fig.7.6. Calculated LA and TA phonon branches along Λ for Mo_6S_8

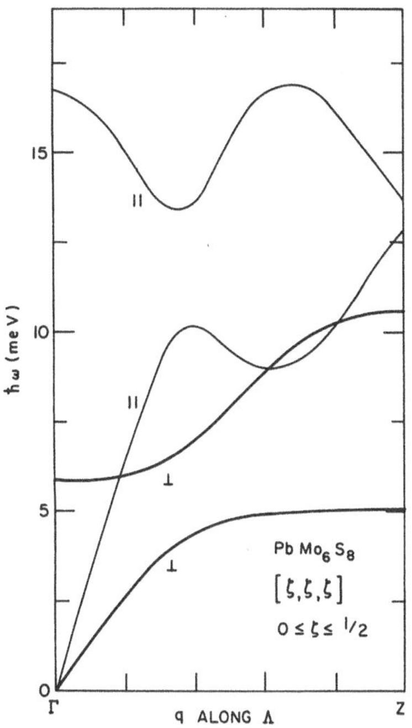

Fig.7.7. Calculated acoustic and Pb-dominated optic phonon branches along Λ for $PbMo_6S_8$

atomic distances and rhombohedral angle assumed in the calculation remained that for $PbMo_6S_8$.) The longitudinal acoustic (LA) and (doubly degenerate) transverse acoustic (TA) branches for this simulated compound and for phonons propagating along the high-symmetry trigonal axis Λ are shown in Fig.7.6. We note that these have a cut-off in the region of ~10.5 meV. Figure 7.7 shows the lowest-lying modes when the Pb atom is inserted. One can see immediately that the TA modes hybridize with transversely polarized M-atom modes producing a pronounced flat region in the vicinity of 5 meV, while the LA mode hybridizes correspondingly with a longitudinally polarized M-atom mode along Λ, but that both modes are fairly dispersive and no observable flat region results. The rhombohedral distortion causes the fairly sizeable splitting between the longitudinally and transversely polarized M-atom modes. It may be seen that the presence of the M atom also accentuates the splitting between the LA and TA branches, as is to be expected, since the LA modes correspond to vibrations along the line joining the M atom to nearest-neighbor S_2-site S atoms, while in the TA modes the M atoms are moving into regions of empty space. Figure 7.8 shows the full set of dispersion curves along Λ as calculated for $PbMo_6S_8$. Mixed in with the set of acoustic and M-atom modes are a set of fairly dispersionless modes centered around ~13 meV. By analyzing the eigenvectors of these modes at $\underline{q} = 0$ (where they are real) it was deduced [7.30] that these modes are associated with predominantly torsional oscillations of the Mo_6S_8 clusters.

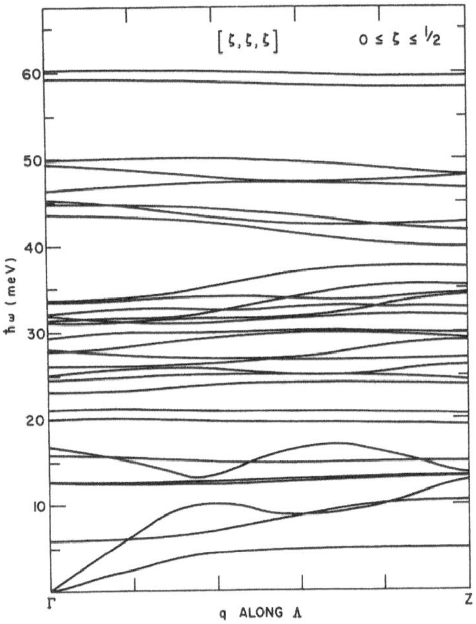

Fig.7.8. Complete set of dispersion curves calculated along Λ for $PbMo_6S_8$

Fig.7.9. Mixed torsional-breathing character of Mo eigenvectors for a 12-meV mode at $\underset{\sim}{q} = 0$ calculated for $PbMo_6S_8$

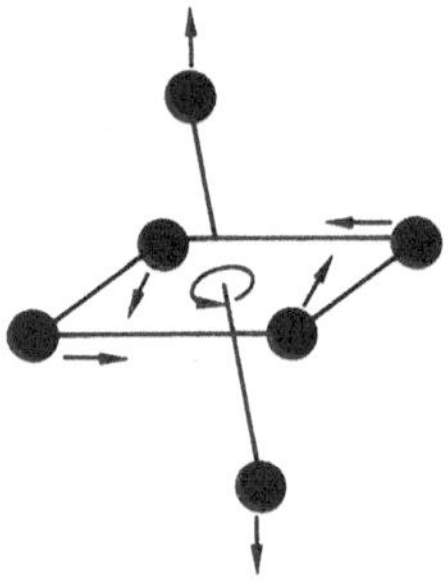

The eigenvector analysis showed that these modes were not purely torsional in nature, however, since there was some "breathing" character as can be seen in Fig. 7.9, i.e., there was some mixing of these modes with "internal" modes of the Mo_6S_8 clusters. Above these modes, without a sizeable gap, lies the set of remaining modes, predominantly internal in character with respect to the Mo_6S_8 clusters.

7.4.3 Calculated Phonon Densities of States

In order to gain further insight into the INS powder data, let us examine the phonon density of states $F(\omega)$ obtained from the above calculations. $F(\omega)$ is defined by

$$F(\omega) = \sum_{\underset{\sim}{q}j} \delta(\omega - \omega_j(\underset{\sim}{q})) \quad , \tag{7.12}$$

and is obtained by sampling $\underset{\sim}{q}$ values over the irreducible section of the first Brillouin zone, which in the case of the Chevrel-phase structures amounts to a volume that is 1/12 the volume of the Brillouin zone. There are well-known methods of performing this sampling so as to reproduce a fairly accurate phonon spectrum [7.32-34], but for the calculations under discussion, where only semiquantitative results were being sought, a much cruder method was used. A uniform mesh of 30 $\underset{\sim}{q}$ values was sampled in the irreducible zone and a histogram of the $F(\omega)$ was constructed. Figure 7.10 shows the resulting histogram for $PbMo_6S_8$, along with those for "Mo_6S_8", "$SnMo_6S_8$", and "Mo_6Se_8". The last three compounds correspond to qualitative *simulations* as pointed out above, since they were obtained from the prototypical $PbMo_6S_8$ force-constant model by simply changing appropriate masses or

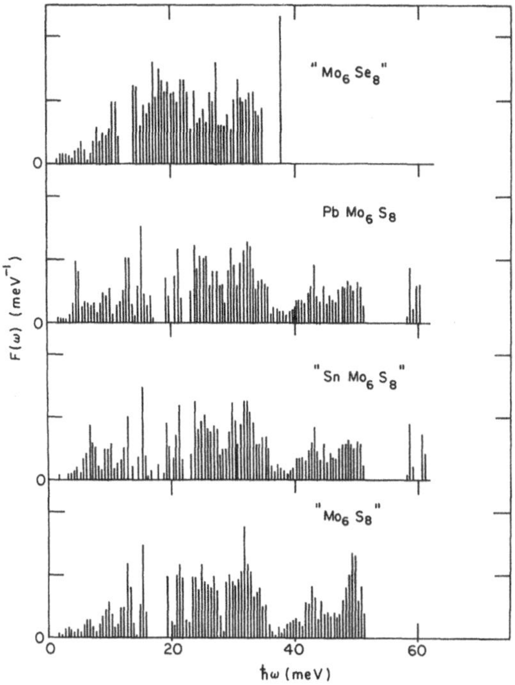

Fig.7.10. Histograms of the phonon spectra $F(\omega)$ based on 30 $\underline{q}$ points sampled in the irreducible zone

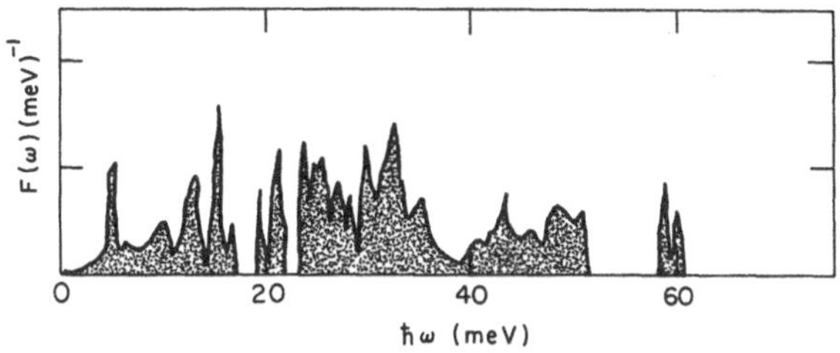

Fig.7.11
$F(\omega)$ for $PbMo_6S_8$ — 650 $\underline{q}$ points

"turning off" the appropriate force constants where necessary. Figure 7.11 shows a more accurate density of states calculated for $PbMo_6S_8$ from a sampling of 650 $\underline{q}$ values in the irreducible zone. Finally Fig.7.12 shows for $PbMo_6S_8$ the phonon spectrum decomposed into Pb, Mo, and S motions according to

$$F_\kappa(\omega) = \sum_{\underline{q}j} [\underline{e}_\kappa^*(\underline{q}j) \cdot \underline{e}_\kappa(\underline{q}j)]\delta(\omega - \omega_j(\underline{q})) \quad , \tag{7.13}$$

where $\underline{e}_\kappa(\underline{q}j)$ is the eigenvector associated with the κth ion. Each $F(\omega)$ was normalized to unity, so that

$$F(\omega) = F_{Pb}(\omega) + 6F_{Mo}(\omega) + 8F_S(\omega) \quad . \tag{7.14}$$

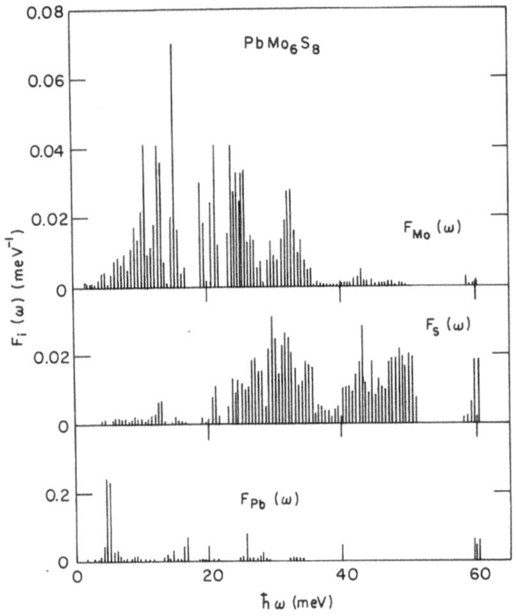

Fig.7.12. F(ω) decomposed for $PbMo_6S_8$ into Pb, Mo, and S vibrational components

7.4.4 Calculated Neutron-Weighted Densities of States Comparisons

In order to examine to what extent the neutron-weighted density of states G(ω) observed in the INS powder data reflects the features of F(ω) as calculated above, G(ω) as given by (7.3,4) was also calculated from the model. Neglecting the rhombohedral distortion, the Debye-Waller factors were calculated by averaging over all directions of the neutron wave vector transfer $\underline{Q}$ relative to the crystal axes;

$$W_\kappa = \frac{1}{3} Q^2 \langle u^2 \rangle_\kappa \quad , \tag{7.15}$$

where $\langle u^2 \rangle_\kappa$ is the mean-square displacement of the atom κ. Q was taken equal to the average value appropriate to the Argonne INS experiments [7.17,21] on $PbMo_6S_8$, and the mean-square displacement (for T = 300 K) was calculated using the usual formula:

$$\langle u^2 \rangle_\kappa = \sum_\alpha \frac{\hbar^2}{NM_\kappa} \sum_{\underline{q}j} e^*_{\kappa\alpha}(\underline{q}j) e_{\kappa\alpha}(\underline{q}j) \frac{1}{\hbar\omega_j(\underline{q})} \left(\frac{1}{2} + \frac{1}{\exp(\beta\omega) - 1} \right) \quad , \tag{7.16}$$

where N is the number of $\underline{q}$ values sampled. Table 7.3 shows the root-mean-square of the displacement along a particular direction for the various atoms in the unit cell at 300 K resulting from this calculation. Figure 7.13 shows the calculated neutron-weighted phonon density of states for $PbMo_6S_8$ and "$SnMo_6S_8$", and Fig.7.14 shows a comparison between the G(ω) calculated for $PbMo_6S_8$ and the G(ω) obtained experimentally by the Argonne group [7.20].

Table 7.3. Root-mean-square displacement component $\langle u_X^2\rangle_\kappa^{1/2}$ calculated at 300 K

κ	$\langle u_X^2\rangle_\kappa^{1/2}$ [Å]
Pb	0.127
Mo[a]	0.080
S[a]	0.081

[a]Value reported is average within unit cell for all atoms of this type.

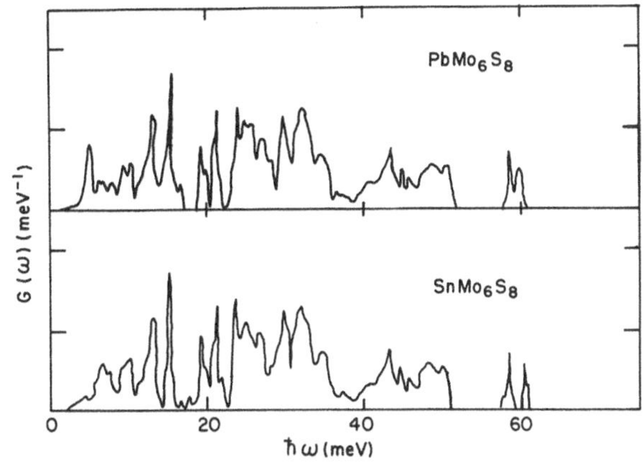

Fig.7.13. Calculated neutron-weighted phonon spectra G(ω)

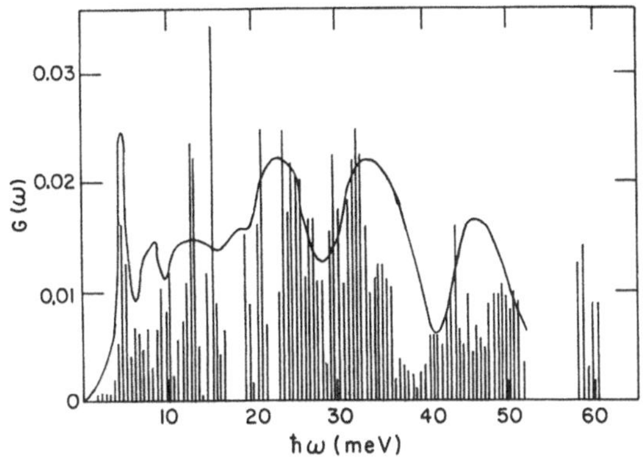

Fig.7.14. Calculated and measured G(ω) for $PbMo_6S_8$

From the above figures, and also the dispersion curves shown in Figs.7.6-8, the following conclusions may be drawn regarding these calculations.

1) There is not much significant distortion of the spectrum in going from F(ω) to G(ω). Thus the INS powder data experiments may be regarded as fairly realistic representations of the true phonon spectrum F(ω).

2) The sharp Einstein-like peak seen at the lowest frequencies in the $MMo_6S_8(Se_8)$ compounds arises from the (doubly degenerate) vibration of the M atoms in the plane *normal* to the trigonal axis, even though this mode is hybridized with the collective TA modes to some extent.

3) The second lowest peak (~ 10 meV) arises from the acoustic external modes of the clusters which are fairly flat in the outer part of the zone.
4) The peak around ~15 meV arises from external modes of the Mo_6S_8 clusters which have a large torsional component. For Mo_6Se_8 clusters this peak is shifted down to ~14.5 meV in the model calculations owing to the larger moments of inertia of the clusters. The measured $PbMo_6Se_8$ INS spectrum, on the other hand, shows this peak at ~12.5 meV, indicating that the force constants are not directly transferable from one compound to another.
5) The "$SnMo_6S_8$" calculations using force constants appropriate to $PbMo_6S_8$ (but with the Sn mass substituted for the Pb mass) yields a Sn-dominated Einstein mode which occurs at too high an energy (7 meV) compared to the experimental value of 5 meV, indicating that the Sn-S interactions are weaker than the Pb-S interactions by a factor of approximately two.
6) For the MMo_6S_8 compounds the structure in $G(\omega)$ above ~20 meV is due to modes which are primarily internal in character, with regard to the cluster. However, since there is not a large gap separating "internal" from "external" modes in these systems, there is mode mixing and the molecular-crystal model is not quantitatively correct. While the structure of the calculated $G(\omega)$ shows qualitatively the correct peak structure for these modes, there appears to be quantitative discrepancies particularly in the region of 20 meV which must be ascribed to a shortcoming of the force-constant model used. For the "Mo_6Se_8" calculation, the internal modes are significantly lower in frequency, reflecting the larger Se mass. This is borne out by the INS data, but again there is only qualitative agreement with experiment.
7) The model calculation indicates that the internal-mode structure is relatively insensitive to the presence of the M atom. As discussed in the previous section, this is not the case experimentally, particularly for the case of the selenides. This indicates that interactions beyond the nearest-neighbor M-S(Se) interactions should be considered when the Mo atom is introduced.

The model calculation thus confirms and clarifies some of the qualitative features discussed in connection with the neutron-measured phonon spectra.

7.4.5 Relationship to Mössbauer Effect

An attempt was made within the framework of the model to evaluate the temperature dependence of the Sn Debye-Waller factor as obtained from the Mössbauer measurements of BOLZ et al. [7.27] and KIMBALL et al. [7.26]. The approach taken [7.30] was to consider the softening of the *external* modes observed in the Argonne experiments [7.21] on $SnMo_6S_8$, and their hybridization with the Sn Einstein-like mode (for which no softening was detected in the Argonne measurements). The result was too small an effect compared to experiment. The more recent neutron measurements shown in Figs.7.4,5 by the Karlsruhe group [7.24] do show a softening of the Sn mode, as discussed in the previous section, but its magnitude again would appear to be too small to account for the observed effect on the Debye-Waller factor; so the latter may be largely due to anharmonic effects in the Sn motion.

7.4.6 Calculated Moments of the Phonon Spectrum

This section contains tabulated moments (Tables 7.4-6) of the phonon spectra for representative Chevrel-phase compounds, taken from the lattice-dynamics calculations of (7.30). The moments provide a valuable summary of the results of the calculations for future reference. For instance, they should be quite useful for the evaluation of any of the standard expressions which arise in a microscopic calculation of the superconducting properties. The moments are defined for $n > -3, \neq 0$, and in the limit $n \to 3^+$ and $n \to 0$, as

$$<\omega^n> = \int \omega^n F(\omega) d\omega \quad . \tag{7.17}$$

The associated Debye temperatures are

$$\Theta_D(n) = \frac{\hbar}{k_B}\left[\left(\frac{n+3}{3}\right)<\omega^n>\right]^{1/n} \quad . \tag{7.18}$$

Site-dependent quantities $<\omega^n>_i$ and $\Theta_i(n)$ can be defined also, by substitution of the eigenvector-weighted phonon density of states $F_i(\omega)$ for $F(\omega)$ in (7.17,18). The zeroth moment commonly is expressed also in its logarithmic-moment form,

$$\omega_g = \exp(<\ln\omega>) = \exp \int (\ln\omega) F(\omega) d\omega$$

$$= \left[\prod_{j=1;q}^{3n} \omega_j(\underline{q})\right]^{1/3nN} \tag{7.19}$$

and

$$\Theta_D(0) = (\hbar/k_B) e^{1/3} \omega_g \quad , \tag{7.20}$$

where N is the number of $\underline{q}$ values sampled in the calculation and ω_g is referred to as the geometric mean frequency.

In the harmonic approximation many of the moments are related directly to experimental observables. The $n = -3$ Debye temperature is that obtained from low-temperature heat-capacity or ultrasonic measurements. The $n = 0$ and 2 Debye temperatures are the high-temperature limiting values obtained from the entropy and the heat capacity, respectively. The $n = -2$ Debye temperature is obtained from high-temperature Debye-Waller-factor measurements, as in X-ray or neutron diffraction or Mössbauer experiments. Tables 7.4,5 contain various Debye temperatures calculated for $PbMo_6S_8$ and Mo_6Se_8, respectively, and Table 7.6 contains $n = 0$ and $n = 2$ Debye temperatures calculated for the Sn and Pb ternary sulfides and the binary chalcogens. The Θ_i values from Table 7.4 for $PbMo_6S_8$ are plotted in Fig. 7.15, which clearly displays that the dominant vibrational character occurs at increasingly higher-frequency ranges for the Pb, Mo and S sites, respectively.

Table 7.4. Debye temperatures associated with n^{th} moments of $F(\omega)$ and $F_i(\omega)$ for $PbMo_6S_8$ as calculated from model

n	$\Theta(n)$ [K]	$\Theta_{Pb}(n)$ [K]	$\Theta_{Mo}(n)$ [K]	$\Theta_S(n)$ [K]
-2	277	115	270	468
-1	336	110	292	531
0	402	119	318	558
1	452	142	342	573
2	486	187	364	583
3	513	248	384	593
4	533	311	403	602
5	551	367	422	610
6	566	413	441	617

Table 7.5. Debye temperatures associated with n^{th} moments of $F(\omega)$ and $F_i(\omega)$ for Mo_6Se_8 as calculated from model

n	$\Theta(n)$ [K]	$\Theta_{Mo}(n)$ [K]	$\Theta_{Se}(n)$ [K]
-2	266	252	277
-1	302	290	312
0	330	324	334
1	347	347	347
2	359	363	356
3	369	376	363
4	377	385	370
5	383	392	376
6	388	398	380

Table 7.6. Debye temperatures associated with logarithmic and second moments of $F(\omega)$ for Chevrel-phase compounds as calculated from model

Compound	Θ_D (n = 0) [K]	Θ_D (n = 2) [K]
Mo_6Se_8	330	359
$PbMo_6S_8$	402	486
$SnMo_6S_8$	410	488
Mo_6S_8	428	484

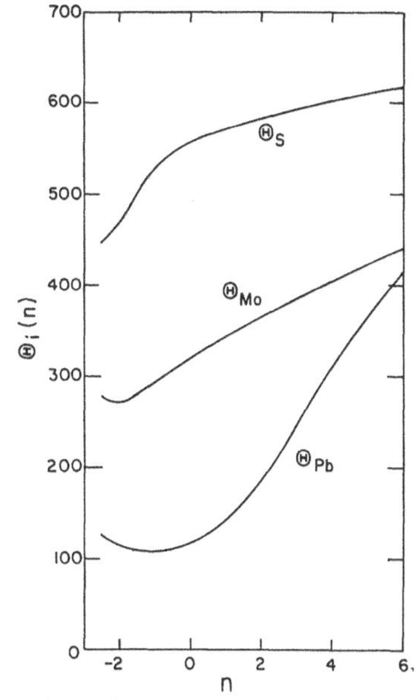

◄ Fig.7.15. Site-dependent Debye temperatures associated with n^{th} moment of the phonon spectrum calculated for $PbMo_6S_8$

▼ Fig.7.16. Temperature dependences of the heat-capacity (Θ^C) and entropy (Θ^S) Debye temperature calculated for $PbMo_6S_8$

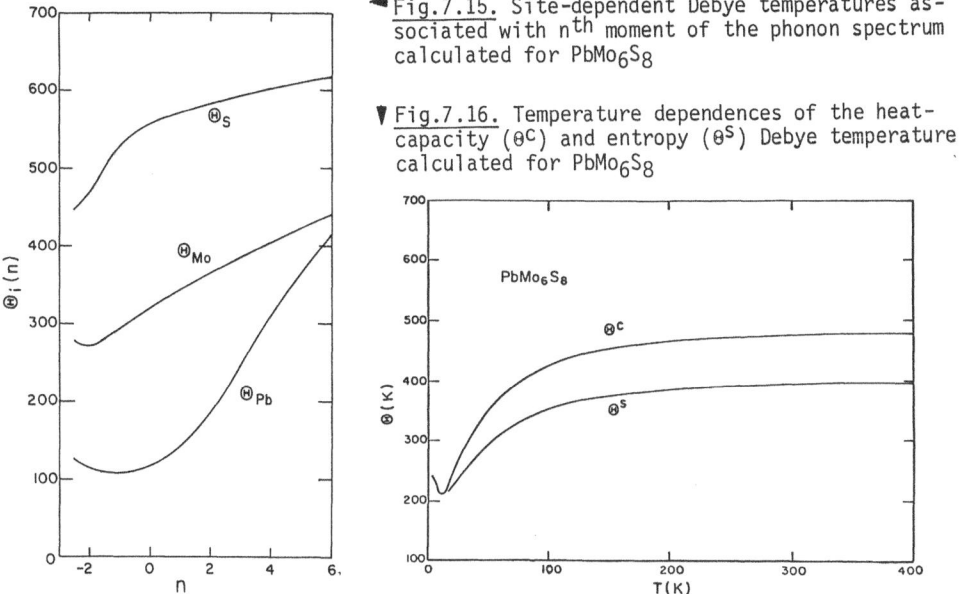

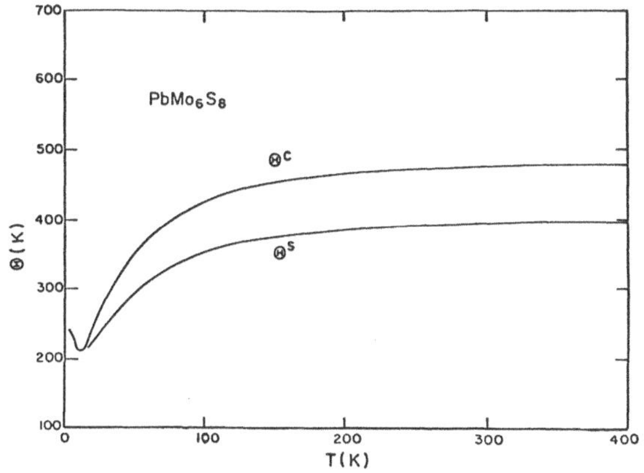

7.4.7 Calculated Lattice Entropy and Heat Capacity

In this section the model calculations are used to calculate thermodynamic properties of interest. In the previous section it was pointed out that in the harmonic approximation the low- and high-temperature limits of the entropy and heat capacity yield pure moments of the phonon spectrum. In the intermediate temperature range the following expression was used to calculate the lattice entropy S_L per gram atom [7.30]:

$$S_L(T) = 3N_Ak_B \int [xn_x - \ln(n_x + 1)]F(\omega)d\omega \quad , \tag{7.21}$$

where N_A is Avogadro's number, $x = \hbar\omega/k_BT$, $n_x = (e^x - 1)^{-1}$ is the Bose-Einstein distribution function, and $F(\omega)$ is normalized to unity. The heat capacity is $C_L(T) = T(dS_L/dT)$. The calculated values of $C_L(T)$ and $S_L(T)$ were converted to $\Theta_D^C(T)$ and $\Theta_D^S(T)$ values, effective Debye temperatures associated with the heat capacity and entropy, respectively, by comparison with thermodynamic Debye-function tables. Figure 7.16, shows the $\Theta_D^C(T)$ and $\Theta_D^S(T)$ curves calculated for $PbMo_6S_8$. Note that the overall shape and magnitude compares favorably with the experimental curve (Fig.7.3) discussed in Sect.7.2 for an early measurement on a ternary Pb sulphide sample [7.18,28] that was nominally reported to be $PbMo_{5.1}S_6$. Also, in [7.30] a generally favorable comparison was made between the ternary Pb sulphide calculation of Fig.7.16 and experimental data for a stoichiometric $SnMo_6S_8$ sample. Figure 7.17 shows a more direct comparison of a binary "Mo_6Se_8" calculation to experiment [7.30]. However, note again that the calculations where adjusted for $PbMo_6S_8$ force constants and geometry. To simulate the binary selenide, the Pb force constant was set to

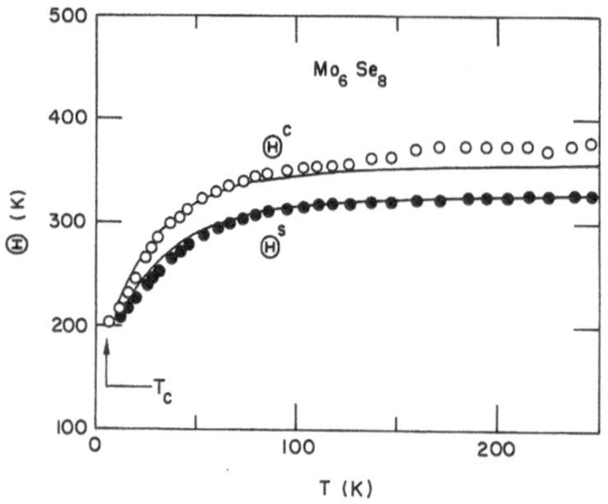

Fig.7.17. Experimental (circles) and calculated (curves) heat-capacity and entropy Debye temperatures for Mo_6Se_8

zero and the appropriate chalcogen mass change was inserted as discussed previously. No other force constant or geometric changes were taken into account.

7.5 Inelastic Neutron Scattering Experiments on Single Crystals

Recently single crystals of Mo_6Se_8 and $Cu_2Mo_6S_8$ have become available so that direct phonon dispersion curve measurements could be attempted with inelastic neutron scattering [7.23]. However, the size and perfection of these crystals considerably restricted the measurements which could be made. The measurements were made at both 300 K and 5 K along directions of high symmetry in the effective *cubic* system (since the rhombohedral distortion is small in these compounds, e.g., the rhomobohedral angle α is 91.58° for Mo_6Se_8 and 95.1° for $Cu_2Mo_6S_8$). It should be noted that the rhombohedral distortion makes, for instance, the [110] and [1$\bar{1}$0] axes nonequivalent. Figures 7.18,19 show the results obtained for these two crystals. The most interesting result is a low-temperature phonon anomaly in the [1$\bar{1}$0] TA branch of Mo_6Se_8 at a $q \simeq 0.5q_{max}$. This is in the frequency region where the low-temperature anomalies in the INS-observed $G(\omega)$ were seen in Mo_6Se_8 powders as discussed in Sect.7.3.2. This would indicate that at low temperatures the softening might extend over fairly large regions of reciprocal space. This type of softening is typical of the anomalies associated with strong electron-phonon interactions in superconducting transition metals [7.29] and is consistent with Mo_6Se_8 being a relatively high-T_c (7.3 K) superconductor. The structure of the measured acoustic branches are consistent with the predictions of the lattice-dynamical-model calculation referred to in Sect.7.4.2. For instance, in $Cu_2Mo_6Se_8$ in the [111] direction, there are indications for hybridization of the TA mode with an optical branch at 8.5 meV, a frequency which corresponds well to the pronounced peak in the 5-K INS-measured $G(\omega)$ spectrum. Interestingly enough, the disorder associated with the occupation of the Cu sites at 300 K does not seem to significantly affect the phonon spectrum.

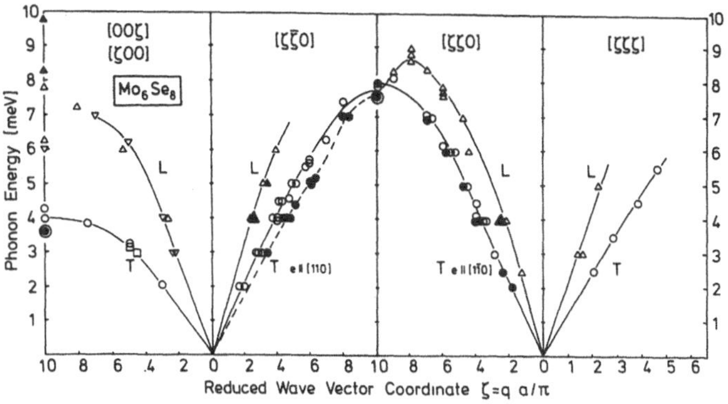

Fig.7.18. Experimental phonon dispersion curve sections at 300 K (open symbols) and 5 K (solid symbols) for Mo_6Se_8 showing a phonon anomaly in the TA [1$\bar{1}$0] branch. Within experimental error no differences between measured results in the [001] (Δ longitudinal, o transverse) and the [100] direction (∇ longitudinal, □ transverse) were observed

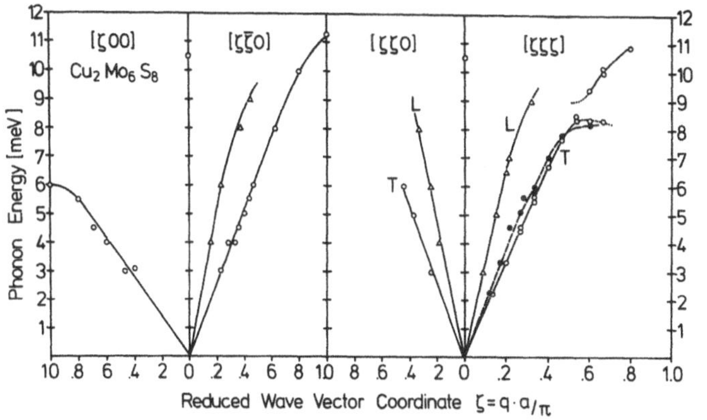

Fig.7.19. Experimental phonon branches for $Cu_2Mo_6S_8$ taken at 300 K (open symbols) and 5 K (solid symbols)

7.6 Relationship of the Phonon Spectrum to the Electron-Phonon Interaction

We now ask what the above results on the phonon spectrum reveal about the electron-phonon interaction in this group of compounds. Microscopic theories of lattice dynamics [7.32,35] show that an essential ingredient in the calculation of the phonon spectrum is a knowledge of the electron-phonon interaction for all modes. Owing to the complexities of the Chevrel-phase crystal structure, such knowledge is at present beyond our grasp. The electronic structure calculations for the Chevrel-phase compounds ([7.11,12] and [Ref.7.13, Chap.6]) reveal a high density of electronic states at the Fermi level predominantly associated with Mo d orbitals and S(Se) p orbitals. According to theories of the electron-phonon coupling

[7.35], therefore we should associate temperature-dependent electron-phonon renormalization of those modes with appreciable Mo or S(Se) atom displacements, i.e., all modes except those associated predominantly with M-atom motion. The low-temperature softening of the Sn modes in $SnMo_6S_8$ thus would be related to the *quartic anharmonicity* of the Sn sites as discussed in Sect.7.2 rather than to strong electron-phonon couplings. This is borne out by the absence of Pb-mode softening in the highest-T_c superconductor $PbMo_6S_8$. The results of SCHWEISS et al. [7.24] show clear evidence that the phonon softening on cooling, which extends throughout the low-frequency (< 20 meV) region of the spectrum, is roughly correlated with T_c. The neutron data at low temperatures does not extend far into the internal mode region, so it is not clear whether there is a low-temperature softening of the internal modes as well.

A rough microscopic calculation for the frequency of a particular class of modes (the torsional-type modes of the Mo_6S_8 cluster) in terms of the electronic band structure was performed by ANDERSEN et al. [7.11] (Chap.6). They showed that the frequency of this mode was affected predominantly by neighboring Mo-Mo atomic distances, for which a short-ranged pair-potential function between the Mo atoms of the form $A[(r/\sigma)^{12} - (r/\sigma)^8]$ could be obtained. Their calculation yielded values of 10 and 15 meV for the torsional mode frequency of Mo_6Se_8 and Mo_6S_8, respectively, in close agreement with the experimental value discussed in the previous sections.

We have seen in this chapter that a framework already exists to characterize $F(\omega)$ in a reasonable fashion. A great challenge then is to relate $F(\omega)$ to the electron-phonon interaction α^2F and to the key superconductivity parameter $\lambda = 2\int \alpha^2F\omega^{-1}d\omega$. This is the subject of the next chapter in this volume. We summarize some highlights here.

If the *resistivity* ρ is governed by phonon-assisted ($\underline{s}$-$\underline{d}$) interband scattering, then the temperature dependence of ρ can act as a useful constraint on the similarities and differences of $F(\omega)$ and α^2F. This is because the resistivity then can be expressed using the Wilson model:

$$\rho(T)\alpha \int \frac{y}{\sinh^2 y}\, \alpha^2_{tr}(\omega)F(\omega)d\omega \quad , \tag{7.22}$$

where $y = \hbar\omega/2k_BT$, and $\alpha^2_{tr}(\omega)$, the transport electron-phonon coupling strength, is very similar to α^2 of superconductivity theory. Equation (7.22) was applied to interpret Chevrel-phase data [7.30], as well as A15 compound data [7.36], and to test various models for α^2_{tr} against the observed behavior for $\rho(T)$. The conclusions reached based on an analysis of Cu and Pb ternary sulfide film data (see [7.30] for a complete discussion) is that "the gross form of α^2F... is related to $F(\omega)$ to an extent similar to that found for other superconducting systems. No strong δ-function-type regions seem to be present in α^2F and controlling T_c. This is in contrast to the early speculation that T_c may be governed by torsional and acoustical branches of $F(\omega)$."

Microscopic calculations of the strength of the electron-phonon coupling constant using the Gaspari-Gyorffy method (Chap.6, [7.11,12], and [Ref.7.13, Chap.6]) yield too low a value compared to that obtained from the experimental T_c values. In particular, these authors (Chap.6, [7.11,12], and [Ref.7.13, Chap.6]) concluded that one must postulate a predominance of low-frequency mode contributions to λ. There now seems to be little evidence for this; perhaps the approximations of the Gaspari-Gyorffy method are breaking down [7.12].

Novel studies of the relationship of $F(\omega)$ to α^2F have been performed at Jülich (see Chap.8 for a complete discussion). *Isotope-effect measurements* for Mo_6Se_8 clearly indicated that modes associated with Se atom displacements are important in determining T_c, as are the Mo modes [7.37]. Also, preliminary *tunneling* investigations performed on Cu and Pb ternary sulfides indicate a particularly strong coupling to the high-frequency (20-50 meV) modes, in addition to coupling to the low-frequency region of $F(\omega)$ [7.38]. These experiments, when taken together with the detailed electronic structure information provided by recent computational studies, are beginning to yield a coherent description of the electron-phonon interaction in this class of superconductors.

7.7 Summary

We have seen in this chapter how the broad features of the phonon spectra of the Chevrel-phase superconductors as revealed by heat capacity, Mössbauer, and powder inelastic neutron scattering experiments can be understood qualitatively, and in some cases even semiquantitatively in terms of a rather simple force-constant model. Obviously this force-constant model needs to be refined and extended further, but this will be possible only after more detailed measurements of phonon dispersion relations on single crystals become available. Such measurements should also yield more detailed information on the anharmonicity associated with the M-atom motions, and the mode softenings associated with strong electron-phonon interactions. In addition, high-resolution low-frequency measurements on high-quality single crystals should yield the spectrum associated with the degrees of freedom associated with the multiple (M_6 or M_{12}) sites associated with the M atom, another topic of considerable interest in these compounds. Finally, there is the challenging problem of a microscopic theory of the phonon spectra in these compounds and a first principles calculation of the electron-phonon coupling associated with the various types of modes. Again, the first steps in this direction have been taken, but much more detailed calculations need to be made. It is hoped that the next few years will show much further progress in these directions.

References

7.1 R. Chevrel, M. Sergent, J. Prigent: J. Solid State Chem. *3*, 515 (1971)
7.2 M. Marezio, P.D. Dernier, J.P. Remeika, E. Corenzwit, B.T. Matthias: Mater. Res. Bull. *8*, 657 (1973)
7.3 Ø. Fischer, A. Treyvaud, R. Chevrel, M. Sergent: Solid State Commun. *17*, 721 (1975)
7.4 B.T. Matthias, M. Marezio, E. Corenzwit, A.S. Cooper, H.E. Barz: Science *175*, 1465 (1972)
7.5 S. Foner: *Superconductivity in d- and f-band Metals*, ed. by D.H. Douglass (Plenum, New York 1976) p.161
7.6 Ø. Fischer, M. Decroux, R. Chevrel, M. Sergent: *Superconductivity in d- and f-band Metals*, ed. by D.H. Douglass (Plenum, New York 1976) p.175
7.7 R. Odermatt, Ø. Fischer, H. Jones, G. Bongi: J. Phys. C*7*, L13 (1974)
7.8 M. Ishikawa, Ø. Fischer: Solid State Commun. *23*, 37 (1977)
7.9 J.W. Lynn, D.E. Moncton, G. Shirane, W. Thomlinson, J. Eckert, R.N. Shelton: J. Appl. Phys. *49*, 1389 (1978)
7.10 R.W. McCallum, D.C. Johnston, R.N. Shelton, W.A. Fertig, M.B. Maple: Solid State Commun. *24*, 501 (1977)
7.11 O.K. Andersen, W. Klose, H. Nohl: Phys. Rev. B*17*, 1209 (1978)
7.12 T. Jarlborg, A.J. Freeman: Phys. Rev. Lett. *44*, 178 (1980)
7.13 Ø. Fischer, M.B. Maple (eds.): *Superconductivity in Ternary Compounds II*, Topics in Current Physics, Vol.(Springer, Berlin, Heidelberg, New York, to be published)
7.14 R. Chevrel, M. Sergent, Ø. Fischer: Mater. Res. Bull. *10*, 1169 (1975)
7.15 R. Yvon: Solid State Commun. *25*, 327 (1978)
7.16 J.D. Jorgensen: Private communication
7.17 S.D. Bader, G.S. Knapp, S.K. Sinha, P. Schweiss, B. Renker: Phys. Rev. Lett. *37*, 344 (1976)
7.18 S.D. Bader, G.S. Knapp, A.T. Aldred: Ferroelectrics *17*, 321 (1988); Ref.7.28
7.19 M.M. Bredov, B.A. Kotov, N.M. Okuneva, V.S. Oskotskii, A.L. Shakh-Bodagov: Sov. Phys. *9*, 214 (1967);
V.S. Oskotskii: Sov. Phys. *9*, 420 (1967)
7.20 F. Gompf, H. Lau, W. Reichardt, J. Salagado: In *Neutron Inelastic Scattering 1972*, ed. by F.M. Markhof (International Atomic Energy Agency, Vienna 1972) p.137
7.21 S.D. Bader, S.K. Sinha, R.N. Shelton: *Superconductivity in d- and f-band Metals*, ed. by D.H. Douglass (Plenum, New York 1976) p.209
7.22 B.P. Schweiss, B. Renker, E. Schneider, W. Reichardt: In *Superconductivity in d- and f-band Metals*, ed. by D.H. Douglass (Plenum, New York 1976) p.189
7.23 W. Marshall, S.M. Lovesey: *Theory of Thermal Neutron Scattering* (Clarendon, Oxford 1971) p.83
7.24 B.P. Schweiss, B. Renker, R. Flükiger: *Proc. of Conf. on Ternary Superconductors*, Lake Geneva, 1980, ed. by G.K. Shenoy (Elsevier-North Holland, Amsterdam 1981) p.29
7.25 H. Mook, S.K. Sinha, F.Y. Fradin: Unpublished
7.26 C.W. Kimball, L. Weber, G. VanLanduyt, F.Y. Fradin, B.D. Dunlap, G.K. Shenoy: Phys. Rev. Lett. *36*, 412 (1976)
7.27 J. Bolz, J. Hauck, F. Pobell: Z. Phys. B*25*, 351 (1976)
7.28 F.Y. Fradin, G.S. Knapp, S.D. Bader, G. Cinader, C.W. Kimball: In *Superconductivity in d- and f-band Metals*, ed. by D.H. Douglass (Plenum, New York 1976) p.297
7.29 M.J.G. Lee, J.M. Perz, E. Fawcett (eds.): *Transition Metals 1977*, Conf. Series No.39 (Inst. Phys., Bristol, London 1978) pp.674-715; Ref.7.35
7.30 S.D. Bader, S.K. Sinha: Phys. Rev. B*18*, 3082 (1978)
7.31 M. Born, K. Huang: *Dynamical Theory of Crystal Lattices* (Clarendon, Oxford 1954)
7.32 G. Venkataraman, L.A. Feldkamp, V.C. Sahni: *Dynamics of Perfect Crystals* (MIT, Cambridge 1975)

7.33 G. Gilat, L.J. Raubenheimer: Phys. Rev. *144*, 390 (1966);
L.J. Raubenheimer, G. Gilat: Phys. Rev. *157*, 586 (1967);
G. Gilat: J. Comput. Phys. *10*, 432 (1972)
7.34 J. Rath, A.J. Freeman: Phys. Rev. B*11*, 2109 (1975)
7.35 S.K. Sinha: In *Dynamical Properties of Solids*, Vol.III, ed. by G.K. Horton, A.A. Maradudin (North Holland, Amsterdam 1980) Chap.I
7.36 G.W. Webb, Z. Fisk, J.J. Engelhardt, S.D. Bader: Phys. Rev. B*15*, 2624 (1977); S.D. Bader, F.Y. Fradin: In *Superconductivity in d- and f-band Metals*, ed. by D.H. Douglass (Plenum, New York 1976) p.567
7.37 F.J. Culetto, F. Pobell: Phys. Rev. Lett. *40*, 1104 (1978)
7.38 U. Poppe, H. Wühl: J. Phys. (Paris) *39*, C6-361 (1978)

8. Electron-Phonon Interaction in Chevrel-Phase Compounds

F. Pobell, D. Rainer, and H. Wühl

With 13 Figures

In this chapter, experiments on the electron-phonon interaction in Chevrel-phase compounds (CPC) and a theoretical discussion of their results will be presented. We will particularly discuss measurements of the isotope effect [8.1-4] and tunneling spectroscopy experiments [8.5,6]. These investigations have been performed in order to obtain information about the strength of the electron-phonon interaction in CPC, and to answer the question of whether there are phonon modes that couple particularly strongly to the electrons in these compounds.

8.1 Introductory Remarks

The effect of the electron-phonon interaction on various physical properties of metals, such as the heat capacity, transport properties, phonon damping, or superconducting properties, is among the most thoroughly theoretically investigated and best-understood fields of solid-state physics. A recent review and an extensive list of references can be found in GRIMVALL's article [8.7]. The main development started when it became clear that an understanding of the superconducting transition temperature needs a quantitatively correct description of electron-phonon effects. Some fortunate circumstances, like the existence of a conveniently small expansion parameter ($\sqrt{m_{el}/M_{ion}}$), allowed a very accurate theory of the effects of the electron-phonon interaction in metals to be developed, with an estimated error in the range of a few percent. This accuracy has been verified, in particular for superconductors where the theory is known as "strong coupling theory of superconductivity."

Extensive theoretical investigations have shown that most low-energy properties of metals are strongly influenced by the electron-phonon interaction and hence carry information about this interaction. In view of these facts, it is at first sight surprising how little experimental information on the various electron-phonon interaction parameters is available for many of the materials of interest. There is a lack of data even for elements like the transition metals V, Pd, Mo, Ta ..., and almost nothing is known about complicated compounds such as the CPC. This unsatisfactory situation is primarily a consequence of serious difficulties in disentangling effects due to the coupling of conduction electrons to phonons from so-called band-structure effects. In general, both classes of effects contribute equally to the measured quantities. The notion "band-structure effects" in conduction electron properties means a collection of phenomena that are best described by "everything else but electron-phonon effects." This includes effects of the rigid (periodic) background potential as well as correlation effects caused by the electron-electron interaction.

The important difference between band-structure and electron-phonon effects is their characteristic energy (frequency) scale. The typical energy for band-structure effects is of the order of one Rydberg and is set by quantities like the conduction bandwidth, the Fermi energy, the plasmon frequency, or typical spin-fluctuation frequencies. Consequently, one needs frequencies of about 10^{15} Hz or temperatures of the order of 10^5 K in order to "break up" the electron-electron correlations and to see "dynamical band-structure effects". In contrast, electron-phonon effects, i.e., the correlations between conduction electrons and lattice motions, start breaking up at frequencies of order ω_D ($\simeq 10^{13}$ Hz) or at the corresponding temperatures θ_D ($\simeq$ 400 K). The pronounced separation in energy is a consequence of the large difference in the masses of electrons and nuclei and is necessary for a reasonable distinction between band-structure effects and electron-phonon effects. This condition of separation is, indeed, well fulfilled for conventional metals and is expected to also hold in Chevrel-phase metals. There is no indication of soft electronic modes in CPC. The typical energy for the conduction band is about 1 eV [8.8-11], which is quite small but still much larger than the typical phonon energy. Like all known superconducting metals, CPC are not on the verge of being itinerant magnets, which excludes soft paramagnons. There is also no evidence for soft plasmons in these compounds. Therefore, it is reasonable to assume that they can be described by the conventional theory of electron-phonon coupling, and we will make this assumption throughout. The separation in energy of band-structure and electron-phonon effects is of particular importance for measurements of electron-phonon interaction parameters.

Electron-phonon induced features in conduction-electron properties are most pronounced in the superconducting state. Hence, the most valuable information is usually obtained from measurements of superconducting properties. In fact, our present – admitted rather limited – knowledge of electron-phonon interaction parameters in CPC comes nearly entirely from measurements in their *superconducting* state. These experiments and their analysis will be discussed in detail in later sections. To our knowledge, the only *normal*-state data in CPC which have been analyzed with regard to electron-phonon parameters are the temperature dependences of the normal-state resistivity and the electronic specific-heat coefficient (Sect.8.2).

Before going into the details of experiments on the electron-phonon coupling, we will mention some of the properties of CPC that will be needed later for an understanding of the results of these experiments.

The general formula of ternary CPC is $M_xMo_6X_8$, where M stands for a large number of metals (or can be absent), X = S, Se or Te, and x = 0 to 4. Many of the properties of CPC are intimately related to their unique (mostly hexagonal rhombohedral) crystal structure [8.12,13]. The main building blocks are the almost-cubic units Mo_6X_8. They – and especially their Mo_6 octahedra – are believed to be essential for many properties of CPC. In the channels between the Mo_6X_8 units, metals M can be inserted. They occupy the origin of the rhombohedral unit cell, or—if M is a small ion—some of 12 sites which are closely grouped around it [8.13]. This crystal structure is the origin of the remarkable electronic and vibrational properties of CPC.

The influence of the electron-phonon coupling in CPC can perhaps be seen by the mode softening observed in neutron scattering [8.14-18] and Mößbauer effect [8.19, 20] experiments, as well as by lattice instabilities [8.21-23]. In addition, at least in some CPC, the structure leads to an anharmonic, large vibrational amplitude of the M atom [8.22]. The latter is quite considerable in $SnMo_6S_8$, for example, as shown by Mössbauer-effect measurements [8.19,20] and X-ray structural analysis [8.24]; a harmonic model is inadequate for a description in this case.

The unique structure of CPC and the results of heat capacity measurements led BADER et al. [8.14,15,25] to introduce a "molecular crystal model" in order to discuss the complex lattice dynamics of MMo_6X_8. This simplifying model allowed grouping the 45 normal modes of MMo_6X_8 into 36 hard internal modes of the Mo_6X_8 unit ($E > 18$ meV) and 9 soft external modes of MMo_6X_8 ($E < 18$ meV). The latter contain 3 acoustical and 3 optical translational modes of M and Mo_6X_8, and 3 torsional modes of the Mo_6X_8 unit.

This model has been used in qualitative discussions of various experimental information on CPC, such as results from specific heat and inelastic neutron scattering experiments [8.14-18,25], and Mössbauer effect [8.19,20] or isotope effect [8.1,2] data. But YVON [8.13] has pointed out that at best the Mo_6 octahedron and not the entire Mo_6X_8 unit can be considered as a "rigid molecule." Furthermore, the generalized phonon density of states obtained from the neutron experiments indicated a strong hybridization of the phonon modes [8.14-18]. The recent lattice dynamical calculations of BADER and SINHA [8.26] showed in detail a strong mixing of external and internal modes, for example, of torsional and breathing modes of the Mo_6 octahedron. They demonstrated that Mo_6X_8 can not be treated as quasi rigid.

In connection with the molecular crystal concept, it was suggested [8.14,15,25] that the external modes, particularly the torsional modes which involve relative displacements of the Mo_6 octahedra, are important for the electron-phonon interaction in CPC. We will demonstrate that this concept is incompatible with results for the isotope effect exponents. This conclusion is supported by the tunneling spectroscopy data and theoretical arguments.

Rather early, FISCHER et al. [8.27,28] suggested that superconductivity of CPC is due to the Mo 4d electrons which are strongly localized within the Mo_6 octahedron. This is confirmed by recent band-structure calculations [8.8-11]. ANDERSEN et al. [8.8] showed that the Fermi level falls into a doubly degenerate band of Mo 4d electrons dominated by E_g levels of an isolated Mo_6 octahedron. The very small bandwidth results from the weakness of intercluster coupling, which is about an order of magnitude weaker than the intracluster coupling [8.8-11]. This implies that the electron-phonon interaction is dominated by intracluster coupling of Mo d states. Modes of a rigid cluster interact with the E_g states only via the weak intercluster coupling and should be unimportant.

Before we discuss in detail the isotope effect and tunneling experiment, we will summarize information on the electron-phonon interaction in CPC which has been obtained from other experiments.

8.2 Experiments on the Electron-Phonon Coupling

Among the normal-state quantities influenced by electron-phonon interaction, it is the electronic specific heat, $C_e = \gamma T$, and the temperature dependence of the electrical resistivity, $\rho(T)$, which have been measured and analyzed. Further information on the coupling strength has been obtained from the temperature dependence of C_e in the superconducting state, from the jump ΔC_e of the specific heat at T_c, from T_c itself, and from EPR and NMR measurements. Measurements of the isotope effect and tunneling experiments contain the most detailed information on the electron-phonon coupling and are discussed separately in Sects.8.4,5. A collection of experimental results on the electronic specific heat coefficient γ, on $\Delta C_e/T_c$, and on the reduced energy gap ratio $2\Delta_0/kT_c$ for some CPC of interest is displayed in Table 8.1.

None of the above-mentioned experiments allow a direct determination of any of the parameters like λ or η ($\lambda = \eta/M\langle\omega^2\rangle$) that are commonly used to characterize the strength of the electron-phonon coupling. They provide, at most, a qualitative estimate of trends in the strength of the coupling. The proper determination of λ (or η) by McMillan's inversion of tunneling spectra is not yet applicable in CPC because of the insufficient accuracy of present tunneling data (Sect.8.4). All alternative methods for generating λ values from experimental data rely on ad-hoc assumptions about other unknown material parameters and should be taken with care. We will, nevertheless, review in the following the λ values obtained by such methods but also mention in each specific case the ad-hoc assumptions needed to pin down λ. The accuracy or reliability of these λ values cannot be assessed.

In the coefficient γ of the electronic specific heat which is proportional to the quasi-particle density of states at the Fermi level, $N_\gamma(E_F)$, the electron-phonon interaction shows up as a factor $1 + \lambda$ enhancing the band density of states

Table 8.1. Collection of experimental results on the electronic specific heat coefficient, the specific heat jump at T_c, and on the reduced energy gap at zero temperature. In [8.5,34] the gap has been determined from tunneling data, whereas in [8.30,31,35,36] a gap is analyzed from thermodynamic, EPR and NMR measurements

	Mo_6Se_8	Mo_6S_8	$PbMo_6S_8$	$Cu_{1.8}Mo_6S_8$	$SnMo_6S_8$	References
γ	44	-	105	-	79	[8.29]
$\left[\frac{mJ}{mole\ K^2}\right]$	47	-	-	-	-	[8.30]
	75	28	125	63	84	[8.31,32]
	21	-	-	-	-	[8.33]
$\Delta C_e/T_c$	70	-	215	-	155	[8.29]
$\left[\frac{mJ}{mole\ K^2}\right]$	107	-	-	-	-	[8.30]
	93	-	160	-	104	[8.31,32]
	41	-	-	-	-	[8.33]
	-	-	4.8	4.2	-	[8.5]
	4.1	-	-	-	-	[8.34]
$\frac{2\Delta_0}{kT_c}$	4.2	-	-	-	-	[8.30]
	3.5	-	3.9	-	-	[8.31]
	-	-	5	-	5	[8.35]
	-	-	3.5	-	-	[8.36]

$N_{BS}(E_F)$. Measurements of γ in CPC are very difficult. Strong deviations of the lattice specific heat from Debye behavior due to the complicated phonon spectrum raise problems when separating the electronic and lattice contributions in the measured heat capacity. The dependence on stoichiometry or amount of impurity phases seems also to contribute to the large scatter in the data. Furthermore, the bare band density of states is not known. Hence, an estimate of λ from γ measurements barely has significance. We take $PbMo_6S_8$ as a typical example. If one extracts $N_{BS}(E_F)$ from susceptibility data, assuming that the measured susceptibility is unenhanced and arises from spin only, the comparison with N_γ gives a value for λ in $PbMo_6S_8$ of about 1.3 [8.37]. Taking into account on the other hand, an exchange interaction of the size obtained for Mo metal in estimating $N_{BS}(E_F)$, the unduly high value $\lambda = 2.5$ has been obtained [8.8].

The most common way of estimating λ is by means of the empirical McMillan formula for T_c. It expresses T_c in terms of λ, the Coulomb parameter μ^*, and an average phonon frequency. The most serious uncertainty in estimating λ from measurements of T_c arises from considerable uncertainties in choosing the average phonon frequency. The most popular guess, a frequency centered around 12 meV (the range of intercluster vibrations in the molecular crystal model), yields $\lambda = 1.2$ for $PbMo_6S_8$ [8.37]. A list of λ's obtained in this way ("poor man's λ") is given in [8.29,37]. With these λ values, and using the relations $\gamma \sim N_{BS}(E_F)(1 + \lambda)$, a strong dependence of λ on the band density of states $N_{BS}(E_F)$ is obtained for CPC

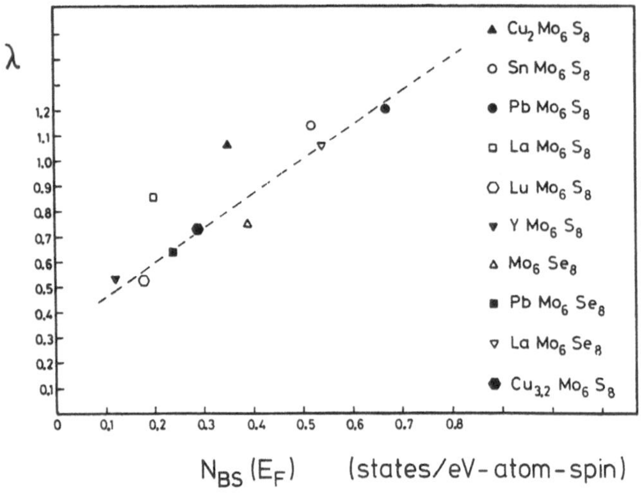

Fig.8.1. Dependence of the "poor man's" electron-phonon coupling constant λ on the bare density of states at the Fermi level for several CPC, from [8.37]

[8.29,37] as plotted in Fig.8.1. Some additional discussion on estimates of λ from T_c's from a more theoretical point of view, but also not free of ad-hoc assumptions, is given in Sect.8.3.

Qualitative information on the strength of the electron-phonon interaction comes from the jump of the electronic specific heat at T_c and the reduced energy gap $2\Delta_0/kT_c$. Measurements on carefully prepared Mo_6Se_8 samples with at most 3% impurity phases showed $\Delta C_e/\gamma T_c = 2.25$ [8.30]. This value is substantially larger than the BCS value 1.43. The enhancement is a pronounced strong-coupling effect which indicates a sizeable coupling to low-frequency phonons and hence an unusually large λ parameter (see also Sect.8.3). In [8.30] the full temperature dependence of C_{es} was analyzed by a phenomenological equation of PADAMSEE et al. [8.38], and a good fit was obtained using a reduced gap of $2\Delta_0/kT_c = 4.2$. This, however, should be compared with the value of 3.5 obtained by ALEKSEEVSKI et al. [8.31] from an analysis of their specific heat data. The most direct information on the gap of Mo_6Se_8 comes from tunneling experiments [8.34]. The authors find a ratio $2\Delta_0/kT_c = 4.1 \pm 0.1$ in agreement with the specific heat data of [8.30]. More data for $2\Delta_0/kT_c$ in CPC, including the tunneling data discussed in Sect.8.5, are given in Table 8.1.

Another source of information on the electron-phonon coupling has been exploited by BADER and SINHA [8.26]. They calculated the electrical resistivity $\rho(T)$ in order to get insight into the coupling of distinct phonon groups to the electrons. If all phonon modes are weighted equally, a dependence of $\rho(T)$ proportional to T^2 instead of the usual T^3 law is obtained at $T < 40$ K. A T^2 law has been observed in sputtered $PbMo_6S_8$ and other CPC [8.39,40]. The authors [8.26] showed the importance of high-frequency modes for the T^2 behavior. The omission of modes associated with the displacements of S atoms or the complete cut-off of all modes with $\omega > 18$ meV led to deviations from the T^2 behavior. Modes associated with displacements of the Pb atom have no decisive influence on the temperature dependence. But on the other side, for sintered and single crystalline $PbMo_6S_8$ samples believed to be free of impurity phases, a linear T dependence has been observed [8.30,40,41].

The strength of the electron-phonon coupling has also been investigated with electron paramagnetic resonance (EPR) [8.35] and nuclear magnetic resonance (NMR) [8.36]. The EPR experiments have been performed on Gd diluted in $SnMo_6S_8$ and $PbMo_6S_8$. From the temperature dependence of the impurity-to-conduction electron relaxation rate, a value for $2\Delta_0/kT_c$ of roughly 5 has been extracted. The temperature dependence of the nuclear spin lattice relaxation rate measured in $PbMo_6S_8$ and $PbMo_6Se_8$ by means of NMR was interpreted by the authors as an indication of an energy gap of only 3.5 kT_c "at both sites of Pb and Se."

Summarizing the experimental evidence, one can say that although no accurate values for λ are yet available, essentially all or at least most experiments indicate a rather strong electron-phonon coupling in CPC, even for the compounds with a relatively low T_c.

8.3 Theoretical Models for the Electron-Phonon Coupling

Presently available experimental data on the electron-phonon interaction in CPC are much too limited to reveal details of the full coupling constant $g(\underline{k}\nu,\underline{k}'\nu';j)$ for the transition of a conduction electron with "momentum" $\underline{k}$ in band ν into a state with momentum $\underline{k}'$ in band ν' under the emission (absorption) of a phonon of momentum $(\underline{k}-\underline{k}')$ and branch-index j. One does not even know the Eliashberg function $\alpha^2F(\omega)$, which contains information on some average over electron-phonon processes at the Fermi surface:

$$\alpha^2F(\omega) = N(E_F) \sum_{\nu,\nu',j} \langle |g(\underline{k}\nu;\underline{k}'\nu';j)|^2 \delta(\omega - \omega_j(\underline{k} - \underline{k}'))\rangle_{FS} \quad , \tag{8.1}$$

a function which determines T_c and other superconducting data. A full determination of $\alpha^2F(\omega)$, which usually comes from a computer inversion of tunneling spectra, is still unattainable for Chevrel-phase superconductors (Sect.8.5). Adding to the present uncertainty is the absence of detailed information on the lattice dynamics $(\omega_j(\underline{q}))$ in these compounds (Chap.7).

In such a situation, any theoretical assistance for an interpretation of physical data should pursue the following strategy.

1) One needs plausible models as working hypotheses whereever direct information is missing (models for the lattice dynamics, for the electron-phonon coupling, etc.).
2) From these models one must calculate observable quantities (e.g., the transition temperature, the electric conductivity, etc.).

3) One must judge whether the results are compatible with experimental findings and whether to keep or discard the underlying models.

Starting from a model for the lattice dynamics and for the electron-phonon coupling, one can calculate superconducting properties such as T_c, the isotope effects, the thermodynamic critical field or other thermodynamic quantities, and the tunneling spectrum including the energy gap. A powerful theory, the strong-coupling theory of superconductivity [8.42], and efficient numerical methods allow a rather easy, fast, and accurate evaluation. Further quantities which are influenced by the electron-phonon coupling, and which can be calculated for given models, are the temperature-dependent resistivity, infrared properties, the Landau damping of phonons, the de Haas-van Alphen effect, the upper critical field (H_{c2}), and others. However, all the latter quantities rely heavily on band-structure models for the electronic quasi particles at the Fermi surface. This adds further uncertainties and might rule out an adequate check of a particular model of the electron-phonon coupling.

The most widely used model for the Eliashberg function $\alpha^2F(\omega)$ is due to McMILLAN [8.43]. It was designed for general superconductors and has a remarkably small number of parameters. McMILLAN assumed that the superconducting parameters are insensitive to the details in the shape of $\alpha^2F(\omega)$. For all superconductors a standard shape for $\alpha^2F(\omega)$ is used, namely, the phonon density of states of Nb with the low-frequency part cut off. The only adjustable parameters in this model are the scale of the phonon frequencies (it is conventionally fixed by adjusting the maximum frequency in $\alpha^2F(\omega)$ to the maximum phonon frequency), the coupling strength (conventionally expressed in terms of λ), and the Coulomb pseudopotential μ^*. The Eliashberg equations plus the Nb-type shape for $\alpha^2F(\omega)$ (the McMillan model) allows a straightforward computer analysis of important data such as the transition temperature, the excitation gap, the heat capacity, or the thermodynamic critical field. Its most widespread use is to estimate the coupling parameter λ from measurements of T_c. We have performed such an analysis [8.44] for several Chevrel-phase superconductors. The results are shown in Figs.8.2,3, where we display λ versus μ_5^* (the index 5 indicates a high-frequency cut-off at five times the maximum phonon frequency). For comparison we have included results from an alternative, but a priori not better justified, guess for the shape of $\alpha^2F(\omega)$. For each compound we have taken the form of its "generalized phonon density of states" $G(\omega)$ of 8.16-18 as model for the corresponding $\alpha^2F(\omega)$ spectrum. The $G(\omega)$ spectra have more spectral weight at low frequencies than the Nb-type spectrum (Fig.8.4), and this leads to a higher λ value at given T_c and μ_5^*. The substantial differences indicate the sensitivity of T_c to the particular shape of $\alpha^2F(\omega)$ and demonstrate the uncertainties in the determination of λ from T_c measurements alone. Nevertheless, the results displayed in Figs.8.2,3 allow an estimate of the λ parameters which should be reasonable, unless the true (and presently unknown) $\alpha^2F(\omega)$ spectra differ significantly from the model spectra.

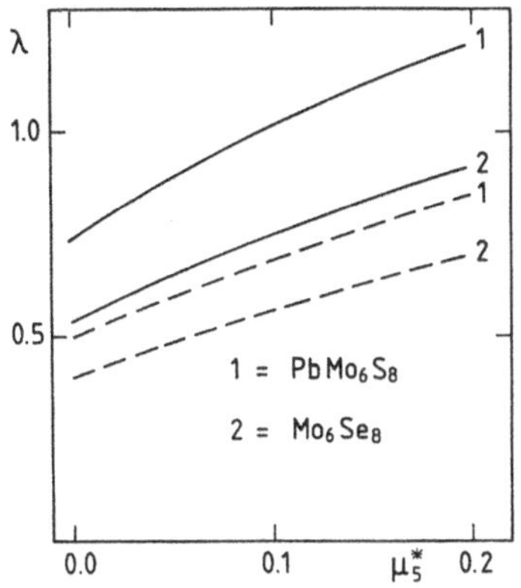

Fig.8.2

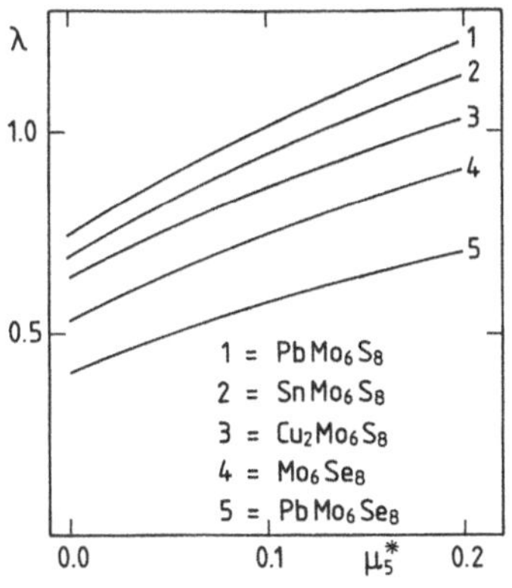

Fig.8.3

Fig.8.2. Calculated λ values versus Coulomb pseudopotential μ_5^* for Mo_6Se_8 and $PbMo_6S_8$ using two ad-hoc guesses for $\alpha^2F(\omega)$. The dashed lines are obtained for McMILLAN's Nb-type spectrum [8.39] (Fig.8.4) and the solid lines are for the generalized phonon density of states (G(ω)) of [8.16-18]. The data demonstrate the sensitivity of λ to the shape of $\alpha^2F(\omega)$. G(ω) represents a spectrum with about equal coupling to all phonons, whereas the coupling to high-frequency phonons dominates in the Nb-type spectrum

Fig.8.3. Calculated λ values versus Coulomb pseudopotential μ_5^* for several Chevrel-phase superconductors. The results are obtained by taking the generalized phonon density of states of [8.16-18] as a model for $\alpha^2F(\omega)$ in the corresponding compounds. The data allow an estimate for λ provided the $\alpha^2F(\omega)$ and the phonon spectra do not differ significantly

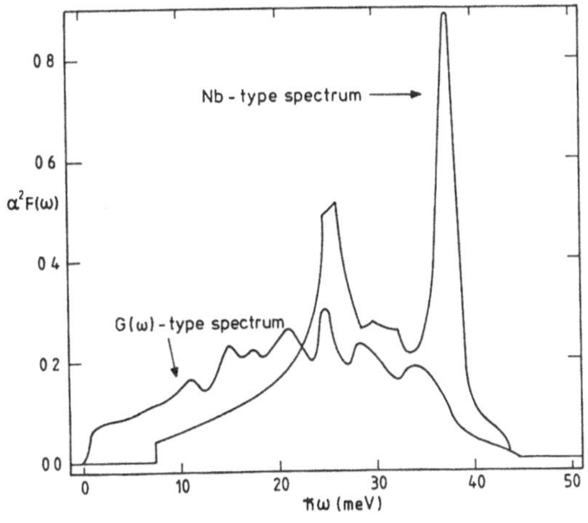

Fig.8.4. Comparison of the Nb-type spectrum (McMILLAN's spectrum [8.39]) and the G(ω)-type spectrum for Mo_6Se_8 [8.16-18]. Both models for $\alpha^2F(\omega)$ yield a T_c of 6.3 K for $\mu_5^* = 0.14$

In addition, we have calculated [8.44] the zero-temperature energy gap and the heat-capacity jump at T_c for two representative materials (Mo_6Se_8, $PbMo_6S_8$). The results are presented in Table 8.2. A comparison with the experimental results of [8.5,30,34] (Table 8.1 and Sect.8.2) indicates the importance of the low-frequency part of $\alpha^2F(\omega)$ for CPC. There is no sizeable spectral weight at low frequencies in the Nb-type spectra, whereas the G(ω)-type spectrum has considerable low-frequency

Table 8.2. Reduced specific-heat jump and reduced zero-temperature energy gap for two representative Chevrel-phase superconductors as obtained from a strong-coupling calculation using the two different models for $\alpha^2F(\omega)$ spectra shown in Fig.8.4 and $\mu_5^* = 0.13$. For a comparison to experimental results see Table 8.1

	$\Delta C_e/\gamma T_c$	$2\Delta_0/kT_c$
Mo_6Se_8 (Nb-type spectrum)	1.51	3.6
Mo_6Se_8 (G(ω)-type spectrum)	1.76	3.8
$PbMo_6S_8$ (Nb-type spectrum)	1.67	3.7
$PbMo_6S_8$ (G(ω)-type spectrum)	2.07	4.1
BCS	1.43	3.5

contributions. As a consequence, the G(ω) models yield more pronounced strong-coupling features and better agreement with experiment. But neither the Nb- nor the G(ω)-type spectra yield satisfactory agreement with current experimental data. One needs a somewhat softer spectrum than G(ω) in order to explain the data of [8.5,34]. Although the shape of the $\alpha^2F(\omega)$ spectra could be altered in an arbitrary number of ways to obtain perfect agreement, the benefit would be negligible, since the few known data cannot determine a reasonably unique $\alpha^2F(\omega)$.

It should also be emphasized that the averaging procedure which leads to $\alpha^2F(\omega)$ washes out details of the electron-phonon coupling which are essential for an interpretation of data such as the isotope effects in superconducting compounds [8.45,46]. For example, $\alpha^2F(\omega)$ contains only information about the total contribution of the coupling from all atoms. But the additional information on the separate contribution of the coupling to the Mo or to the X atoms is an important ingredient for the explanation of partial isotope effects (Sect.8.4). Hence, a more complete analysis of experimental data needs more detailed models for the electron-phonon coupling; two such models have recently been suggested for Chevrel-phase compounds and will be discussed in the following.

The first approach is based on the "molecular-crystal model" for the lattice dynamics. It has been introduced by BADER et al. [8.14,15,25], who in addition postulated the dominance of a coupling of conduction electrons to soft external modes of rigid Mo_6X_8 units. This appealingly simple model for the electron-phonon coupling was neither compatible with the observed isotope effect in Mo_6Se_8 [8.1] (Sect.8.4) nor did its basis, the clear separation of external and internal modes [8.26], withstand a more detailed lattice dynamical calculation. We shall not, therefore, extensivly discuss this model but refer the interested reader to [8.1,2,14,15,25,26,45] and to Sect.8.4.

All present evidence favors an alternative model, the "model of weakly coupled cluster states." It is a spin-off from band-structure calculations by MATTHEIS and FONG [8.9], ANDERSEN et al. [8.8], and BULLETT [8.10]. Even though the approaches of these groups are quite different, they all indicated a very typical and unique feature of the conduction band in Chevrel-phase metals. To a good approximation, the most important states at the Fermi energy originate from the 4d atomic levels of Mo, and the conduction-electron states can be constructed by first combining the 6×5 atomic levels of a Mo_6 cluster into 30 cluster orbitals. Intracluster interactions lead to a splitting of the cluster levels of about 10 eV. Cluster orbitals from different cells interact only weakly and form tight-binding bands. The weak intercluster interaction results in narrow bands (about 0.5 eV wide) with a high density of states.

This picture for the conduction band has very definite implications for the electron-phonon coupling. It predicts a much stronger coupling of the conduction electrons to internal deformations of a Mo_6 cluster than to the relative displacement of two neighboring clusters. The relative size of the couplings to the two distinct types of lattice displacement can be estimated by the ratio of intercluster to intracluster splitting energies, of the order 0.1. Hence, one may safely neglect intercluster contributions to the electron-phonon coupling, and gain significant simplifications. For example, the high (octahedral) symmetry of an isolated Mo_6 cluster may be utilized fully because the electron-phonon coupling is unaffected by neighboring cells. Strong selection rules and a small number of electron-phonon coupling parameters result. A conduction band formed from cluster states of E_g type, as has been suggested in [8.8], needs only two coupling parameters (g_1, g_2). They determine the coupling to A_{1g}-type deformations of the Mo_6 octahedron (breathing modes) and to the doubly degenerate E_g-type deformations (cigar modes). Hence, the Hamiltonian for the electron-phonon interaction has the simple two-parameter form:

$$
\begin{aligned}
H_{e\text{-}ph} = \sum_\sigma \sum_i \Big[& g_1 Q_{iA} (C^+_{i1\sigma} C_{i1\sigma} + C^+_{i2\sigma} C_{i2\sigma}) \\
& + g_2 Q_{iE1} (C^+_{i1\sigma} C_{i1\sigma} - C^+_{i2\sigma} C_{i2\sigma}) \Big] \\
& - g_2 Q_{iE2} (C^+_{i1\sigma} C_{i2\sigma} + C^+_{i2\sigma} C_{i1\sigma}) \quad . \qquad (8.2)
\end{aligned}
$$

$C^+_{i1\sigma}$, $C^+_{i2\sigma}$ are creation operators for the two cluster states of E_g symmetry in cell i (Wannier states), and Q_{iA}, Q_{iE1}, Q_{iE2} are the operators for A_{1g}- and E_g-type deformations of the Mo_6 octahedron in cell i.

This model for the electron-phonon coupling must be joined with the band-structure model (from which it is derived) and a model for the lattice dynamics in order to allow a complete description of superconducting properties of CPC. The weakest part of this scheme is the lattice dynamics, for which detailed experimental information or reliable models are not yet available. It is therefore still open to future investigations to check whether the model of weakly coupled cluster states provides an adequate description of Chevrel-phase superconductors. The only check presently available is a calculation of the isotope effects [8.46], which gave reasonable agreement with recent data on Mo_6Se_8 (Sect.8.4). It would be encouraging for further theoretical work on CPC if this surprisingly simple model for a complicated system is confirmed by future experiments.

8.4 Isotope Effect of T_c in Mo_6Se_8 and $SnMo_6S_8$

8.4.1 The Isotope Effect

The mass dependence of T_c is the essential experimental information that electron pairing in superconductors occurs via exchange of phonons [8.47,48]. If the phonon-induced interaction dominates, then the so-called isotope-effect exponent β in the equation $T_c\ M^\beta$ = constant should have a value near 0.5. For a multicomponent metal, the latter relation has to be generalized to

$$d\ \log T_c = -\sum_r \beta_r\ d\ \log M_r \qquad (8.3)$$

with

$$\beta = \sum_r \beta_r \quad .$$

The partial-isotope-effect exponents β_r are the response coefficients of T_c to small changes of the masses M_r of the constituents of the compound. Other than T_c or β, the coefficients β_r can distinguish between contributions from individual constituents.

This was the basic idea behind the experiments on Mo_6Se_8 and $SnMo_6S_8$ [8.1-4] to be discussed in the following. Of course, those experiments can only give information on the contribution of particular phonon modes to β_r and T_c if one applies a model to the lattice dynamics of CPC. Therefore the results of these experiments allow—at least partially—the question about the influence of the various phonon modes on T_c to be answered.

Besides the conventionally considered direct isotope effect, i.e., a mass dependence of the superconducting transition temperature T_c which comes from the mass dependence of the phonons at fixed spring constants, there exist indirect isotope effects. They arise from the mass dependence of the zero-point vibrations of the lattice [8.49-51] and their influence on electronic properties and spring constants. The indirect effects must be subtracted from the experimental results before a comparison to theoretical results for the direct isotope effect is possible.

Usually, these indirect effects are small because they are proportional to some power of the ratio of zero-point displacement to lattice constant. Only in exotic cases, like Pd-H(D) [8.52], is the isotope effect dominated by indirect mechanisms.

For the Pd-H(D) system, this is probably due to the strongly anharmonic motion of the H(D) ion [8.50]. For an order of magnitude estimate, we scale the large and inverse indirect exponent of Pd-H(D) with the ratio of the mean-square displacements, and find $-0.1 \lesssim \beta^{ind} < 0$ for CPC.

Another indirect effect can result from the mass dependence of the unit-cell volume [8.49]. This effect can be expressed as

$$\beta^{ind,V} = -\frac{d \log T_c}{d \log V} \frac{d \log V}{d \log M} \quad . \tag{8.4}$$

It was found to be about + 0.1 for Mo metal, which, to our knowledge, is the only case where it has been determined [8.49].

Using the pressure dependence of T_c and the compressibility κ, the exponent $\beta^{ind,V}$ can be written as

$$\beta^{ind,V} = \frac{1}{kT_c}\left(\frac{dT_c}{dp}\right)_0 \frac{d \log V}{d \log M} \quad . \tag{8.5}$$

Unfortunately, the value of d log V/d log M is not known for Chevrel-phase superconductors, but from the sign of $(dT_c/dp)_0$ [8.53], we can conclude that $\beta^{ind,V}$ is positive for Mo_6Se_8 and $SnMo_6S_8$. Here, we have assumed the usual negative sign for d log V/d log M, of course. As discussed in the next section, it was found that the lattice parameters x of the hexagonal cell of Mo_6Se_8 ($SnMo_6S_8$), which are of the order of 10 Å, remained constant to 0.005 Å (0.006 Å) at room temperature when the isotopic mass was changed by 8% (7%) [8.2]. If we take this value, the value for $(dT_c/dp)_0$ from [8.53], and the value for the compressibility of CPC from [8.54], we find $0 < \beta^{ind,V} \lesssim 0.1$ for the isotopic volume exponent for CPC. But let us repeat that the values Δx have been determined at room temperature, and they are the values for the parameters of the hexagonal cell of CPC and not for a particular constituent, such as Sn.

We have presented these estimates of indirect-isotope-effect exponents to confirm our belief that they are of order ±0.1 or smaller in CPC.

8.4.2 Sample Preparation

The binary system Mo_6Se_8 is well suited for measurements of partial-isotope-effect exponents because both atoms have a substantial number of stable isotopes and their mass can independently be varied by about 8%. For an investigation of the influence of the M atom on T_c, the ternary compound $SnMo_6S_8$ seems to be a good candidate for the same reason. Of course, the expected variation of T_c will only be a few percent. Usually, the transition width of Chevrel-phase superconductors, and the T_c scattering of nominally identical samples are at least a few tenths of a degree. Therefore, the main experimental problem for measuring isotope effects is the synthesis of compounds with a reproducible and sharp transition temperature.

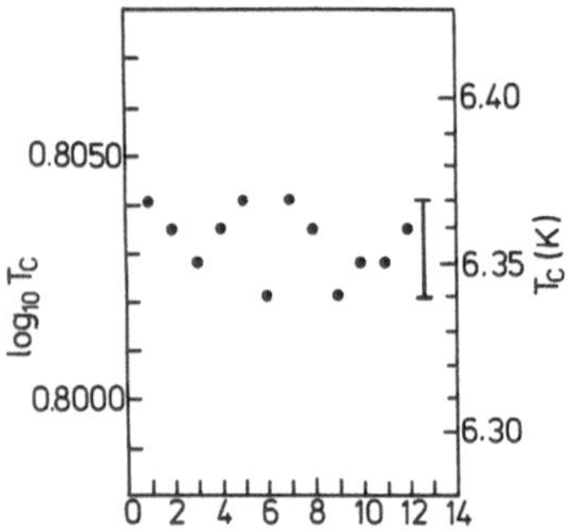

Fig.8.5. Scattering of T_c of 12 nominally identical Mo_6Se_8 samples synthesized in one reaction cycle; from [8.2]

For the experiments of CULETTO and POBELL [8.1-3], the optimal preparation conditions for obtaining single-phase samples with appropriate transitions have been found empirically using natural Mo, Se, S, and Sn as starting materials. Starting compositions $Mo_6Se_{7.6}$ and $Sn_xMo_6S_{7.6}$ were taken[1], because the stoichiometric composition leads to a $MoSe_2$ contamination [8.24,55]. The preparation procedure allowed simultaneous production of up to 12 samples under identical conditions in one reaction cycle. The resulting samples containing the natural isotope composition have been used to study the T_c scatter of nominally identical substances. In the case of $Mo_6Se_{7.6}$, the maximum T_c scatter was $\Delta T_c = 0.015$ K, as shown in Fig.8.5. The scatter of T_c for samples from different reaction cycles was a few hundredths of a Kelvin for this compound. A careful specific heat study on these samples showed that they contain at most 3% $MoSe_2$ and Mo [8.30]; X-ray analysis did not indicate any impurity phases to within the resolution of several percent.

In the production of $Sn_xMo_6S_{7.6}$ with x = 0.9 to 1.2, a strong dependence of T_c on x was detected. Absence of impurity phases and the smallest scatter of T_c were observed for samples with the smallest T_c, which occurred for x = 1.2 [8.2,24]. The dependence of T_c on x is the reason for the rather large scatter of up to $\Delta T_c \sim 0.14$ K for Sn compounds from one reaction cycle, and of $\Delta T_c \sim 0.3$ K for samples from different reaction cycles. The transition widths were 0.07 K.

For the preparation of the isotope samples [8.56], the same procedure was followed. The lattice parameters of the hexagonal cells evaluated from X-ray diffraction diagrams were constant to within ±0.005 Å ($\bar{a}$ = 9.56 Å, $\bar{c}$ = 11.17 Å) for $Mo_6Se_{7.6}$, and to within ±0.006 Å ($\bar{a}$ = 9.16 Å, $\bar{c}$ = 11.34 Å) for $Sn_{1.2}Mo_6S_{7.6}$ [8.2]. All samples were single phase (to within the X-ray detection limit) except for $^{100}Mo_6Se_{7.6}$ which contained a barely detectable amount of an unidentified second phase. The superconducting transitions were measured inductively.

1 It was not investigated whether the appropriate formula for the reaction products is $Mo_6Se_{7.6}$ ($Sn_{1.2}Mo_6S_{7.6}$) or $Mo_{6.3}Se_8$ ($Sn_{1.3}Mo_{6.3}S_8$), or whether the reaction product deviated from the starting composition. In [8.24], indications were found that reaction of $Sn_{1.2}Mo_{6.35}S_8$ to a single crystal results in a stoichiometric compound. In this article we will often use the stoichiometric formula for convenience.

ALEKSEEVSKI and NIZHANKOVSKII [8.4] have also investigated ${}^{s}Sn^{t}Mo_6S_8$ with s = 112 and 124, and t = 92 and 100. The main difference in their sample preparation procedure was an additional annealing at 1000°C for up to 100 h. For $SnMo_6S_8$ samples with natural isotopic composition and prepared in one run, they report a maximum deviation of ±0.07 K from the mean value of T_c, in agreement with our results [8.2, 3]. On the other hand, T_c values for samples from different runs deviated by up to 2 K, from 11.68 K to 13.80 K for ${}^{124}SnMo_6S_8$, which is substantially larger compared to our deviation of 0.3 K. The transition was measured resistively.

8.4.3 Results for the Isotope-Effect Exponent

Figure 8.6 shows the transition curves of the series $Mo_6{}^{m}Se_{7.6}$ and of ${}^{n}Mo_6Se_{7.6}$ from [8.1-3]. A very similar dependence of T_c on isotopic mass is clearly visible in both cases. Figure 8.7 shows plots of log T_c versus log M, where M is the Mo or Se mass. The experimental data are fitted to the relation T_c = const. $M^{-\beta}$ using a least-squares fit procedure [8.2]. Using the above-mentioned value for the T_c scatter of identical samples ($\Delta T_c = \pm 0.015$ K) to calculate the standard deviation of β, the authors obtained for the isotope-effect exponents

$$\beta_{Mo} = 0.27 \pm 0.04 \quad ,$$
$$\beta_{Se} = 0.27 \pm 0.05 \quad . \qquad (8.6)$$

Because of the small transition widths, the exponents are independent of the definition of T_c.

In [8.4] a value of $\beta_{Mo} = 0.48 \pm 0.1$ was obtained from T_c measurements on $Sn^{92}Mo_6S_8$ and $Sn^{100}Mo_6S_8$.

Figure 8.8 shows a plot of log T_c versus log M_{Sn} for $Sn_{1.2}Mo_6S_{7.6}$ obtained from three production cycles [8.2,3]. Because of the larger scatter of T_c, the mean values of T_c of the samples from the three runs have been set equal; Fig.8.8 shows therefore only the deviation from this mean value of 11.70 K. The data may suggest an inverse partial-isotope-effect exponent β_{Sn} of order -0.1. Because of the scatter of the data points and the scatter of T_c of nominally identical samples, and because Fig.8.8 already includes some statistics (it only shows deviations from the mean value), the authors [8.2,3] did not claim to have seen a dependence of T_c on the Sn mass; a limit $|\beta_{Sn}| \le 0.05$ was given [8.2].

ALEKSEEVSKII and NIZHANKOVSKII [8.4] have observed a very large negative isotope-effect exponent β_{Sn} which, however, depended on the annealing procedure applied to the samples. For seven series of ${}^{s}SnMo_6S_8$, each with s = 112 and 124, that were subjected to annealing, a mean value of $\beta_{Sn} = -0.23 \pm 0.04$ was obtained [8.4].

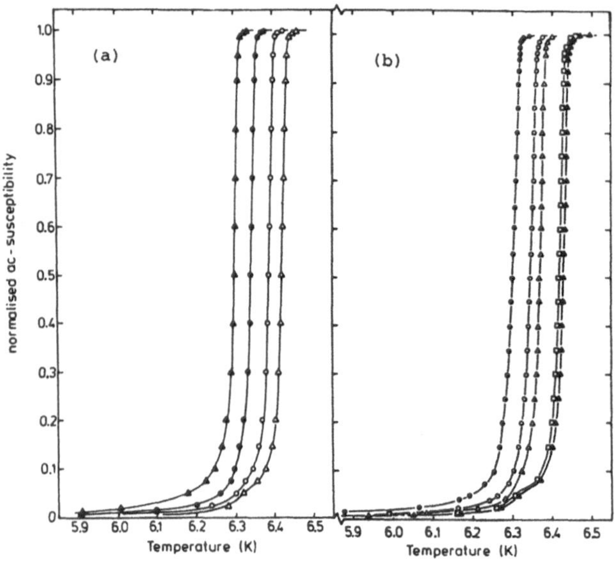

Fig.8.6a,b. Inductively measured superconductive transitions of (a) $Mo_6{}^{m}Se_{7.6}$ and (b) ${}^{n}Mo_6Se_{7.6}$. The nominal isotopic masses are m=82, 80, 78, 76, and n=100, 98, 96, 94, 92 from left to right; from [8.1-3]

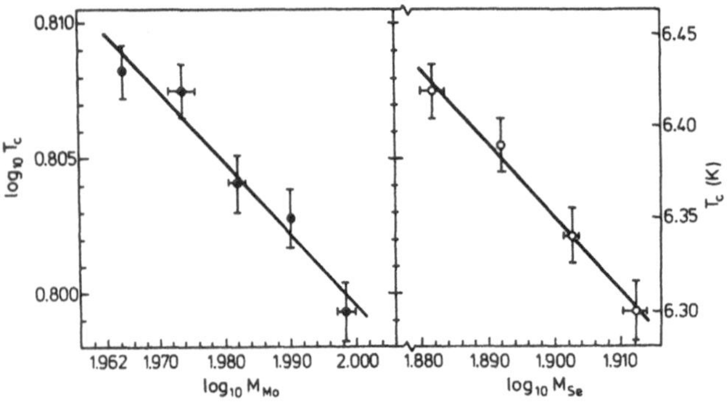

Fig.8.7. Double-logarithmic plot of T_c versus isotopic mass M of Mo and Se in Mo_6Se_8. The lines have slopes of -0.27; from [8.1-3]

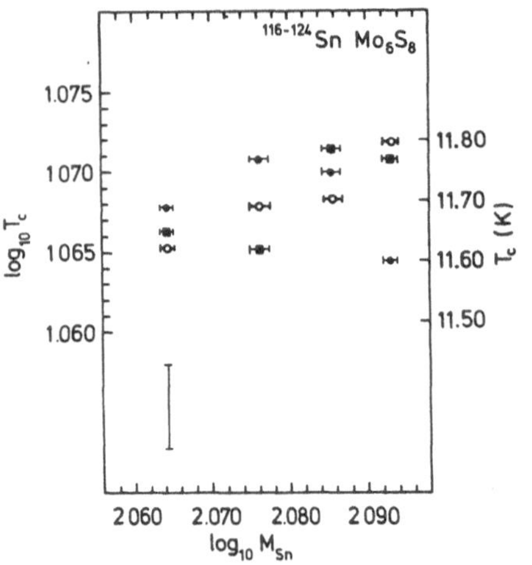

Fig.8.8. Double-logarithmic plot of T_c versus isotopic mass of Sn in ${}^{r}SnMo_6S_8$ for r = 116, 119, 122, and 124. The vertical error bar shows the T_c scatter of nominally identical samples from one reaction cycle. Because the scatter for samples from different reaction cycles is about a factor of two larger, the mean T_c values for the three series have been set equal in this plot; from [8.2,3]

8.4.4 Discussion of the Measured Isotope Effect

a) *Mo_6Se_8*

There are two immediate conclusions available from the data on Mo_6Se_8. First, the result $\beta_{Mo} + \beta_{Se} \sim 0.5$ indicates that the "exotic" properties of CPC do not result from "exotic" interactions; their superconductivity and the electron pairing is dominantly determined by the conventional electron-phonon coupling.

The observation of a total exponent of + 0.54, which means slightly larger than 0.5, might be due to indirect contributions mentioned in Sect.8.4.1. With the values $\lambda \simeq 0.8$ [8.30] and $\mu^* \simeq 0.13$, one would expect a direct total isotope exponent $\beta_{Mo} + \beta_{Se} = 0.45$ [8.45], and therefore $\beta^{ind} \simeq 0.1$ in agreement with our estimates in Sect.8.4.1. In the absence of any further information, we make the plausible assumption that the small indirect contributions are equally distributed between Mo and Se. The second basic information then, the result $\beta_{Mo} \simeq \beta_{Se}$, demonstrates that vibrations of the 6 Mo and the 8 Se atoms contribute about equally to the transition temperature. This is the essential information which we now will use in a more detailed discussion.

A first attempt by the authors of [8.1,2] to interpret their results was based on the molecular crystal model of BADER et al. [8.14,15,25], i.e., on the separation of the 42 phonon modes per unit cell of Mo_6Se_8 into three groups: 3 acoustic translational plus 3 torsional modes of the "rigid" Mo_6Se_8 units (external modes), and 36 internal modes. Each of the three types of modes is characterized by frequencies $\bar{\omega}_{trans}$, $\bar{\omega}_{tors}$, $\bar{\omega}_{int}$, whose mass dependences—which cause the isotope effect—are given by

$$- d \log \bar{\omega}_i^{-2} = c^i_{Mo} \, d \log M_{Mo} + c^i_{Se} \, d \log M_{Se} \quad . \tag{8.7}$$

The coefficients c^i_{Mo} and c^i_{Se} describe how much the 6 Mo or the 8 Se atoms contribute to a particular mode. The Se atoms contribute much more strongly to torsional modes of a Mo_6Se_8 unit than the Mo atoms. This follows from the picture of torsional modes as rigid librations of Mo_6Se_8 units. A detailed analysis of the molecular crystal model yields c^i_{Se}/c^i_{Mo} values of 1.1, 3.3, and 1.2 for the translational, torsional, and internal modes, respectively [8.1,2]. For a discussion of the isotope effect the above phonon model must be supplemented by a model for the coupling of conduction electrons to the three groups of phonon modes. An early conjecture [8.14,15,25] attributed the superconducting properties of CPC to a dominant coupling of conduction electrons to the soft external modes, particularly to the torsional modes of rigid Mo_6Se_8 units. This could be ruled out by the observed isotope effects. A coupling to torsional modes alone leads to a ratio β_{Se}/β_{Mo} of 3.3, and equal coupling to only translational and torsional modes gives a ratio of 1.6 [8.1,2]. This must be compared with the experimental result $\beta_{Se}/\beta_{Mo} = 1.0 \pm 0.3$. If one sticks to the molecular crystal model for the lattice dynamics, the only sensible way to explain the partial isotope effects is to assume that the internal modes couple at least as strongly as the external modes to conduction electrons [8.1,45]. This lead the authors of [8.1,2] to reject the model of "dominant coupling to soft torsional modes" of [8.14,15,25]. The doubts about the model were supported by subsequent lattice dynamical calculations of BADER and SINHA [8.26], which call into question the molecular crystal model for phonons. This removes the basis for the electron-phonon coupling model discussed above and makes a revised interpretation of the isotope effect necessary.

In order to check if the improved phonon model of [8.26] is compatible with the isotope effects in Mo_6Se_8, one has to add to it a model for the electron-phonon coupling and to calculate the isotope effect. This has been done by CULETTO and RAINER [8.45,46], who used the model of "weakly coupled cluster states" described in Sect.8.3. They evaluated the isotope effect on the basis of standard theory of superconductivity [8.42]. This amounts to a calculation of the phonon propagator which enters the kernel of the Eliashberg equations and a subsequent computer solution of the linearized Eliashberg equations. The authors neglected correlations of atomic motions in different cells. This approximation did not influence the calculated isotope effect appreciably, as could be checked by using different embeddings in calculating the lattice modes of a single cell. The agreement of the theoretical results, which are shown in Fig.8.9, with experimental data (8.6) is good when both the theoretical and experimental uncertainties are taken into account [8.45,46]. This result should not be considered as a proof for the applied models but rather as a check for their consistency with the experiments. Any other models have to undergo this check of consistency with the rather stringent requirement of $\beta_{Mo} \simeq \beta_{Se}$.

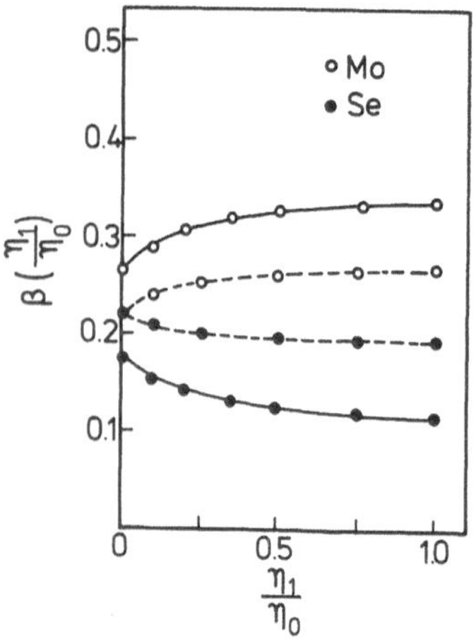

Fig.8.9. Theoretical results for the partial isotope effect exponents in Mo_6Se_8 in the locator model of [8.46] for various ratios of the electronic coupling constants η_1/η_0. The full and dashed lines represent the results of rigid background or periodic boundary conditions, respectively. The overall coupling strength is adjusted to yield a T_c = 6.3 K; from [8.46]

b) *$SnMo_6S_8$*

A small *direct* isotope effect might be expected for the M atom due to the hybridization of its modes with "internal cluster modes" and/or due to the weak overlap of the Mo d orbitals with the Sn atoms. The small value for β_{Sn} reported in [8.2,3] seems to indicate the minor importance of the Sn modes for superconductivity of $SnMo_6S_8$.

If there is a partial isotope effect of Sn due to *indirect* effects, then the result of [8.2,3] is in reasonable agreement with the upper limits calculated in Sect.8.4.1. On the other hand, ALEKSEEVSKII and NIZHANKOVSKII reported the rather large inverse partial-isotope exponent of $\beta_{Sn} = -0.23 \pm 0.04$. (The error estimate seems to be surprisingly small with regard to the scatter of T_c in [8.4]). This result fits into an earlier phenomenological observation by ALEKSEEVSKII of the dependence of T_c on the mass of the M atom in MMo_6S_8 for M atoms from the same group in the periodic table [8.32]. The authors of [8.4] relate their observation to the strongly anharmonic motion of the Sn atoms in CPC, which has been indicated by Mössbauer-effect measurements [8.19,20] and X-ray structure analysis [8.24], and which is due to the asymmetric compressed S surrounding the Sn atoms. This would invoke the same origin, a strong anharmonicity, for an inverse indirect isotope effect as in the case of Pd-H(D) [8.50,51]. The limit, which has been estimated in Sect.8.4.1 for the isotopic volume effect, was calculated with an upper limit for the change of the lattice parameter of the total hexagonal unit cell of $SnMo_6S_8$ due to a mass change at room temperature [8.2]. The behavior of the Sn atom at low temperature may be different. Furthermore, there might be contributions from other indirect effects.

An anomalous indirect Sn isotope effect in $SnMo_6S_8$ can therefore not be excluded at present and might fit into the "Chevrel-phase picture" as another of their exotic properties. More experiments have to be performed to solve the question of the partial isotope effect of the M atom in CPC.

8.5 Tunneling Spectroscopy on $Cu_{1.8}Mo_6S_8$ and $PbMo_6S_8$

8.5.1 Tunneling Experiment

Tunneling spectroscopy offers the only practicable possibility to evaluate the Eliashberg function $\alpha^2F(\omega)$ from experimental data. This function contains all information needed for the calculation of thermodynamical properties, such as T_c, Δ_0, and H_c, of a superconductor. $\alpha^2F(\omega)$ can uniquely reflect the importance of distinct phonon frequencies for superconductivity.

The tunneling density of states, which is proportional to the first derivative dI/dV of the current-voltage characteristic of a normal-metal/insulator/superconductor tunnel junction, is used to calculate $\alpha^2F(\omega)$ by inversion of the Eliashberg equations [8.58]. This procedure demands high-quality junctions, which can be obtained for nontransition metals. The properties of the high-T_c superconductors, binary and ternary compounds, are often sensitive to deviations from stoichiometry, impurities, and lattice disorder. Surfaces are particularly affected. In addition, most of these compounds form bad, ill-defined tunneling barriers by oxidation. Therefore, the short sampling depth of tunneling electrons resulting from a small Fermi velocity in these compounds with high electronic density of states makes it difficult to probe bulk properties. For $Cu_{1.8}Mo_6S_8$ and $PbMo_6S_8$, POPPE and WÜHL [8.5] succeeded in preparing junctions with sufficiently well-defined interfaces. They used bulk samples in order to avoid surface problems of evaporated or sputtered films. The authors could measure quantitatively the energy gap Δ_0 and observe phonon-induced structures in the tunneling characteristics. The latter results only allow a qualitative analysis and are not yet good enough for a numerical evaluation of $\alpha^2F(\omega)$. Sharp drops in dI/dV or minima in the second derivative d^2I/dV^2 are taken as a rough outline for the behavior of $\alpha^2F(\omega)$.

8.5.2 Junction Preparation

Starting from bulk materials, two methods were used for junction fabrication. $Cu_{1.8}Mo_6S_8$ was probed by means of point contacts. The sample was prepared by reaction of the powdered elements at 1200°C and subsequent melting at 1850°C in a sealed Mo crucible. Clean and smooth surfaces appeared on the polycrystalline sample (T_c = 10.2 K) after pealing off a thin Mo layer. The tunneling contact was made by pressing an oxidized Al tip or an etched Zn-doped GaAs tip onto the sample surface very carefully with a piezoelectric drive. These junctions were very sensitive to mechanical vibrations, especially at high voltage.

Better junction quality was obtained by depositing artificial barriers onto single crystals of $PbMo_6S_8$. The cubic crystals with edge lengths of 0.3 to 0.4 mm were prepared in a solid-state diffusion process. The crystals had T_c's of 11.3 to 12 K and a transition width of about 0.3 K [8.6]. The tunneling area was masked with GE varnish. Granular Al deposited at 77 K in presence of oxygen and oxidized at room temperature served as tunneling barrier. The junctions were completed with Al films.

8.5.3 Results and Discussions

The first derivative dI/dV typical for $Cu_{1.8}Mo_6S_8$ is plotted in Fig.8.10. Besides the finite measuring temperature, imperfect tunneling barrier and inhomogeneous pressure distribution at the point contact area contribute to the smearing of the gap structure. Because of the strong pressure dependence of T_c in $Cu_{1.8}Mo_6S_8$ [8.53], T_c deduced from the vanishing of the energy gap varied from 10.2 to 12.5 K. A rough estimate of contact pressure and the observed increase of T_c with

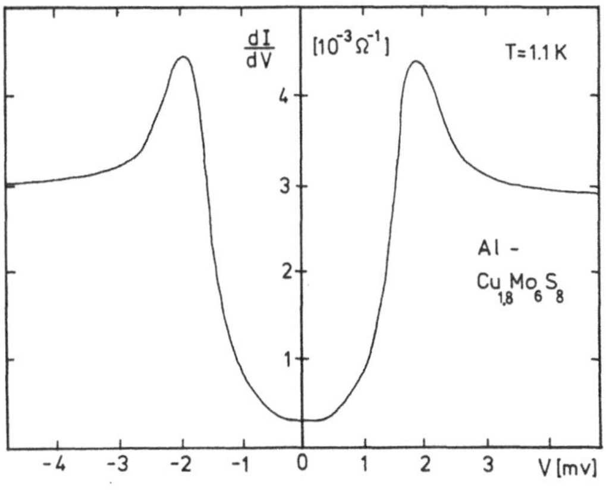

Fig.8.10. Differential conductance dI/dV versus voltage of a $Cu_{1.8}Mo_6S_8$-Al point-contact junction at 1.1 K; from [8.5,6]

pressure, which is in agreement with data in the literature [8.53], proved that the point contact did not impair the superconducting properties. As a measure of junction quality, the ratio of junction impedances (R_s/R_n) measured at zero bias in the superconducting and normal state came up to 30. The energy gap Δ_0 was taken from the energy of the maximum in dI/dV which had been reduced by 7% to allow for the temperature effect at 1.2 K. The procedure of taking the gap at the energy where $dI/dV = R_n^{-1}$ [8.58] would yield too small values, as the gap smearing is dominated by other than thermal effects. Gaps between 1.8 and 2.3 meV were obtained. The quantity $2\Delta_0/kT_c$, for which the actual T_c at the point contact [$\Delta(T_c) = 0$] was measured in each experiment, ranged from 3.9 to 4.2.

The corresponding data for $PbMo_6S_8$ (Fig.8.11) obtained from sandwich junctions typically were $\Delta_0 = 2.4$ meV and $2\Delta_0/kT_c = 4.8$. For a few junctions, values for $2\Delta_0/kT_c$ of up to 5.5 were observed. Values of up to 4.2 were obtained by taking Δ_0 at $dI/dV = R_n^{-1}$. All values for $2\Delta_0/kT_c$ exceed significantly the weak coupling BCS value of 3.5 and indicate the very strong electron-phonon coupling in the CPC that were investigated. Strong-coupling behavior, $2\Delta_0/kT_c = 4.1$, has also been observed for single crystals of binary Mo_6Se_8, which were investigated by a vacuum tunneling technique [8.34].

Due to the strong coupling, it was possible to observe structures in dI/dV and d^2I/dV^2 corresponding to peaks in $\alpha^2F(\omega)$ for most of the junctions made of $Cu_{1.8}Mo_6S_8$ or $PbMo_6S_8$, in spite of the small sampling depth of tunneling electrons in CPC. But instabilities in the tunneling barriers often gave rise to excessive noise, especially at high energies, and then prevented the analysis of the data. Structures at energies $\omega \lesssim 20$ meV could be reproduced for more than 30 junctions, whereas at $\omega \gtrsim 20$ meV, phonon-induced structures could only be established for 4 sandwich junctions with $PbMo_6S_8$.

The best information on $\alpha^2F(\omega)$ at $\omega < 20$ meV were obtained for $Cu_{1.8}Mo_6S_8$ using point contacts with oxidized Al tips. In Fig.8.12a four traces of d^2I/dV^2 from different samples demonstrate the reproducibility. The magnitude of the dips in

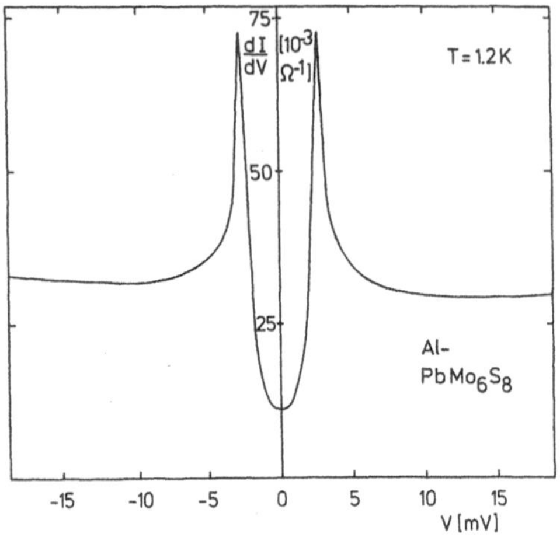

Fig.8.11. Differential conductance dI/dV versus voltage of a $PbMo_6S_8/AlO_x$(12 Å)/Al sandwich junction at 1.2 K; from [8.5,6]

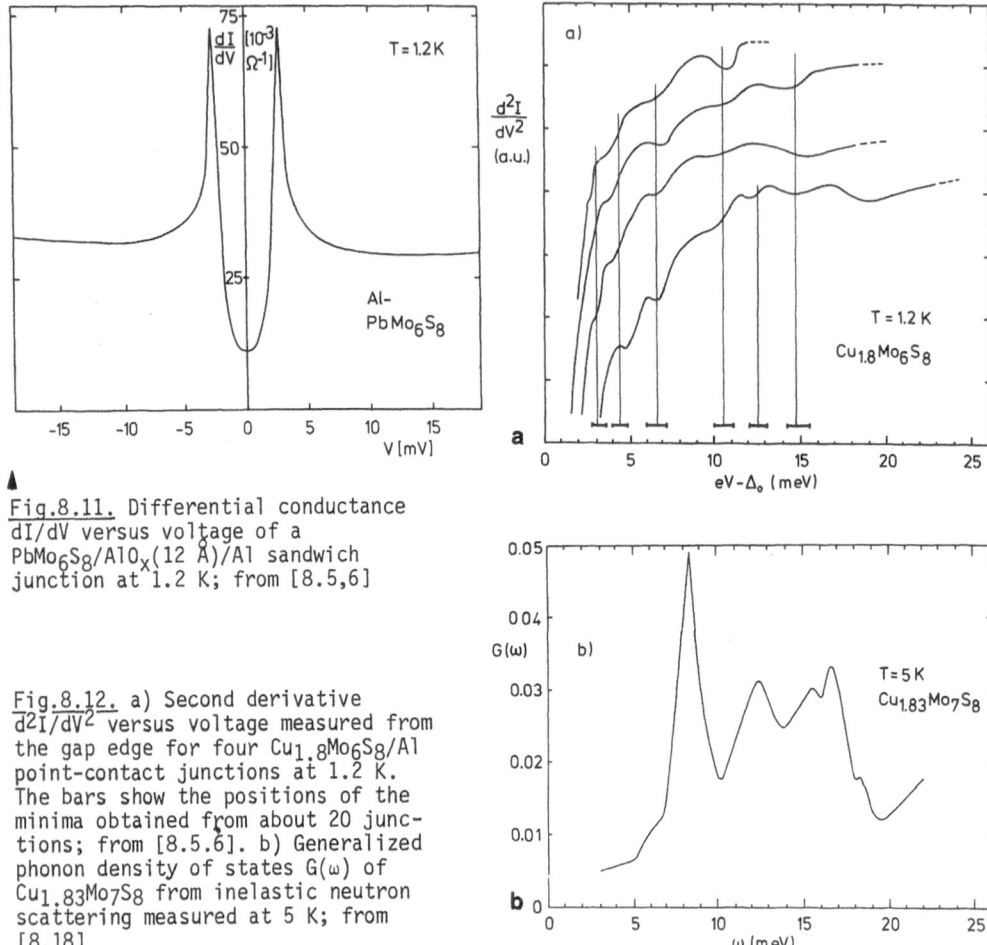

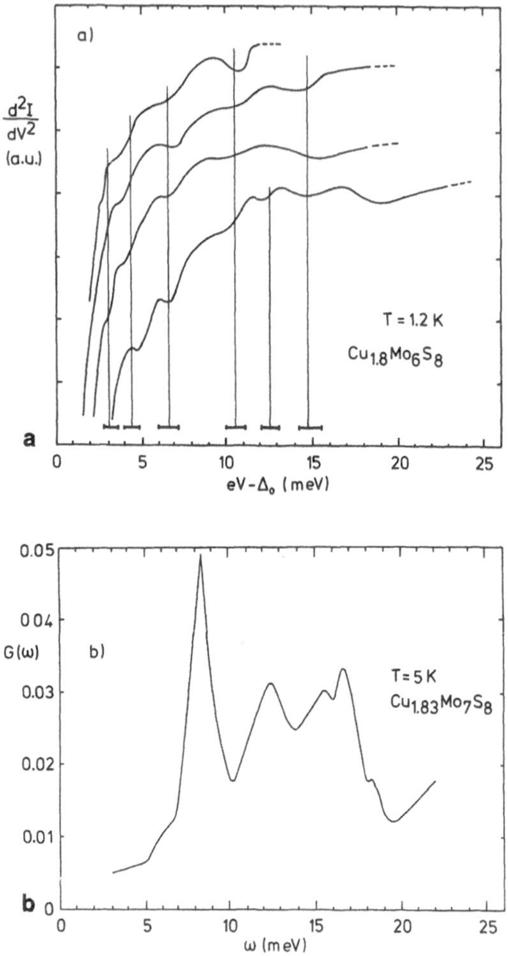

Fig.8.12. a) Second derivative d^2I/dV^2 versus voltage measured from the gap edge for four $Cu_{1.8}Mo_6S_8$/Al point-contact junctions at 1.2 K. The bars show the positions of the minima obtained from about 20 junctions; from [8.5.6]. b) Generalized phonon density of states $G(\omega)$ of $Cu_{1.83}Mo_7S_8$ from inelastic neutron scattering measured at 5 K; from [8.18]

d^2I/dV^2 varies depending on the quality of the junctions. The bars mark the variation of the position in energy of the minima obtained from 20 junctions. The influence of the contact pressure on phonon frequencies is expected to be of minor importance. For comparison, the low-energy part of the generalized phonon density of states $G(\omega)$ obtained by neutron scattering experiments at 5 K [8.18] are shown in Fig.8.12b. At most energies where characteristic phonons appear in $G(\omega)$, minima in d^2I/dV^2 were observed. Qualitative agreement between both results exists at energies of about 4.5, 6.5, 12.5, and 15 meV. The phonon peak at 12.5 meV is reflected as a well-pronounced minimum in only one of the second derivatives shown in Fig.8.12a, but it has been observed in 4 other experiments that are not shown. The existence of a phonon peak at 3 meV could not be confirmed, since elastic neutron scattering dominates at such low energies. Disagreement exists at 8.5 and 10.5 meV. The very strong Einstein-like phonon peak at 8.5 meV is faced with a very weak structure observed only occasionally in d^2I/dV^2. On the other hand, the structure in d^2I/dV^2 at 10.5 meV has no counterpart in $G(\omega)$.

$Cu_{1.8}Mo_6S_8$ transforms at 269 K into a triclinic low-temperature modification in which the Cu atoms condense into pairs [8.13]. The pronounced Einstein-like peak which appears at low temperatures in $G(\omega)$ at 8.5 meV can be attributed to at least one of the optical modes associated with Cu-atom displacements. The absence of comparably strong structure in d^2I/dV^2 at 8.5 meV indicates weak coupling to this mode. The results, however, also imply that there is a special mode at 10.5 meV which is not distinguished in $G(\omega)$, but contributes to $\alpha^2F(\omega)$.

Measurements at high energies were considerably improved using $PbMo_6S_8/AlO_x/Al$ sandwich junctions. The differential conductances of a junction measured in the superconducting and normal state are shown in Fig.8.13. The strong suppression of the conductance in the superconducting state below that of the normal state at $\omega > 6$ meV implies coupling to a broad band of phonons. Peaks in $\alpha^2F(\omega)$ expected from these results are marked by arrows. Maxima in $G(\omega)$ are indicated on the abscissa. The agreement in the energy positions is fairly good.

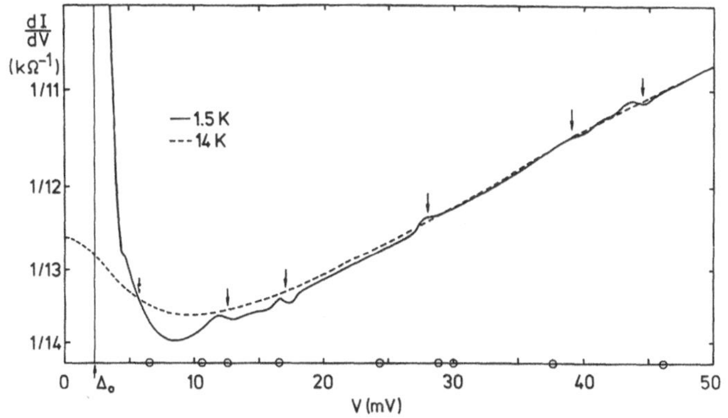

Fig.8.13. Differential conductance dI/dV of a $PbMo_6S_8/AlO_x/Al$ sandwich junction versus voltage measured at 1.5 K ($<T_c$) and 14 K ($>T_c$). The arrows indicate the position of structure in dI/dV and thus in $\alpha^2F(\omega)$. Circles indicate the energies of characteristic maxima in the phonon density of states $G(\omega)$ counted from the gap edge Δ_0 [8.18]; from [8.5,6]

The structure in dI/dV of $PbMo_6S_8$ junctions at high energies amounts to about 1% of the junction conductance. Since the height of a step is caused by a peak in $\alpha^2F(\omega)$ weighted with ω^{-2} [8.42], the pronounced steps at $\omega > 20$ meV indicate a very strong coupling to phonons with high energies.

In the energy range of vibrations attributed to the displacement of the Pb atom ($\omega \simeq 4$ meV), it is conspicuous that the peak in $\alpha^2F(\omega)$ lies 0.5 to 1 meV below that of $G(\omega)$. This discrepancy may also hint to a different coupling of modes as in the case of $Cu_{1.8}Mo_6S_8$. For a quantitative analysis one has to bear in mind that the samples used in the neutron scattering [8.18] and tunneling experiment [8.5,6] had different nominal compositions.

Band-structure calculations [8.8-10] have shown that superconductivity is primarily associated with the d electrons of the Mo_6 octahedra, and the electron-phonon interaction should then be mainly determined by coupling of internal modes of the Mo_6 octahedra to the Mo d electrons. Displacements of the M atoms as well as of the Mo_6X_8 units may therefore contribute to superconductivity only to the extent that these modes are hybridized with internal modes of the Mo_6 octahedra. But the energies where hybridization takes place do not in general coincide with those of Brillouin-zone boundary modes. Phonons with a relatively small density of states may experience a high electron-phonon coupling leading to an enhanced peak in $\alpha^2F(\omega)$. This effect has possibly been observed at 10.5 meV in $Cu_{1.8}Mo_6S_8$.

Tunneling experiments indicated a weak coupling of conduction electrons to optical modes of M atoms. This implies that the M atoms influence superconductivity mainly in an indirect way by charge transfer to the Mo_6X_8 units which reduces the distortion of the Mo_6 octahedra and alters the electron density of states at the Fermi level [8.13]. A comparison of the properties of ternary compounds with those of the binary Mo_6Se_8 shows that the optical modes of the M atoms indeed may not be crucial for the electron-phonon coupling strength. They are absent in Mo_6Se_8, but nevertheless, Mo_6Se_8 shows a high value of $2\Delta_0/kT_c$ (Sect.8.2) and pronounced strong coupling effects. It seems that another set of low-frequency modes is responsible for strong-coupling features in binary as well as in ternary CPC.

8.6 Conclusion

In the preceding sections we have summarized our present knowledge about the electron-phonon interaction in Chevrel-phase superconductors. Even though some information is available, the field is still in its infancy. Tunneling spectroscopy, in general the most informative probe, is not accurate enough to give any quantitative results, except for the size of the energy gap. Conclusive isotope-effect data are only available for Mo_6Se_8, and their interpretation is limited by our ignorance of indirect isotope effects. Any theoretical analysis suffers from a lack of information on the lattice dynamics of CPC. This negative list could be continued. The situation is not much different for most other groups of superconducting compounds.

Nevertheless, we consider this article as a reasonable basis and encouragement for further research, and we want to summarize its main and probably most reliable results. The energy-gap data from tunneling experiments on Mo_6Se_8, $Cu_{1.8}Mo_6S_8$, and $PbMo_6S_8$ clearly demonstrate the strong electron-phonon coupling in CPC, in agreement with the rather large jump in the specific heat of Mo_6Se_8. These data hint at a special coupling to low-frequency phonons. The calculated values for the energy gap and for the jump of the specific heat at T_c can only be made compatible with experimental results if a particularly strong coupling to low-frequency modes is assumed. This raises the question about the particular character of these active low-frequency modes. One might think of optical modes of heavy M atoms. This possibility seems to be ruled out by tunneling data—we think also by the isotope-effect results—and by the fact that strong-coupling anomalies are also found in Mo_6Se_8, which has no M atom. Another suggestion refers to molecular-crystal-type librational modes. This hypothesis cannot explain recent isotope-effect data. The data for Mo_6Se_8 have shown that only a set of phonon modes can be of importance for superconductivity in CPC to which the Mo and the Se atoms contribute about equally (even though the decisive electronic states are probably dominated by Mo electrons). Band-structure theory claims that the most important conduction electrons in CPC react predominantly to internal deformations of the Mo octahedra, which excludes a reasonably strong coupling to librational and other external modes of molecular-type units.

All these requirements together lead to a rather stringent characterization of the important phonon modes for superconductivity in CPC. The modes should be soft, be coupled strongly to deformations of the Mo octahedra, and should be equally strongly affected by the Mo and Se masses. Do we have such modes (which might be specified as "*soft internal modes*") in CPC? To date there is no direct proof for their existence. One has even good reason to doubt the importance of soft modes for CPC superconductivity. Tunneling data indicate that the superconducting electrons couple especially strongly to high-frequency modes. The assumption of a strong coupling to *hard internal modes* is compatible with the isotope effects and with theoretical ideas based on band-structure calculations, but cannot explain the pronounced strong-coupling effects in the energy gap and the specific heat. Further experimental data or even a confirmation or improvement of present data is needed in order to resolve the striking discrepancies. One might speculate that these discrepancies are intrinsic and indicate a nonconventional type of superconductivity in CPC, e.g., a system of Josephson-coupled "superconducting molecules" as suggested by REVZEN and RON [8.58]. In our opinion, present uncertainties in the experimental data do not yet justify this fascinating but barely justified alternative. On the contrary, the good and nearly quantitative agreement of CPC superconductivity with the strong-coupling model of superconductors indicates that all remaining puzzles may find their solution within this conventional model. It might, however, be important for a quantitative understanding to consider anisotropy or multiband effects [Ref.8.59, Chap.3]. They are neglected throughout in our analysis, which employs a dirty-limit theory.

Acknowledgement. We are very grateful to Dr. F.J. Culetto and Dr. U. Poppe for their substantial contributions to the work summarized in this article and for many helpful discussions.

References

8.1 F.J. Culetto, F. Pobell: Phys. Rev. Lett. *40*, 1104 (1978)
8.2 F.J. Culetto: Ph. D. Thesis, JÜL-Report 1587, Kernforschungsanlage Jülich, West Germany (1979)
8.3 F. Pobell: In *Proc. Int. Conf. on Ternary Superconductors*, Lake Geneva, Wisconsin (1980), ed. by G.K. Shenoy, B.D. Dunlap, F.Y. Fradin (North Holland, New York 1981) p.35
8.4 N.E. Alekseevskii, V.I. Nizhankovskii: JETP Lett. *31*, 58 (1980)
8.5 U. Poppe, H. Wühl: J. Phys. (Paris) *39*, C6-361 (1978); J. Low. Temp. Phys. *43*, 371 (1981)
8.6 U. Poppe: Ph. D. Thesis, JÜL-Report 1635, Kernforschungsanlage Jülich, West Germany (1980)
8.7 G. Grimvall: Physica Scripta *14*, 63 (1976)
8.8 O.K. Andersen, W. Klose, H. Nohl: Phys. Rev. B*17*, 1209 (1978)
8.9 L.F. Mattheis, C.Y. Fonq: Phys. Rev. B*15*, 1760 (1977)
8.10 D.W. Bullett: Phys. Rev. Lett. *39*, 664 (1977)
8.11 T. Jarlborg, A.J. Freeman: Phys. Rev. Lett. *44*, 178 (1980); A.J. Freeman, T. Jarlborg: In *Proc. Int. Conf. on Ternary Superconductors*; Lake Geneva, Wisconsin (1980), ed. by G.K. Shenoy, B.D. Dunlap, F.Y. Fradin (North Holland, New York 1981) p.59
8.12 M. Marezio, P.D. Dernier, J.P. Remeika, E. Corenzwit, B.T. Matthias: Mat. Res. Bull. *8*, 657 (1973); O. Bars, J. Guillevic, J. Grandjean: J. Solid State Chem. *6*, 48 (1973); R. Chevrel, M. Sergent, J. Prigent: Mater. Res. Bull. *9*, 1487 (1974)
8.13 K. Yvon: In *Current Topics in Materials Science*, Vol.II, ed. by E. Kaldis (North Holland, Amsterdam 1979) p.53
8.14 S.D. Bader, S.K. Sinha, R.N. Shelton: In *Superconductivity in d- and f-Band Metals*, ed. by H.D. Douglass (Plenum, New York 1976) p.209
8.15 S.D. Bader, G.S. Knapp, S.K. Sinha, P. Schweiß, B. Renker: Phys. Rev. Lett. *37*, 344 (1978)
8.16 B.P. Schweiss, B. Renker, E. Schneider, W. Reichardt: In *Superconductivity of d- and f-Band Metals*, ed. by H.D. Douglass (Plenum, New York 1976); B.P. Schweiß, B. Renker: Progr. Report of the Teilinstitut Nukleare Festkörperphysik, KfK 2357 (1976)
8.17 B.P. Schweiss, B. Renker, J.B. Suck: J. de Physique *39*, C6-356 (1978)
8.18 B.P. Schweiss, B. Renker, R. Flükiger: In *Proc. Int. Conf. on Ternary Superconductors*, Lake Geneva, Wisconsin (1980), ed. by G.K. Shenoy, B.D. Dunlap, F.Y. Fradin (North Holland, New York 1981) p.29 We thank Drs. B. Renker and B.P. Schweiss for supplying us with their most recent neutron scattering results
8.19 C.W. Kimball, L. Weber, G.v. Landuyt, F.Y. Fradin, B.D. Dunlap, G.K. Shenoy: Phys. Rev. Lett. *36*, 412 (1976)
8.20 J. Bolz, J. Hauck, F. Pobell: Z. Phys. B*25*, 351 (1976)
8.21 A.C. Lawson: Mater. Res. Bull. *7*, 733 (1972); D.C. Johnston, R.N. Shelton: J. Low Temp. Phys. *26*, 561 (1977)
8.22 K. Yvon: Solid State Commun. *25*, 327 (1978)
8.23 D.C. Johnston: R.N. Shelton, J.J. Bugaj: Solid State Commun. *21*, 949 (1977); R. Flükiger, A. Junod, R. Baillif, P. Spitzli, A. Treyvaud, A. Paoli, H. Devantay, J. Muller: Solid State Commun. *23*, 699 (1977)
8.24 R. Chevrel, C. Rossel, M. Sergent: J. Less Common Metals *72*, 31 (1980)
8.25 S.D. Bader, G.S. Knapp, A.T. Aldred: Ferroelectrics *17*, 321 (1977)
8.26 S.D. Bader, S.K. Sinha: Phys. Rev. B*18*, 3082 (1978)
8.27 Ø. Fischer: Colloques Int. CNRS *242*, 79 (1974); *Proc. 14th Int. Conf. Low Temp. Phys.*, Vol.V, ed. by M. Krusius, M. Vuorio (Otaniemi, 1975) p.172
8.28 Ø. Fischer, H. Jones, G. Bongi, M. Serqent, R. Chevrel: J. Phys. C*7*, L450 (1974); Ø. Fischer, A. Treyvaud, R. Chevrel, M. Sergent: Solid State Commun. *17*, 721 (1975)

8.29 F.Y. Fradin, G.S. Knapp, S.D. Bader, G. Cinader, C.W. Kimball: In *Superconductivity in d- and f-Band Metals*, ed. by D.H. Douglass (Plenum, New York 1976) p.297
8.30 K.P. Nerz, U. Poppe, F. Pobell, M. Weger, H. Wühl: In *Superconductivity in d- and f-Band Metals*, ed. by H. Suhl, M.B. Maple (Academic, New York 1980) p.501; K.P. Nerz: Ph. D. Thesis, JÜL-Spez 30, Kernforschungsanlage Jülich, West Germany (1979)
8.31 N.E. Alekseevskii, G. Wolf, N.M. Dobrovolskii, C. Hohlfeld: J. Low Temp. Phys. *38*, 253 (1980); J. Low Temp. Phys. *40*, 479 (1980)
8.32 N.E. Alekseevskii: Cryogenics *26*, 257 (1980)
8.33 R.W. McCallum, L.D. Woolf, R.N. Shelton, M.B. Maple: J. Phys. (Paris) *39*, C6-359 (1978)
8.34 U. Poppe, H. Schröder, F. Pobell: Verh. Deutsche Phys. Ges. *3*, 503 (1981); Unpublished results
8.35 R. Odermatt, M. Hardiman, J. van Meijel: Solid State Commun. *32*, 1227 (1979)
8.36 N. Sano, T. Taniguchi, K. Asayama: Solid State Commun. *33*, 419 (1980)
8.37 Ø. Fischer: Appl. Phys. *16*, 1 (1978)
8.38 H. Padamsee, J.E. Neighbor, C.A. Shiffman: J. Low Temp. Phys. *12*, 387 (1973)
8.39 K. Kitazawa, T. Matsuura, S. Tanaka: In *Proc. Int. Conf. on Ternary Superconductors*, Lake Geneva, Wisconsin (1980), ed. by G.K. Shenoy, B.D. Dunlap, F.Y. Fradin (North Holland, New York 1981) p.83
8.40 J.A. Woollam, S.A. Alterovitz: Solid State Commun. *27*, 571 (1978); Phys. Rev. B*19*, 749 (1979)
8.41 R. Flükiger, R. Baillif, E. Walker: Mater. Res. Bull. *13*, 743 (1978)
8.42 D.J. Scalapino: In *Superconductivity*, ed. by R.D. Parks (Marcel Dekker, New York 1969) p.449
8.43 W.L. McMillan: Phys. Rev. *167*, 331 (1968)
8.44 We have used the same computer program as Jong-Chul Park, J.E. Neighbor, C.A. Shiffman: Phys. Lett. *50*A, 9 (1974)
8.45 F.J. Culetto, D. Rainer: JÜL-Report 1504, Kernforschungsanlage Jülich, West Germany (1978)
8.46 D. Rainer, F.J. Culetto: Phys. Rev. B*19*, 2540 (1979)
8.47 H. Fröhlich: Phys. Rev. *79*, 845 (1950)
8.48 J. Bardeen: Phys. Rev. *80*, 567 (1950)
8.49 T. Nakajima, T. Fukamachi, O. Terasaki, S. Hosoya: J. Low Temp. Phys. *27*, 245 (1977)
8.50 B.N. Ganguly: Z. Phys. B*22*, 127 (1975); Phys. Rev. B*14*, 3848 (1976)
8.51 R.J. Miller, C.B. Satterthwaite: Phys. Rev. Lett. *34*, 144 (1975)
8.52 B. Stritzker, W. Buckel: Z. Phys. *257*, 1 (1972)
8.53 R.N. Shelton, A.C. Lawson, D.C. Johnston: Mater. Res. Bull. *10*, 297 (1975); R.N. Shelton: In *Superconductivity in d- and f-Band Metals*, ed. by D.H. Douglass (Plenum, New York 1976) p.137
For $Sn_{1.2}Mo_6S_8$ we take $dT_c/dp \simeq 1.10^{-4}$ K/bar, which corresponds to the sample with a T_c closer to the one of our sample
8.54 A.W. Webb, R.N. Shelton: J. Phys. F *8*, 261 (1978)
8.55 M. Sergent, R. Chevrel, C. Rossel, Ø. Fischer: J. Less Common. Met. *58*, 179 (1978)
8.56 The isotopes were obtained from Rohstoff Einfuhr GmbH, Düsseldorf; their mean masses were 76.15, 77.94, 79.91, 81.68 for Se; 92.17, 94.11, 95.92, 97.75, and 99.61 for Mo; and 115.99, 119.18, 121.78, and 123.86 for Sn
8.57 W.L. McMillan, J.M. Rowell: In *Superconductivity*, ed. by R.D. Parks (Dekker, New York 1969) p.561
8.58 M. Revzen, A. Ron: Private communication (1978)
8.59 Ø. Fischer, M.B. Maple (eds.): *Superconductivity in Ternary Compounds II*, Topics in Current Physics, Vol.34 (Springer, Berlin, Heidelberg, New York, to be published)

Subject Index

Excitons

Editor: **K. Cho**

1979. 118 figures, 8 tables. XI, 274 pages
(Topics in Current Physics, Volume 14)
ISBN 3-540-09567-5

Contents:

K. Cho: Introduction. – *K. Cho:* Structure of
Excitons. – *P. J. Dean, D. C. Herbert:* Bound
Excitons in Semiconductors. – *B. Fischer,
J. Lagois:* Surface Exciton Polaritons. –
P. Y. Yu: Study of Excitons and Exciton-
Phonon Interactions by Resonant Raman and
Brillouin Spectroscopies.

Glassy Metals I

Ionic Structure, Electronic Transport, and
Crystallization

Editors: **H. Beck, H.-J. Güntherodt**

1981. 119 figures. XIV, 267 pages
(Topics in Applied Physics, Volume 46)
ISBN 3-540-10440-2

Contents:
H. Beck, H.-J. Güntherodt: Introduction. –
P. Duwez: Metallic Glasses – Historical Back-
ground. – *T. Egami:* Structural Study by
Energy Dispersive X-Ray Diffraction. –
J. Wong: EXAFS Studies of Metallic Glasses. –
A. P. Malozemoff: Brillouin Light Scattering
from Metallic Glasses. – *J. Hafner:* Theory of
the Structure, Stability, and Dynamics of
Simple-Metal Glasses. – *P. J. Cote, L. V. Meisel:*
Electrical Transport in Glassy Metals. –
J. L. Black: Low-Energy Excitations in Metallic
Glasses. – *W. L. Johnson:* Superconductivity in
Metallic Glasses. – *U. Köster, U. Herold:*
Crystallization of Metallic Glasses.

Structural Phase Transitions

Editors: **K. A. Müller, H. Thomas**

1981. 61 figures. XI, 190 pages
(Topics in Current Physics, Volume 23)
ISBN 3-540-10329-5

Contents:
K. A. Müller: Introduction. – *P. A. Fleury,
K. Lyons:* Optical Studies of Structural Phase
Transitions. – *B. Dorner:* Investigation of
Structural Phase Transformations by Inelastic
Neutron Scattering. – *B. Lüthi, W. Rehwald:*
Ultrasonic Studies Near Structural Phase
Transitions.

**S. V. Vonsovsky, Y. A. Izyumov,
E. Z. Kurmaev**

Superconductivity of Transition Metals

Their Alloys and Compounds

Translated from the Russian by E. H. Brandt
and A. P. Zavarnitsyn

1982. 182 figures. XIII, 512 pages
(Springer Series in Solid-State Sciences,
Volume 27)
ISBN 3-540-11382-7

Contents:
Introduction. – The Theory of Strong Coup-
ling Superconductors. – Superconductivity
and Magnetism. – Superconductivity in
Transition Metals. – Superconductivity in
Transition-Metal Alloys. – Compounds with
A-15 Structure. – Other Compounds Based
on Transition Metals. – High-Temperature
Superconductors and Lattice Instability of
Compounds. – Radiation Effects on Super-
conductors. – References. – Subject Index.

Springer-Verlag Berlin Heidelberg New York